T0321919

Research and Practice in Social Skills Training

Research and Practice in Social Skills Training

Edited by

Alan S. Bellack
Clinical Psychology Center
Pittsburgh, Pennsylvania

and

Michel Hersen
Western Psychiatric Institute and Clinic
Pittsburgh, Pennsylvania

Plenum Press · New York and London

Library of Congress Cataloging in Publication Data

Main entry under title:

Research and practice in social skill training.

Includes index.

CONTENTS: Trower, P. Fundamentals of interpersonal behaviour.–Kazdin, A. E. Sociopsychological factors in psychopathology.–Bellack, A. S. Behavioral assessment of social skills.–Rinn, R. C. and Markle, A. Modification of social deficits in children. [etc.]

1. Psychology, Pathological–Social aspects. 2. Social psychiatry. 3. Interpersonal relations. I. Bellack, Alan S. II. Hersen, Michel.

RC455.R47 158′.2 79-12118

ISBN 0-306-40233-5

A Division of Plenum Publishing Corporation
227 West 17th Street, New York, N.Y. 10011

Printed in the United States of America

To Barbara, Jonathan, and Adam

To Jonathan

Contributors

Alan S. Bellack • Department of Psychology, University of Pittsburgh, Pittsburgh, Pennsylvania

Gary R. Birchler • Mental Health Clinic, Veterans Administration Medical Center, and University of California School of Medicine at San Diego, La Jolla, California

James P. Curran • Mental Hygiene Clinic, Veterans Administration Medical Center, and Brown Medical School, Providence, Rhode Island

Kelly J. Egan • Department of Psychology, University of Washington, Seattle, Washington

John P. Galassi • Counseling Psychology, School of Education, University of North Carolina, Chapel Hill, North Carolina

Merna Dee Galassi • Office of Developmental Counseling, Meredith College, Raleigh, North Carolina

Michel Hersen • Department of Psychiatry, Western Psychiatric Institute and Clinic, University of Pittsburgh School of Medicine, Pittsburgh, Pennsylvania

Alan E. Kazdin • Department of Psychology, Pennsylvania State University, University Park, Pennsylvania

Marsha M. Linehan • Department of Psychology, University of Washington, Seattle, Washington

Allan Markle • Private Practice, Northbrook, Illinois

Roger C. Rinn • Private Practice, Huntsville, Alabama

Peter Trower • Department of Psychology, Hollymoor Hospital, Northfield, Birmingham, England

Preface

It is perhaps trite to refer to human beings as social animals, but nevertheless it is true. A substantial portion of our lives is spent in interactions with other people. Moreover, the nature, quality, and quantity of those interactions have a tremendous impact on behavior, mood, and the adequacy of adjustment. Faulty interpersonal relationship patterns have reliably been associated with a wide variety of behavioral-psychological dysfunctions ranging from simple loneliness to schizophrenia.

Most "traditional" analyses of interpersonal failures have viewed them as consequences or by-products of other difficulties, such as anxiety, depression, intrapsychic conflict, or thought disorder. Consequently, remediational efforts have rarely been directed to interpersonal behavior *per se*. Rather, it has been expected that interpersonal relationships would improve when the source disorder was eliminated. While this model does account for some interpersonal dysfunctions (e.g., social anxiety can inhibit interpersonal behavior), it is not adequate to account for the vast majority of interpersonal difficulties. In fact, in many cases those difficulties either are independent of or underlie other dysfunctions (e.g., repeated social failure may produce depression or social anxiety).

What then is the source of interpersonal problems and how can they be alleviated? From a behavioral perspective, the basis of interpersonal behavior is a set of learned abilities: social skills. People must learn how to interact in the same manner that they learn to swim, play the violin, or play bridge. Interpersonal failures result when faulty learning histories leave the individual with social skill deficits. Consequently, remediation of the difficulty entails relatively straightforward training on how to make appropriate responses: social skills training.

Scores of studies have been conducted in the past several years in an effort to examine the behavioral conception and to develop and validate effective training procedures. Considerable success has been reported, and social skills training is now frequently touted as one of the most significant achievements of behavior therapy. However, similar exuberant claims for some other behavioral approaches (e.g., token economies) have subsequently been shown to be somewhat premature. The social skills literature is now sufficiently extensive to permit (and justify) a comprehensive, critical appraisal. The purpose of this volume is to present such an evaluation, to discuss the state of the art, identify strengths and weaknesses, and point out directions in which future research should proceed.

The reader will not find the book to contain a raft of self-serving congratulations for a job well done. Rather, the contributors generally have been quite objective and realistically self-critical. Numerous problems, unanswered questions, and faulty conclusions have been identified. We certainly do not have all of the answers to either analysis of social behavior or treatment. Nevertheless, the appraisal is, generally, quite positive. Initial claims of success appear to have been exaggerated, but not unfounded.

Alan S. Bellack
Michel Hersen

Pittsburgh, Pennsylvania

Contents

PART ONE: GENERAL ISSUES

PART TWO: TREATMENT

PART THREE: METHODOLOGICAL ISSUES

PART ONE

GENERAL ISSUES

CHAPTER 1

Fundamentals of Interpersonal Behavior: A Social-Psychological Perspective

Peter Trower

INTRODUCTION

The pace of development of social skills training over the past few years has hardly allowed for the orderly assimilation of scientific knowledge from social psychology and other disciplines. Research in these disciplines remains a relatively untapped source (Kopel & Arkowitz, 1975) and it is the purpose of the present chapter to attempt a review and integration of some aspects of the literature which have actual or potential application in social skills training. Subsequent chapters show how some of these topics can be developed in practice.

Topics and Approaches

Social psychologists and researchers in related specialities, such as sociology, social anthropology, human ethology, and sociolinguistics, pursue quite diverse areas of interpersonal behavior. Some examine the minutiae of the behavioral stream; some analyze the cognitive, motivational, and perceptual processes underlying behavior; some examine social and cultural rule systems that govern behavioral sequences; others take situations and contexts as the focus of analysis. Some kind of theoretical framework is clearly needed to organize and integrate the

Peter Trower • Department of Psychology, Hollymoor Hospital, Northfield, Birmingham, B31 5EX, England.

findings and relate them to the topic of social skills. There is no shortage of approaches to choose from, and indeed the theories are as diverse as the topics. In one topic alone—interpersonal attraction—there are no less than 14 identifiable theories (Duck, 1977). However, there is one group of conceptually related theories which, despite criticism by a number of radical behaviorists, nonetheless find more favor with other critics (e.g., Harré & Secord, 1972) and appear to have broad explanatory and predictive utility for the area of social skills. These are the feedback-loop theories which include the social skills model (Argyle & Kendon, 1967), aspects of social learning theory (Bandura, 1977), Kanfer's closed-loop model of self-control (Kanfer & Karoly, 1972), and the TOTE model (Miller, Galanter, & Pribram, 1960). Of these, the first two directly address the acquisition and production of skilled social behavior. These theories integrate cognitive and behavioral components and also accommodate important developments such as interactional psychology (Endler & Magnusson, 1976) and cognitive social psychology (Triandis, 1977), as well as the self-control (Mahoney, 1974) and cognitive therapies (Meichenbaum, 1977).

The social skills model—which has been influential in British research on social skills—provides an illustrative case. It is based on the analogy of serial motor skills and assumes that the individual has goals or targets which he seeks in order to obtain rewards. Goal achievement is dependent on skilled behavior which involves a continuous cycle of monitoring and modifying performance in the light of feedback. Failure in skill is defined as a breakdown or impairment at some point in the cycle, which results in failure to achieve targets, leading to negative outcomes and abnormal behavior patterns.

Although differing in important detail, several of the feedback-loop theories appear to emphasize similar component processes, the central ones being a perceptual component for observing and receiving feedback from the environment, a cognitive (or translation) component for making judgments and decisions about response choices, and a performance component for carrying out actual behavioral sequences. The central process is the feedback loop, an operation which Bandura (1977) terms reciprocal determinism. A thorough treatment of these theories is beyond the scope of the present review, and discussion will be confined to these main component processes and related phenomena.

BASIC PROCESSES IN SOCIAL SKILL

Perception

In order to learn and produce skilled behavior the individual needs (1) information on the state of affairs in the immediate situation and (2) information concerning outcomes of possible strategies, gained from either trial-and-error experience or observation of modeled displays. Information of the second kind is retained in memory for guiding future performance (Bandura, 1977). It would seem to be a reasonable assumption that the perception of such information should be relevant and accurate if performance is to be skilled.

The individual can attend to only a fraction of the available information, and, therefore, what he selects should be relevant to the target sought. In first-encounter situations the most salient information is the most obvious—static physical features and dynamic or behavioral cues. The information selected will also depend on such factors as its frequency, recency, primacy (occuring earlier rather than later in a sequence), and reinforcement value. During observational learning the most salient conditions which govern learning will include the model's affective valence (rewarding properties), similarity to the observer, and so on. Generally speaking, *A* will attend to cues which he believes have utility in predicting outcomes, e.g., *B*'s future behavior.

It is assumed that such cues must not only be salient but accurately perceived if they are to have predictive value in informing people about the probable effects of actions and events. For example, it would seem obvious that people need to decode social signals accurately in order to know how to deal with the other and predict his behavior. However, stimuli are actively selected, interpreted, coded, and otherwise processed, in the course of which errors of perception may arise, particularly in the social environment which often presents highly ambiguous phenomena. Research shows that accuracy in interpersonal judgments is variable (Warr & Knapper, 1968). Individuals may fail to monitor the immediate environment or do so intermittently; fail to register significant events; possess limited, superficial, or ambiguous information—all conditions which encourage projection or distortion through stereotypes, personal beliefs, and other perceptual processes. Many of these processes have been identified and some are now described.

Labeling. Individuals employ conceptual labels to classify and categorize others. The function of this labeling process is to reduce complex information (which would otherwise be overwhelming) to simpler concepts which have predictive value. Dependence on global perceptual labels is greatest in unfamiliar and ambiguous situations where information is limited. In such situations people attend to and interpret or "decode" the most obvious cues and make inferences from them about the attributes of others. These social cues include static physical features such as physiognomy of the face, body build, skin color, and clothes, and dynamic cues such as vocal characteristics, nonverbal and verbal behavior (Cook, 1971).

One kind of perceptual label is the social stereotype, which derives from a belief which the individual acquires from his social/cultural group. Once a person is identified as belonging to a stereotyped group (e.g., all blacks), a standard set of attributes is applied to him, irrespective of his individual characteristics which are ignored or overlooked.

Some groups, including mental patients, the physically handicapped, and even the unattractive, are stigmatized through stereotyping. Goffman (1963) has described the damaging effects of stigma. Some believe that the process of labeling itself produces the deviance, although this has been challenged, for example by Gove (1970). We shall see later, however, that there is strong evidence that labeling is a "self-fulfilling prophecy."

Some stereotypic beliefs derive not from the culture but from the individual's personal experiences. Byrne, McDonald, and Mikawa (1963) report that past history can lead to a belief that others are mainly punishing, leading to interaction avoidance. Experiments show that perceptions are affected by failure experiences (Mischel, Ebbeson, & Zeiss, 1973), emotional state (e.g., Feshback & Singer, 1957), and mood. Perception also varies as a function of cognitive style. Robbins (1975) found that "dogmatic" subjects formed lasting impressions on little evidence, and would tend to explain away inconsistent information. Leonard (1976) showed that cognitively complex judges were more able to discriminate similar from dissimilar others.

Self-Fulfilling Prophecies. Although perceptual judgments are often found to be inaccurate, it seems to be accepted among some psychologists that at least some stereotypes have a kernel of truth. More important, the evidence shows that some judgments end up as true by their effect on behavior. Indeed, it is in this sense that some perceptual biases have their most serious consequences.

Paranoid interpretations of social cues may lead an individual to behave with hostility, which will in turn provoke hostility in others. This vicious circle becomes autonomous and self-sustaining in that such interpretations become a form of self-fulfilling prophecy (Rosenthal, 1973). For example, Jones and Panitch (1971) showed that when an individual believes another person dislikes him, his consequent behavior can actually produce feelings of disliking in the other person—even when the original belief was totally incorrect. This false definition of the situation evokes behavior which makes the false belief into a true one.

Snyder, Tanke, and Berscheid (1977) similarly found evidence of the self-fulfilling nature of the attractiveness stereotype. This belief basically is that "what is beautiful is good" (Dion, Berscheid, & Walster, 1972). Snyder *et al.* (1977) gave their male subjects false information on the attractiveness of their unseen female telephone partners, which was sufficient to arouse an erroneous attraction stereotype of the partner.

The men then proceeded to behave in a way congruent with their belief, in turn producing and reinforcing behavior in the partner which was also congruent with the stereotype.

Attribution Tendencies. Stereotypes and other beliefs are a person's intuitive theories about others, and social behavior and other stimuli form his data. His social competence presumably depends on the accuracy and adequacy of his hypotheses, evidence, and inferences. Yet research has now established that attribution errors are common and widespread. Many of these are well reviewed by Ross (1977), from whom the following examples are mainly drawn.

A claim, challenged by Ross, is that people have a motivated "ego defensive" bias to attribute "successes" to their own efforts, abilities, or dispositions while attributing "failures" to luck, task difficulty, or other external factors.

There is much stronger evidence to show that people systematically attribute too much to the person (internal cause) and too little to the situation (external cause) (Jones & Nisbett, 1971). This is a variant of the discovery that the average individual is a "trait theorist," believing people behave consistently according to their personalities (internal cause) although the evidence shows that people behave variably according to the situation (Mischel, 1968), or, more exactly, an interaction of the two (Endler & Magnusson, 1976). The exception to this is that the individual attributes his own responses to situational constraints (Jones & Nisbett, 1971) unless *he* is the focus of attention (Duval & Hensley, 1976).

Ross reports evidence of a "false consensus" bias by which people see their own behavior and judgments as common and normal but alternative ones as relatively uncommon and deviant. Subjects believed that most of their peers would behave the same way as they would.

Ross also reports that judges fail to take into account the constraints of people's social roles when assessing others and so draw inaccurate inferences about role-advantaged and role-disadvantaged others. An example of this may occur in mental hospitals, where staff have status, power, and credibility, the patients none of these, yet both staff and patients tend to see these as qualities of persons rather than role-conferred advantages.

Nisbett and his associates (Nisbett, Borgida, Crandall, & Reed, 1976) showed in a series of studies that people do not take into account consensus information (i.e., what the majority do) when making causal judgments, but instead tend to generalize from the particular. Depressed subjects were not influenced by knowing that most other people were depressed by the same situation. It appears that concrete and vivid stimuli are more easily perceived and remembered.

Initial judgments about the self or others are difficult to reverse and may survive in the face of totally disconfirming evidence (e.g., Ross, Lepper, & Hubbard, 1975). Perceivers further distort evidence by selectively and uncritically attending to confirming evidence, and questioning and in other ways attempting to discredit contradictory evidence. Depressives can maintain their pessimistic perspective by attending to negative rather than positive feedback (Wener & Rehm, 1975). It was shown earlier that this can develop into a self-fulfilling prophecy in interpersonal behavior.

Perception and Competence. Given the extent of perceptual and attribution error of the average person, can it be true that social competence depends on perceptual accuracy? Several studies have shown that effective encoders are also effective decoders; that is, they have a general communication ability (e.g., Levy, 1964). However, other research reaches a contrary conclusion. Cunningham (1977) found that high scorers on "neuroticism" were actually superior in accuracy of decoding emotions compared to skilled emotion senders. He explains this by saying that neurotics are motivated to be more vigilant because, presumably they are more vulnerable. Cunningham cites previous findings that sensitivity in person perception was associated with social passivity and withdrawal. Lanzetta and Kleck (1970) also found sending and receiving ability to be negatively related, and Young (1977) found no relationship between social competence and accurate empathic understanding. Young argues that successful interaction does not necessarily require empathic understanding but rather an understanding of the social rules, such as rules of synchrony and ritual, and rules of situations (see below). In the final analysis, these may be the type of cues that are most crucial in social interactions.

Cognition

Problem Solving. Feedback-loop theories usually postulate a cognitive stage between perception and performance. Thus, faced with significant incoming information about the immediate situation, the individual has to decide what, if anything, to do. According to the social skills model, individuals acquire a central store of "translation processes" which prescribe the action to be taken for any given perceptual information (Argyle, 1969). In other words, the individual should have a store of information to call on to guide action, and the memory codes of modeled events could perhaps serve this function. However, this assumes a fairly direct and simple relationship between perception and performance, whereas the relationship is usually more complex and problematic. Indeed, the main point of having a cognitive stage in the model is the necessity for having a mechanism for problem solving.

Bandura (1977) argues that "the functional value of thought rests on the close correspondence between the symbolic system and external events, so that the former can be substituted for the latter" (p. 172). He also argues that symbols representing events, cognitive operations, and relationships, being infinitely easier to manipulate than their physical counterparts, greatly increase the flexibility and power of cognitive problem solving. A number of investigators have explored the stages of problem solving (e.g., Goldfried & Goldfried, 1975; Spivack, Platt, & Shure, 1976). For èxample, Spivack *et al.* (1976) distinguish the following: means–end thinking—the ability to identify and plan the steps required to reach a goal; alternative thinking—the capacity to generate alternative solutions; consequential thinking—evaluating particular solutions as to their probable effectiveness; causal thinking—understanding the causes and effects of one's behavior. The amount of problem solving required varies with the type of interaction, unstructured discourse like casual conversations requiring a great deal, structured discourse very little.

A body of research shows that some behavior is under the guidance and control of such thought patterns in the form of verbal symbols or imagery, and that such patterns can be manipulated with therapeutic effect (e.g., Meichenbaum, 1977).

Individuals of course vary in the way that they process information and solve (or fail to solve) problems. Harvey, Hunt, and Schroder (1961) distinguish cognitively simple (or concrete-thinking) from cognitively complex individuals, in that the former show few differentiations, have an inability to be neutral about issues, do not introspect, are egocentric and nonanalytical. Similarly, Witkin and his associates distinguished the field-dependent, who are concrete and conforming, from field-independent, who are the reverse. Witkin and Goodenough (1977) show that field-dependent people are more socially oriented and more socially skilled, while the latter are more autonomous, have more skill in cognitive analysis and structuring. This makes it unlikely that there is any simple relationship between cognitive complexity and social skill.

Sequencing. The social skills model postulates that in the early stages of learning a skill a new response has to be selected and evaluated for every new piece of perceptual information or feedback, but that once learned these molecular units are strung together to form a unit or sequence which is automatically run off as a complete molar unit. Conscious monitoring, therefore, is decreased and occurs only at the transition points between habitual chains. Since consciousness is a limited-capacity device, social transactions take place on the basis of habit "simply because of the cognitive strain which would be involved were each element of behavior to be preceded by conscious deliberations" (Thorngate, 1976, p. 33).

The above cognitive organization of elements into larger units is well documented. Miller *et al.* (1960) describe plans or TOTE units as being composed of a hierarchy of subplans that are themselves TOTE units. Hayes–Roth (1977) expounds a "knowledge-assembly" theory in which an assembly of cogits (the smallest unit of information) is strengthened by frequent activation and association into a single, integrated unit, a process she terms unitization. She reports that the theory is consistent with a large body of experimental results.

Performance

If responses are selected and organized at a cognitive level there should be similarity between thought and behavior patterns. Cautela and Baron (1977) and others put forward the view that there is a continuity or "homogeneity" between overt (public) and covert (private) events, and that there is a mutual interaction between the two. One would therefore expect to find the same hierarchical organization at both cognitive and behavioral levels, with molar units composed of molecular elements. This is true of the structure of grammar in language, and Birdwhistell (1966) assumes a similar structure for "kinesics" in which microelements or kinemes combine to form kinemorph units. The principle applies to a great deal of social behavior, as shown later. It is also assumed that larger units are built up by learning and, once learned, are run off automatically without conscious monitoring of the component parts. The rest of this section will be devoted to a description of the social elements and higher order units and the rules which combine them.

A final point on component processes. It is assumed in the social skills model that all levels are goal directed—the aim is to keep on course and achieve the desired goal as efficiently as possible, and this emphasizes the cybernetic analogy of the model. The goal influences what is relevant and salient at the perceptual level, and at the cognitive and performance levels the goal dictates the hierarchical structure. It will be shown, for example, that in discourse the hierarchical structure grows out of the speaker's purpose for talking—where to begin, how to proceed, what to make prominent, and where and how to end.

ORGANIZATION OF FACE-TO-FACE INTERACTION

It was suggested above that social behavior is hierarchically organized. At the lower, molecular level there are single elements, and at the higher, molar level are units or sequences comprised of displays or strings of elements (Argyle, 1969). For example, in face-to-face interaction there is a constant two-way exchange of glances, gestures, vocalizations, and other social signals which are not generally monitored individually, but which form higher order chunks (like episodes of conversation) which are consciously monitored. The individual is presumed to have a store or repertoire of elements which he retrieves and assembles into sequences in order to communicate (encode) messages to another individual, who attempts to understand (decode) the intended (and sometimes unintended) meaning.

THE ELEMENTS

Elements can be classified in various ways: according to the channel used, e.g., visual (nonverbal) or auditory (vocal, verbal); according to the part of the body involved, e.g., gaze, vocal; according to their informative importance and function, e.g., indexical cues, synchronizing cues. The reader is referred to Argyle (1975), Knapp, (1978), and Morris (1978) for comprehensive reviews of nonverbal behavior, and Clark and Clark (1977, Chapters 3 and 6) for speech. This account will be confined to a brief overview.

Nonverbal

Face. The face has the greatest encoding capacity for nonverbal information. Ekman and Friesen (1975) point out the wide range of discriminable stimulus patterns—more than 10,000 facial actions; the potential speed of expressiveness—from a second or two to a split second, and its prominence and visibility. It is also the most closely monitored region, both for external (reaction from other) and internal (conscious awareness) feedback. Ekman (1978) describes four "facial sign vehicles"—static, such as bony structure; slow, such as aging; rapid, such as changes in muscle tonus; and artificial, such as cosmetics. He also lists 18 types of information conveyed. The first group concerns stable properties of the person, such as personal identity, kinship, race and gender, temperament, personality, beauty, sexual attractiveness, and disease. The next is emotion and mood, expressed in pure form or complex controlled displays, such as blends or fragments of full expressions (see below). Another is the group of rapid movements which accompany speech and are involved in conversation management, and could be termed facial gestures, such as emblems, adaptors, illustrators, and regulators, the last of these being involved in listener feedback.

Gaze. Brief encounters can be opened, maintained, and closed by the eyes alone (Argyle & Cook, 1976). This is partly because of their unique function as a simultaneous channel and signal; gaze synchronizes conversation by collecting (monitoring) and sending information (feedback) at the same time. Gaze is also important purely as a signal of emotions and attitudes. In interpersonal attitudes gaze may serve to express preference or power simply by length, direction, and pattern of looks (Exline, 1974), and the intensity of emotion is revealed by amount of looking, more looking intensifying some emotions like anger, less looking intensifying others, such as shame and embarrassment. Amount of looking is also associated with perceived personality, more looking being more positively perceived than less.

Gesture. Drawing on earlier work, Ekman and Friesen (1969) developed a taxonomy of gestures which has been widely accepted and used. *Emblems* have specific meaning much as words, e.g., head nods for agreement, clapping for approval. *Illustrators* are closely linked to speech, and include pointing, showing spatial relationships, showing tempo or rhythm, emphasizing, showing a bodily action, drawing "pictographs," and showing a direction of thought. Researchers have explored how illustrators operate in synchrony between interactors. *Affect* displays to some extent reveal type of emotion but mainly reveal intensity. *Regulators* play an important role in conversation management. *Adaptors* are of three types: self-, object-, and other-directed. Many adaptors "leak" emotions like anxiety and aggression.

Posture. Posture is a significant communicator of interpersonal attitudes, and Mehrabian (1972) has distinguished two main styles—"immediacy," which relates to the warm/cold dimension and has the effect of increasing or decreasing distance and visibility between people; and "relaxation," which relates to the dominant/submissive dimension and is characterized by tension or relaxation of the musculature. The intensity of emotion is also shown by the relative tension or relaxation. Finally, posture plays some part in speech in that posture changes mark off larger sections of interaction.

Spatial Behavior. Spatial behavior consists of proximity, orientation, territorial behavior, and movement in a physical setting. Proximity is simply the distance between people, and Hall (1966) suggested four zones: intimate (18 in), personal (18 in to 4 ft), social-consultative (4–12 ft), and public (12 ft and above). Orientation is the angle at which one person faces another, and is a cue to the intimacy or formality of the relationship. Territoriality refers to the significance of owned or claimed space and the use made of it to achieve certain effects, such as status.

Bodily Contact. Touching is the most intimate form of communication. It is severely restricted in adult life, especially in "noncontact" cultures such as the United Kingdom, and allowed under certain socially defined circumstances, which often affect and transform the meaning. Heslin (1974) suggested a taxonomy of touching, reminiscent of Hall's zones: functional-professional, as in tailor to customer, doctor to patient; social-polite, as in shaking hands; friendship-warmth, as in arm touching, hand to shoulder; love-intimacy, as in touching cheek or full embrace, and sexual arousal. Convention can formalize an otherwise intimate kind of touch, as the kiss in greeting, and the full embrace in some sports, and cross culturally where the conventions are different, such as in hand holding.

Physical Appearance. Manipulable aspects of appearance include styles of dress and hair, cosmetics and artifacts. They are manipulated to send information about the personality, status, and group membership of the wearer, as well as sexual availability, aggressiveness, and other interpersonal attitudes. Nonmanipulable aspects of appearance are body build, color, and height. Research shows some correlations between these appearances and personality and attitudes, e.g., ectomorphs (thin and bony) as quiet and tense; endomorphs (round and fat) as warm hearted, agreeable, and dependent; and mesomorphs (muscular and athletic) as adventurous and self-reliant (Wells & Siegel, 1961).

Nonverbal Aspects of Speech. The distinction is made between speech-related sounds, such as pauses, stress, and timing, which affect the meaning of utterances, and paralinguistic sounds which convey emotions and personality characteristics. Speech-related sounds include prosodic signals, synchronizing signals, and speech disturbances. For example, short pauses give emphasis, while longer ones signal grammatical junctures. Emotions are revealed by speed, loudness, pitch, speech disturbances, and voice quality. Other types of information revealed by paralinguistic signals include interpersonal attitudes, personality, social class, race and cultural group, age, gender, and personal identity. Mehrabian (1972) reports that tone of voice contributed much more than the contents of speech to impressions of interpersonal attitudes.

Verbal

Linguists divide speech into three parts—the speech *act* (doing something by uttering a sentence), the propositional content (conveying a proposition or belief), and the thematic content (fitting the topic into the ongoing discourse) (Clark & Clark, 1977). We are here concerned with speech acts since they are elements concerned with action. Speech acts have a certain force in having an effect on the listener, e.g., an assertion, a request, or a warning. Searle (1975a) distinguishes five categories of speech acts as follows:

1. *Representatives.* The speaker conveys his belief that something is true. Examples are asserting, swearing, suggesting, and hinting.

2. *Directives.* The speaker attempts to get the listener to do something, the two main ways being requests and questions. Other examples are ordering, commanding, begging, and pleading.

3. *Commissives.* The speaker is committing himself to some future course of action, the main way being a promise. Other examples are vows, pledges, contracts, and guarantees.

4. *Expressives.* The speaker is expressing his "psychological state"—how good or how bad he feels about an event. For example, he apologizes, thanks, congratulates, welcomes, or deplores.

5. *Declarations.* The speaker brings about a new state of affairs by the very uttering of the words, for example: "You're fired," "I resign," "I sentence you. . . . "

Each of the above categories requires something different of the listener. Of the five categories, the first two are the most common and best studied. There are also several other uses of language in social interaction not listed here because of space.

THE SEQUENCES

Elements rarely occur in isolation but are embedded in spatial displays (e.g., facial expression, gesture, posture, etc., in expressing anger) or temporal sequences (a greeting) which have a meaning of their own. The form of these combinations depends on social rules and norms, the immediate situation, the individual's plans, competence, etc. The social skills model suggests how some of these sequences are built by means of the feedback loop, although others may be innate (emotional expression in the face) or acquired "ready-assembled" by means of modeling.

Apart from innate influences there are two main sources that influence the pattern of social behavior sequences: (1) the environment (or situation) and (2) the person. The final outcome is a function of an interaction of the two—the person in the situation. This approach, the P (person) $\times$ S (situation) model, has been well explored by interactional psychologists (Endler & Magnusson, 1976) and will be adopted to organize the material below.

The Social Environment

The social environment contains a system of rules and conventions (or contingencies) which influence and shape (control) the behavior of "normal" individuals in characteristic ways. This structure ranges from culture-wide or even universal social rule systems to local, situation-specific conventions. Goffman (1961) graphically describes this rule system as the "membrane" surrounding all encounters and giving them their social reality. Individuals break these rules at their peril: "To be awkward, or unkempt, to talk or move wrongly, is to be a dangerous giant, a destroyer of worlds. As every psychotic and comic ought to know, any accurately improper move can poke through the thin sleeve of immediate reality" (Goffman, 1961, p. 72).

The propensity of ordinary people to conform to rules, norms, and explicit orders has been strikingly demonstrated by the classic experiments of Asch (1952), Milgram (1974), and many others. Snyder (1974) showed that more competent people are distinguished by the extent to which they monitor the situation, which includes, presumably, cues as to the norms of behavior. Below we briefly describe some of the rules and norms which regulate and prescribe social behavior in a wide variety of situations.

Rules of Conversation. Conversation usually follows an intricate pattern such as that described by Kendon (1973). Speaker and listener orient their bodies toward one another. They repeatedly scan each other; from time to time their eyes meet. Their bodily movements are coordinated, as is their speech. Even before they start to interact, they are preparing to do so by synchronous head, eye, and hand movements, with one picking up the rhythm of the other. They synchronize with equal precision the initiation and termination of the encounter. What are the rules that enable this to be achieved?

Conversation usually involves taking turns at speaking and listening. As Chapple (1970) puts it, "given two persons A and B, only four states are possible: (1) A acting, B inactive; (2) A inactive, B acting; (3) A and B both acting; and (4) A and B both inactive" (p. 631). Chapple describes the first two as synchronous, the latter two as asynchronous. Asynchronies of this kind—interruptions and silences—cause stress, for example, "the mother not responding to the child, the wife constantly dominating (interrupting) the husband" (Chapple, 1970, p. 631). Sacks (unpublished manuscript, cited in Coulthard, 1977) suggests there is a basic norm in American English conversation—"at least and not more than one party talks at a time," and there should not be undue silences except in familiar encounters, since, as Goffman (1955) says, "undue lulls come to be potential signs of having nothing in common or of being insufficiently self-possessed to create something to say, and hence must be avoided. Similarly interruptions and inattentiveness may convey disrespect and must be avoided . . . " (p. 227). Investigators show that conversationalists typically work to eliminate asynchronies, by yielding the floor or ending quickly, by speaking earlier, asking questions, or incorporating silences into their speaking turn (Coulthard, 1977). Many studies find that conversationalists manage a remarkably smooth synchronization of their speaking turns. "Normal" participants monitor the other's performance, as shown by patterns of gaze, the speaker looking periodically, such as at the ends of phrases, the listener looking in long glances, with short glances away (Argyle & Cook, 1976). Participants are attending to various cues emitted by the others, such as readiness to switch speaker roles.

One way people can deal with the synchronizing problem is for the speaker to nominate the person to speak next; this is the main system in formal situations like lecturing and interviewing. In more spontaneous conversation, the next speaker can identify points of possible completion (e.g., a completed idea) and has the technical capacity to select a precise spot to start his own talk "no later" than the exact appropriate moment (Jefferson, 1973). In cases of asynchrony there are rules for who speaks next, e.g., during interruptions the one who yields the floor takes the next turn.

There are a variety of rules for dealing with asynchronies. In interruptions one quickly yields the floor but then gets the next turn. A speaker (just ending) may transform the ensuing silence into a pause by continuing, or by a *post completor* which is either a question or a repeat (Schegloff & Sacks, 1973). Alternatively, the next speaker may begin with "uh" or an audible intake of breath while formulating his words. Duncan and Fiske (1977) showed that speakers have several cues with which to offer the turn: an intonation-marked clause (rising or falling pitch) on the last syllable or two; a stereotyped expression termed a "sociocentric sequence" such as "But, uh . . . " "or something," and "you know"; the completion of a grammatical clause involving a subject–predicate combination; a drawl on the final or last stressed syllable; a decrease in pitch or loudness, namely a "lowering of the voice" or "trailing off," on a sociocentric sequence; the termination of a hand gesture or relaxation of tensed hand position. Kendon (1967) found that the speaker signaled the end of his turn by a prolonged gaze. If this terminal gaze was not given there was a long pause before the other replied. (The terminal gaze signal has not been found in all studies, e.g., Beattie, 1978.)

Intending speakers were found to signal this intention by looking away just before or just as they begin to talk and/or by the initiation of a gesture and by speech preparatory movements. In some situations speech was preceded by a sharp drawing-in of breath. Intending speakers may speak more loudly (Meltzer, Moriss, & Hayes, 1971) or may "speed up" the speaker by rapid head nods and verbalizations of pseudo-agreement (Knapp, 1978).

During a turn a speaker seeks feedback without wishing to relinquish the floor and so gives a "within-turn signal" (Duncan & Fiske 1977) consisting of a shift in head direction toward the partner and completion of a grammatical clause. The listener gives "back-channel" responses at these points, such as brief verbalizations ("Mhmm"), completion of speaker's sentence, request for clarification, brief restatement (of preceding thought), head nods and shakes. However, the speaker is vulnerable at these points. Ferguson (1975) discovered that almost a third of interruptions occur following filled pauses within a phrase such as "um," "er," and "y'know," while Beattie (1977) found most floor switches occurred during unfilled pauses. Strategies for keeping the floor include using an *utterance incompletor* like "but," "and," or "however," which serves to make a complete utterance into an incomplete one, or to prefix an *incomplete marker* like "if," "since." Coulthard (1977) says that speakers reject interruptions also if they speak more loudly, more quickly, and in a higher pitch. Duncan and Fiske (1977) found that starting or continuing a gesture was almost totally successful in suppressing a turn attempt.

Sometimes a listener will refuse a turn offer by continuing with back-channel cues, remaining silent, or producing a *possible preclosing* such as "alright," "okay."

Brazil (1975) has shown that vocal pitch alone is a powerful signal for affecting the way the listener responds. An utterance ending in high key is perceived as "incomplete" and this constrains the listener to respond in some way, whereas an utterance ending in low key suggests completeness and finality, having the opposite effect on the listener. Mid-key sets up no particular expectations. It is well known that terminal changes in pitch and loudness, as well as kinesic behaviors, serve to mark syntactic junctures in the conversation stream (see below).

Sacks (unpublished manuscript, cited in Coulthard, 1977) observes that conversation consists of at least two turns. However, most conversation is more complex than this, and Sinclair, Forsyth, Coulthard, and Ashby (1972) describe a rank scale model for the classroom, and, they believe, for most discourse, which is like a grammatical model in that a unit of one rank, say a word, is made up of one or more units of the rank below—morphemes. The ranks, from smallest to largest, are act, move, exchange, transaction, and lesson. A transaction (equivalent to a topic), for example, is made up of a sequence of exchanges beginning with a boundary exchange or frame like "well," "right," "by the way," followed by a pause followed by a *focus* ". . . the other day I . . ." i.e., a change of topic, followed by a series of informing, directing, or eliciting exchanges. Exchanges in turn consist of moves of which the classic example is the question–answer sequence. Sacks and his colleagues (e.g., Sacks, Schegloff, & Jefferson, 1974) isolate a class of sequences of this type called *adjacency pairs* which consist of two moves by two speakers which are ordered and related. The first part (question) may select next speaker and always selects next action (answer) and thus sets up a *transition relevance.* A question automatically reserves the right to talk again after an answer is given, and this may be a further question—a *chaining* rule which allows for an indefinitely long sequence of the type in doctor–patient interviews.

Scheflen (1964) suggests that speakers signal kinesically the boundaries of such discourse units, which he names differently. For example, the point (equivalent to exchange) is marked by changes in head position, hand movement, and eyelid positions. The position (transaction) is marked by a gross postural shift of at least half the body, and the presentation (interaction) is marked by a complete change in location.

One of the most interesting units is the transaction, since this usually involves a familiar cluster of problems having to do with topic: change of topic, topic conflict, stories and topical coherence discussed by Sacks in unpublished lecture notes (1967–71, cited in Coulthard, 1977). For example, an intending storyteller must negotiate the suspension of the usual turn-taking system and does this with a *floor seeker* or *story preface.* Once told, others are likely to reciprocate, but their stories must be topically related and highly prespecified, and breaking this rule may lead to the member's being sanctioned in some way.

Kent, Davis, and Shapiro (1978) have extended some of the above work experimentally. They propose that conversation rests in part on the ability of actors to provide one another with continual instruction in how to proceed, and in part on consensual knowledge of the social world and its interactional rules.

Many of the basic rules of discourse are probably universal in structure, e.g., a turn system would be operating in all cultures, but the details will vary greatly. La France and Mayo (1976) found that black Americans gazed more while speaking while Caucasians gazed more while listening. Lomax (1975) reports many variations between American Indian, "West Euro-American," and black African, such as characteristic differences in speech length, timing, drawl and pitch, pauses, interruptions, and so on.

Social Routines. Rules of discourse vary in the amount of freedom or spontaneity they allow conversationalists. The rules so far discussed allow considerable freedom. However, some forms of behavior are so essential to face-to-face interaction that their form is very precisely prescribed—down to the very words and order of words. These standard routines have symbolic rather than literal meaning, form the basis of polite "etiquette," serve to initiate, terminate, confirm, restore, and change social interaction, are widely recognized and used, and most important, are obligatory in certain situations.

These routines are examples of speech acts with "illocutionary force," referred to in the section on verbal elements. Their form is restricted because they must satisfy certain "felicity conditions," i.e., rules which constitute their meaning and regulate their use (Searle, 1975b). What is the social function of these routines? Goffman (1972) puts it this way: "when individuals come into one another's immediate presence, territories of the self bring to the scene a vast filigree of trip wires which individuals are uniquely equipped to trip over. This ensures that circumstances will constantly produce potentially offensive configurations that were not foreseen or were foreseen but undesired" (p. 135). A similar and much earlier idea was put forward by Malinowski (Ogden & Richards, 1923), who coined the term "phatic communion," i.e., communion achieved through speech, to describe the function of these sequences. This idea has recently been developed by others, for example Laver (1975), to whose account we return later.

Goffman, (1972) distinguishes two types of social routines which he calls *supportive interchanges* and *remedial interchanges*. Greetings and partings are routines of the former kind with special importance, forming "ritual brackets surrounding periods of heightened access." The greeting takes many different forms in different cultures but the basic structure is believed to be universal. A detailed account of one Western greeting has been given by Kendon and Ferber (1973) in which there occur four phases as follows:

1. Distant salutation—two people sight each other, followed by face-to-face orientation and mutual glance, followed by eyebrow flash, smile, open-palm greeting gesture with head toss, etc.
2. Approach and preparation—gaze cut off, head dipped, self-grooming, approach begun.
3. Close salutation—halt, face-to-face position taken up, stereotyped utterances exchanged, usually with bodily contact—handshake, embrace, or kiss.
4. Attachment phase—less stereotyped conversation, establishing identity and status, inquiring about activities, purpose of visit, etc.

There are several types of greeting, of which Goffman describes three: the *passing greeting* used simply to maintain minimum social contact, the normal *full greeting* (as described by Kendon and Ferber, 1973), and a nongreeting or *"civil inattention"* used when passing strangers in public places. The parting is similarly sequenced, and an account is given by Knapp, Hart, Friedrich, and Shulman (1973). There are other supportive interchanges, which have the function of confirming and supporting relationships, either symbolically, with compliments, praise, congratulations, encouragement, and sympathy, or tangibly, with gifts, loans or actions.

Moves to initiate or change interactions are in danger of becoming ensnared in the "filigree of trip wires," and routines to avoid or repair such damage—i.e., "remedial" routines—are clearly needed. One of these is the *request,* which consists of a series of moves and responses: a preparatory statement, request, and justification is followed either by a compliant response including agreement and more-or-less willingness, or a refusal containing an excuse or apology; a second move of appreciation (if successful) is followed by a response of minimization.

An *account* denies, explains, or excuses an act in at least four ways, ranging from full to partial diminution of blame. *Apologies* accept partial blame and follow a certain sequence: the apology itself; an acknowledgment of rule breaking; an account, disclosure of embarrassment, guilt, etc.; promise of good behavior. The response may be relief or a demand for a further account. Apologies and accounts are routines for the offender to make amends and to give reassurance about future rule-following behavior. *Assertive* routines are used by the offended person to assert his rights—to point out a transgression and threaten sanction if the offender fails to make amends. One type of assertive routine is as follows: an implicit request to "prime" the offender by drawing attention to the offense; an explicit request pointing out the offense and insistence on recompense or apology; an explicit demand to restore the situation or apologize with a threat of sanction if this is not done. Probably most assertive routines go no further than the first, priming stage, which can be done in many subtle ways, such as withholding the expected reciprocal response to a supportive routine (e.g., failing to return a greeting thereby informing that something "is wrong").

The notion of rule is used above in a prescriptive and regulative sense (Collett, 1977) which implies that infringements may occur and that sanctions and other forms of social behavior may be brought to bear to bring the rule breaker into line. Goffman (1972) says sanctions themselves are norms about norms—techniques for ensuring conformance that are themselves approved. Scheflen (1972) distinguishes between communication and metacommunication, the latter referring to those elements of behavior—be they verbal, paralinguistic, kinesic, and so on—which serve to regulate and monitor the communication *per se.* One form of metacommunication is "procedural and judgmental." It consists of monitors or warnings of deviation, instructions to hold enactment to the usual format. Thus, frowns or other behaviors of censure may be directed at two people who are sitting too close together; sometimes such a censure emerges from the background and becomes a topic in its own right, an argument about the propriety of a way of behaving. It is not surprising that there exist a number of special routines, some of which Goffman has drawn attention to ("remedial interchanges"), which form the basis of many assertive training programs.

Situation Rules and Other Components. The rules of discourse are as basic to conversation as grammar is to language. Many of these rules are believed to be universal or at least pancultural. However, many rules apply to smaller groups—a culture, subculture, or still smaller group (family, club), or even a dyad. Situations are often uniquely bounded by their own rule systems. Barker and Wright (1955) recorded no less than 800 "behavior settings" in a small midwestern town (Kansas). Barker and his co-workers have developed the concept of the behavior setting as a stable, extraindividual unit with great coercive power over the behavior that occurs within, shown by the fact that behavior differs markedly from setting to setting and also because people display very similar behavior in the same setting (Gump, 1971). Gump distinguishes three variables, using a school recess as an example: nonbehavioral factors—space, ground, bounded by fence; objects—like swings, trees; standing behavior patterns—children play with equipment in a pattern which is stable over generations. These behavior patterns are also affected by physical forces, physiological process, social forces, and imitation tendencies of individuals. It is also pointed out that settings select persons and vice versa.

Following in a similar tradition, Argyle (1977) analyzes situations into essential components as follows:

1. *Special moves.* Certain moves are relevant and important (as are standing behavior patterns), e.g., passing food, pouring wine at a dinner party, asking and answering questions in an interview.
2. *Goals.* Each situation has a purpose, e.g., to strike a bargain, to teach (and learn), to make something, to be romantic. One situation may have multiple goals, some of which are formal, others informal.
3. *Rules.* Each situation is defined by a set of rules, which govern the type and sequence of moves, e.g., a dinner party has rules about time, order of courses, dress, topics of conversation, seating, etc.
4. *Roles.* Some roles are essential to the situation, e.g., dinner parties must have hosts, guests, and sometimes servants, as well as male/female roles, senior/less senior.
5. *Pieces.* Situations involve special environmental settings or props, e.g., a seminar requires blackboard, slides, projector, lecture notes, etc.
6. *Concepts.* There is a special terminology for many situations, such as "resistance" and "transference" in psychodynamic psychotherapy.

The structure of situations makes them predictable and people know how to behave so long as they know the rules and other components. The more formal a situation is, the more restricted is the behavior within it, as occurs in the social routines discussed earlier.

Social Ecology. One element of the situation mentioned earlier is the physical environment, and one topic of research has been the effects of environment on social behavior.

One group of studies examined the impact of temperature and crowding. Baron and Lawton (1972) found that high ambient temperatures (90°F) facilitate aggression, especially if aggressive models are present. Griffitt and Veitch (1971) similarly found that hot and crowded conditions produced negative interpersonal responses and more negative feelings about a stranger. Freedman (1975) found that crowding increases whatever emotion is being experienced, be it fear and antagonism or excitement and friendliness. Milgram (1970) argues that cities tend to produce norms of noninvolvement, resulting in withdrawal of simple courtesies, less helpfulness, and diffusion of responsibility.

Another group of studies has examined the effects of architecture and the arrangement of space. Osmond (1956) distinguished between *sociofugal* space which discourages human interaction and tends to keep people apart (e.g., waiting rooms of bus stations, seating arrangements in churches, and many mental hospital day rooms), and *sociopetal* space, which has the opposite influence. Such a distinction helps explain the results of the classic study by Festinger, Schacter, and Back (1950) in which friendships were shown to be affected by distance between houses and the direction in which a house faced. Residents whose houses faced the street had less than half as many friends as those whose houses faced the court area. People who lived near entrances and exits of stairways and near mailboxes had more friends than those who lived away from these spaces. Clarke (1952) found that people tended to marry those who lived near, went to the same school, or worked in the same place.

The Person

Some aspects of behavior are mainly attributable to personality, sex, age, and other "person" variables. Each of these person variables is associated with characteristic behavior patterns, for example "working class" with a style of dress, language, accent, and behavioral conventions. It is presumed that person-dependent behavior is comparatively stable across situations. No attempt will be made to give an exhaustive account of this issue, but rather it is intended to highlight the principles and most important variables.

Personality. Since Mischel's influential review on the empirical weakness of personality theory (1968), more sophisticated analyses have clarified different types of cross-situational consistency for which there is good evidence (Magnusson & Endler, 1977).

Perhaps the most extensively researched dimensions of personality are those of extraversion and adjustment. A number of studies have distinguished extraverts from others in behavioral terms. Extraverts tend to speak first and for a greater proportion of the time (e.g., Rutter, Morley, & Graham, 1972), and with a louder voice (Scherer, 1974). Extraverts have been found to look at another more often when speaking (Kendon & Cook, 1969), and to look more frequently (Rutter *et al.*, 1972). However, Campbell and Rushton (1978) found that they looked *less,* which is readily interpretable in terms of their greater amount of talking. There is reasonable evidence that sociability and extraversion correlate with self-disclosure (Cozby, 1973). Absence of hesitation pauses and a generally faster speech rate gave impressions of extraversion, competence, and likability (Brown, Strong, & Rencher, 1973), and people who use many gesture illustrators are judged to be outgoing and sociable (Ekman & Friesen, 1975).

Turning to adjustment, Campbell and Rushton (1978) found that students perceived as poorly adjusted touched themselves more while speaking and listening, used few communicative gestures, and paused more often while speaking. Self-reported neurotics also averted their gaze more. Cozby (1973) found the literature inconclusive as regards self-disclosure and adjustment, but suggests that poorly adjusted people will disclose too much or too little. Derlega and Chaikin (1976) found high-disclosing men and low-disclosing women to be poorly adjusted (well-adjusted men and women show the opposite pattern).

Interpersonal Attitudes. One of the main functions of nonverbal as opposed to verbal communication is to communicate interpersonal attitudes. Argyle and his colleagues (Argyle, Alkema, & Gilmour, 1972; Argyle, Salter, Nicholson, Williams, & Burgess, 1970) found that nonverbal signals had a more powerful effect than verbal ones in the communication of attitudes on two dimensions: friendly-unfriendly and dominant-submissive. These two dimensions were chosen because they represent the two main interpersonal styles which have emerged from a number of investigations. One of the earliest was by William James (1932), who found the following main postures: approach—an attentive posture with forward lean; withdrawal—a negative posture of drawing back or turning away; expansion—a proud posture with expanded chest, erect or backward-leaning trunk, erect head, and raised shoulders; contraction—a depressed, downcast posture of forward-leaning trunk, bowed head, dropping shoulders, and a sunken chest. Studies by

Mehrabian and others have extended this work. Mehrabian (1972) describes a group of "immediacy" cues associated with an affiliative style and which increase physical proximity and mutual sensory stimulation—bodily contact, closer proximity, direct orientation, forward lean, mutual gaze, and so on. Clore, Wiggins, and Itkin (1975) and Scheflen (1965) have explored this dimension in courtship behavior. Argyle and Dean (1965) offered an equilibrium model of intimacy which argued that increased intimacy in one element, say gaze, would be balanced by decreased intimacy in another, say proximity, so that interacting pairs would maintain a comfortable equilibrium.

Turning to the dominant-submissive dimension, Mehrabian (1972) and Eisler, Miller, and Hersen (1973) have identified a set of intercorrelated cues for a dominant style including asymmetrical limb positions, sideways lean and/or reclining position (seated), relaxed hands and neck, lengthy speech, loud volume, greater affect appropriate to the situation, and less smiling. However, others have found, contrary to expectations, that vocal loudness does not correlate with dominance (Aronovitch, 1976), although pitch level does (Scherer, 1977) within a narrow frequency range. Rogers and Jones (1975) found that dominant partners held the floor for about twice as long as their partners, were about half again as likely to attempt interruptions, and held a slight edge in interruption successes.

Emotions. The face and voice are the most important communicators of the content of emotion, while gaze, gesture, and posture communicate the intensity.

From their cross-cultural work on expression of emotion in the face, Ekman (1971) put forward their "neurocultural theory." They state that there are seven basic expressions or "primary affect displays": surprise, fear, anger, disgust, sadness, happiness, and interest (or neutral). These expressions are said to be innate and universal in man insofar as they were recognized reliably in seven widely differing cultures. The configurations of these expressions are striking and compelling. To give just one example, surprise is universally expressed by raised and curved brow, horizontal wrinkles across the forehead, eyelids fully opened and sclera showing above the iris and often below, jaw dropped open but no tension of the mouth. These primary expressions are not often seen in their pure form, being modified by "cultural display rules" so that expressions are intensified, deintensified, neutralized, or masked. A common occurrence is the "blend" of two or more expressions. Individuals develop facial styles, for instance "the withholder," where expression of feelings is totally inhibited. The control of emotion expression leads to "leakage" in microfacial and micromomentary movements or imperfectly performed simulations, or characteristic movements in the hands or feet which are less well monitored.

Turning to the voice, Beier and Zautra (1972, cited by Knapp, 1978) sought to find universality of meaning for emotions expressed vocally. They found accuracy across three cultures for the recognition of the following: angry, sad, flirtatious, fearful, and indifferent. Others, however, have found the relationship between voice characteristics (e.g., speed, loudness, pitch, quality) and emotional state much more complex than for facial expression. The same vocal characteristic may be a feature of several emotions, e.g., moderate pitch variations in anger, boredom, disgust, and fear (Scherer, 1974).

Demographic Variables. Age, sex, race, social class, and other demographic and sociological variables have been studied for their effect on behavior. We take just one of these to illustrate—gender differences.

Numerous studies highlight gender differences in behavior. Some show women to be more affiliative: they are visually more interactive (Exline, 1974), prefer closer proximity (Liebman, 1970), smile more (Mackey, 1976), self-disclose more (Cozby, 1973), and show more empathy (Hoffman, 1977). Women also appear to be better at nonverbal encoding (Buck, Miller, & Caul, 1974). Buck *et al.* (1974) interpret this as showing that men are under socialization pressures to inhibit expression and that women are under a reverse pressure—to exhibit it. This account would also explain why women are more intimate self-disclosers.

While women are more expressive and affiliative, men are more assertive. Davis (1978) showed that men took the initiative in conversation, selected topics, and dictated the pace at which intimacy increased. When women do assert themselves, the pattern is quite different from that used by men. Summarizing status and power gestures between men and women, Henley (1977) reports that men adopt a relaxed posture, closer position, use touch, tend either to stare or ignore, show "informal demeanor," conceal emotions, and do not smile. Women have a tense posture, distant position, avoid touching, avert eyes but watchful, adopt a "circumspect demeanor," show emotion, and smile.

The Person in the Situation

All aspects of social behavior are inextricably linked, so that individuals obey conversation rules at the same time as self-presenting, expressing attitudes, and so on. The outcome is a function not just of situation or person variables, but an interaction of the two. Endler and Magnusson (1976) argue that the "empirical results support an interactional view of behavior, in which actual behavior is determined by a continuous and multidirectional interaction between person and situation variables" (p. 968). This is certainly consistent with the feedback-loop models.

The person brings certain behavioral tendencies to the situation while the situation prescribes and allows certain patterns of behavior. The process of influence between the two is reciprocal—the person alters the situation as well as the situation altering the behavior of the person. The person is commonly unaware of his influence on the situation and this can lead to a self-fulfilling prophecy in which *B*'s defensive reaction to *A*'s hostility is misattributed by *A* to *B*'s personality.

In addition, persons and situations vary in their "consistency" and this influences outcome. For example, situations range from unstructured dialogues, like casual conversations in which person variables will exert greatest influence, to highly structured situations, like ceremonies where situation variables exert greatest effect. Similarly, persons range from high "self-monitors" who vary across situations to low self-monitors, such as neurotics and some mental patients who are more consistent across situations (Moos, 1968; Snyder & Monson, 1975). Furthermore, some elements of behavior are more dependent on persons, others on situations. Person-dependent behaviors include smiling, gaze, scratching, and addictive behavior like smoking, while situation-dependent behaviors include amount of talk and attention feedback and amount of intimate information revealed (Argyle, 1976).

An individual responds not only to properties of the situation, as described earlier, but also to the personalities, social class, and other behavioral elements of persons in the situation. Given their individual differences, two strangers interacting have a "coordination problem" which they solve by means of conventions. They have to negotiate an episode of interaction, agree when to start and stop it, and on what type of interaction it will be (i.e., the purpose, task, or goal), agree on the level of intimacy and comparative status, synchronize their responses, and so on. For example, the "adjacency pairs" rule will operate to signal invitations and willingness to cooperate, coordinate turns, and instruct each other how to proceed. Person *A*'s question invites *B* to join the conversation, turns over the conversation to him, and instructs him to respond within a circumscribed range of topics (Clark & Clark, 1977, pp. 228–229; Kent *et al.*, 1978). It will be shown later how this works in practice, by reciprocation, accommodation, and convergence of behaviors.

People do not lose their individuality by accommodating and synchronizing to the other. The system of sociolinguistic rules is permissive rather than coercive in that it prescribes limits but provides options, e.g., to take a speaking turn or not (Duncan & Fiske, 1977). Duncan argues that the individual uses the options available within the appropriate conventions to pursue his goals, which may be to dominate the other, win an argument, present himself in a particular way. Rogers and Jones (1975) show where in a cooperative task between equals the dominant partner will use conventions of turn taking to talk more. Scherer (1977) suggests that dominant partners will habitually use continuation or turn-claiming signals such as increased vocal amplitude, lack of hesitations, continuation gestures, and so on. Clearly, individuals develop specialized skills in this way which become a recognized feature of their style as stable habit patterns.

While they may thus express their individuality, people nonetheless repress behavior which they believe does not conform to the social rules, for example the expression of emotions which are not socially sanctioned, like anger and fear. As shown earlier, people control the expression of such emotion in the face by the use of cultural display rules. The individual in the situation is constantly engaged in "self-presentation" (Goffman, 1956), that is, the management of impressions by conscious or habitual manipulation of verbal and nonverbal information—the expression of some aspects, the concealing of others. Ekman and Friesen (1969) say that deception is attempted either by inhibiting (e.g., cutting off entirely), or by simulating (e.g., simulating confidence to conceal nervousness). They also suggest that individuals will spend most effort inhibiting or dissimulating with the face (as reflected in such expressions as "maintaining face") which has the best sending capacity and is the most visible region, medium effort controlling the hands, and least monitoring the legs and feet. Within this framework they suggest a number of clues to leakage and deception, including imperfectly performed facial simulations and microexpressions which are quickly controlled, hand "adaptors" directed at self, other, or an object, and leg/foot movements such as "aggressive foot kicks, flirtatious leg displays, autoerotic or soothing leg squeezing, abortive restless flight movements" as well as tension, frequent shifts of leg posture, and repetitious movements.

In subsequent research, Ekman, Friesen, and Scherer (1976) substantiated some of these ideas and also found that student nurses, asked to conceal negative feelings, used fewer illustrative hand movements and higher voice pitch. Knapp, Hart, and Dennis (1974) found concealers looked at others less, and Mehrabian (1972) found less nodding, more speech errors, slower speaking rates, and less direct body orientations.

THE STAGES OF FRIENDSHIP

Social interaction progresses in stages. Individuals meet and have successive encounters in which their behavior mutually undergoes change. This area of social psychology, known variously as friendship formation or interpersonal attraction, is of central importance to social skills training.

Among the many theories in this area, perhaps the most relevant from the present standpoint are the stage and filter theories (Duck, 1977). This view, as presented by Duck, is basically that individuals proceed through various stages in a relationship from formality to intimacy, at each stage exchanging quite different information, making different inferences, and also applying criteria to filter potential partners before allowing deeper (and riskier) levels of intimacy to develop. This approach also emphasizes the importance of conventional behavioral steps which people must follow if they are to succeed in forming friends. The process is graphically described by Altman (1974) as *social penetration,* and he uses the "onionskin" model which encompasses the important notions of breadth, depth, and layers. Duck (1977) describes the process as follows:

> It is assumed that individuals take certain personal and cultural beliefs into any interaction; that they respond first to the outward appearance of each other; then to behavior, gesture and the like; then to the content of conversation like attitudes; and then to personality structure or content. At each point, it is assumed, individuals use these bits of evidence in the service of model building, i.e., they make initially shaky but progressively more stable and constellated models of each other's personality as their acquaintance proceeds. (p. 73)

One of the important principles operating throughout the friendship process is similarity. Numerous studies arguably show that all aspects of similarity—in appearance, attitude, background, and to some extent personality—are associated with friendship development. One reason postulated for people's seeking similar others is their "need for effectance" (Byrne & Clore, 1967), i.e., the need that is satisfied by evaluating and finding consensual validation for one's view of the world and, by further implication, support for one's personality.

The Background of Friendship

There are a number of background phenomena which facilitate or inhibit friendship. First, propinquity, or mere exposure, enhances liking, even in negative contexts (Saegert, Swap, & Zajonc, 1973). McGovern, Arkowitz, and Gilmore (1975) found that practice dating alone was effective, and *as* effective as practice dating combined with social skills training. We have also seen the effect of building design and arrangement of living space. Second, friendship is affected by such individual differences as personality, cognitive and perceptual style, cultural background, and so on. A third variable is reputation, which precedes acquaintance and influences perception (Kelley, 1950).

The Stages of Friendship

Against these and other background features one can analyze how individuals meeting face to face may either remain strangers or progress by stages to acquaintanceship or friendship. We shall omit the parallel stages of courtship and love for space reasons. Several authors have offered rather different stages or levels (Altman, 1974; Levinger, 1974; Murstein, 1977). For simplicity, the stages will be described here as 1, 2, and 3, although, given the state of knowledge, these only represent approximations.

Stage 1. This stage corresponds to Murstein's stimulus stage (in his Stimulus–Value–Role Theory; Murstein, 1977) and Levinger's unilateral (one-sided) level (Levinger, 1974) in which others are "seen-from-a-distance and judged on external characteristics. Appearance cues, although notoriously unreliable, provide the only evidence (apart from reputation) of the other's potential, and therefore play a crucial role in deciding the fate of the relationship. The other may be filtered out if he has low-stimulus attractiveness, and his real potential then becomes irrelevant. One such cue is physical attractiveness, which is believed to derive at least some of its power from the belief that external quality indicates internal quality or quality of personality, hence the belief that attractive-looking people are warmer and more responsive, sensitive, kind, interesting, strong, poised, modest, sociable, and outgoing, get more prestigious jobs, make more competent spouses, and have happier marriages (Dion *et al.*, 1972; other studies show the "what is beautiful is good" stereotype to be less than universal). Secord (1958) found that certain faces that departed from the all-American ideal were disliked and distrusted. Gibbins (1969) found that people would make quite specific predictions about girls wearing certain types of clothes, such as about their sexual morals. Other studies have shown relations with perceived intelligence (high foreheads and spectacles), leadership and competence (height, body build), and social status (clothes, accent).

Stage 2. This stage refers to initial contact, and has parallels with Murstein's value-comparison stage and Levinger's surface-contact level. As individuals interact, more information becomes available through behavior, such as vocal characteristics, gaze patterns, posture, gestures, and other elements. It is also at this point that culturally defined sequences of progression toward friendship come into operation, and skipping steps and other rule deviations may lead to filtering out. First encounters are formal and ritualized, beginning with greetings and introductions, followed by strictly reciprocal routine exchanges about peripheral matters like the weather (depending on local convention), with attitude expressions of "liking" conventionalized in a modern etiquette of social smiles and touch (handshakes).

A key feature of this reliance upon such conventions and social routines is: (a) to provide structure for strangers who have no shared history or knowledge, which thus enables them to interact and undertake a preliminary exploration; (b) to provide a system of etiquette or "phatic communion" to "lubricate . . . and ease the potentially awkward tension of the early moments of the encounter" (Laver, 1975, p. 218).

Opening phrases in an encounter have these special purposes. For example, the English phrase "nice day" is not intended to convey information but to establish social communion and to initiate an exchange of moves which are as well known to the participants as an actor's lines. Laver (1975) says their function is to defuse the potential hostility of silence in situations where speech is conventionally anticipated, and quotes Malinowski's graphic comment: "another man's silence is not a reassuring factor, but on the contrary, something alarming and dangerous. The breaking of silence, the communion of words, is the first act to establish links of fellowship. . . . " (Ogden & Richards, 1923).

Another feature of the opening phase (between strangers) is an exchange of "indexical" information, including self-presentation, which enables participants to explore their way to a working consensus. Berger, Gardner, Parks, Schulman, & Miller (1976) report that when strangers are asked to get acquainted, the first minute or two is dominated by question asking (about 4/minute) about background and demographic information, e.g., "What do you do?" This tails off to less than one question a minute by the 13th minute, and the type of question also changes—a person is asked to *explain* his feelings about topics such as public affairs, religion, and nonintimate family and personal information.

In the early stages a norm of reciprocity seems to operate which in this context would, for example, imply being prepared to answer a similar question to that which one is asking. Cherulnik, Neely, Flanagan and Zachan (1978) found that skilled subjects shared speaking time more equally than unskilled subjects. Davis (1978) found that cross-sex dyads matched the rate at which they increased the intimacy of their disclosures and also matched their total intimacy, despite the apparent reluctance of the female subjects to disclose at the levels chosen by their partners. This study also shows that in cross-sex encounters it is the male who is the principal architect of the acquaintance process.

Berger *et al.* (1976) studied the effects of rule breaking in initial encounters. They found that people who violate norms of information sequencing (e.g., topics too intimate too early), reciprocity, and appropriate compliment (paying too many compliments) were judged as less attractive and less "mentally healthy" than more conforming subjects. Jones and Gordon (1972) found it was not only intimacy of disclosure that was rule governed. Their results suggest that it is "unattractive" to disclose good fortune early in an encounter but it was equally bad to disclose being a victim of bad fortune. Berger *et al.* (1976) say that norm violations affect the perception of communication attractiveness and impede the pursuit of interpersonal knowledge. A norm violater is seen as an unpredictable person. We have seen that interactors implicitly define what they consider appropriate conduct. When someone violates a conversational norm, the other feels that his definition of the situation has been rejected.

Stage 3. As relationships develop from acquaintanceship to friendship a number of changes in behavior occur, of which two will be emphasized here: the first is increasing affiliation or intimacy; the second is mutual accommodation.

Affiliation: A number of investigators emphasize increasing accessibility combined with less formality and use of convention. Some authors have argued that there would be an increase in intimacy and a decrease in reciprocity as the social process continued. Morton (1978) found both these mechanisms to be operating, with married couples choosing more intimate discussion topics but being less reciprocal, apparently not needing the structure and guidelines of stringent reciprocity. Further evidence comes from the interaction style of couples, which is characterized by more interruptions, rapid-fire exchanges, and simultaneous talking.

Levinger (1974) emphasizes the importance of self-disclosure in enabling the development of shared awareness and knowledge, which is a key feature of stable friendships. Sociologists Scott and Lyman (1968) argue that the linguistic style of interaction becomes more casual because of this shared background. They suggest five linguistic "styles" progressing from: (1) frozen style, as between strangers who are separated by an irremovable barrier; (2) formal style as between people in rigidly defined statuses, e.g., in a bureaucratic organization; (3) consultative style, as between people who have no shared background knowledge; (4) casual style, exemplified by peers, in-group members, and insiders who use "ellipsis," i.e., omissions and slang, and where background information is taken for granted; and finally to (5) the intimate style manifested by spouses or very close friends where single sounds or words or jargon are used to communicate whole ideas. Gumperz (1972) also asserts "communication through style shifting, special intonation, special in-group terminologies and topical selection basically relies on metaphor, and is heavily dependent on shared background knowledge" (p. 220). However, even total strangers have shared background knowledge if they are from the same culture, and this greatly facilitates interaction (Kent *et al.*, 1978).

Affiliation processes are by no means solely mediated by language.

The power of nonverbal communication in the expression of interpersonal attitudes has already been described and it is at the present stage in friendship that such cues would be operating. These cues would include proximity, touch, orientation, forward lean, and other "immediacy" cues (Mehrabian, 1972), patterns of looking, and facial expression, use of space such as seating positions, and so on.

Accommodation: Another feature of this stage is the accommodation or synchronizing of behavior over time. A number of workers have found increased similarity in speech length and synchrony in other speech and nonverbal behavior patterns between patients and psychotherapists, spouses, and other groups, e.g., astronauts and ground communicators (Kendon, 1967; Lennard & Bernstein, 1960; Matarazzo & Wiens, 1967). Condon and Ogston (1967), in presenting their work on interactional synchrony (e.g., movement mirroring by the listener), suggest that synchrony may assist in the identification of the quality of a relationship—rapport and degree of intimate interpersonal knowledge. Charny (1966) observed an increasing relationship between postural congruence and rapport in psychotherapy. Convergence in other types of behavior has been reported, for example, of vocal loudness (Natale, 1975).

Little is known about exactly how people pass from one stage to another in the acquaintanceship process. There must be key sequences which function as transition points. These transition sequences are probably highly structured speech acts in the form, for example, of adjacency pairs which would be used to negotiate such transitions. We have seen earlier how people use such linguistic conventions to achieve a goal, to control or woo the other, for example. Combined with the continuous mutual reinforcement and shaping of desired behavior, one can see how individuals come to develop complementary roles which, as Murstein (1977) states, is an essential feature of the final stage.

Finally, we should note that many of the features of friendship formation described here are taken from Western cultures and are by no means universal. Barnlund (1975), for example, showed that Japanese prefer more regulated communication than Americans and differ sharply in their levels of self-disclosure and physical contact, being far more reserved, even with close friends.

CONCLUSION

Social psychology has grown dramatically in recent years, and the output of research in some areas, as reflected in books, journal articles, and college courses, has been little short of explosive. Of all therapeutic procedures, social skills training stands to gain the most from this growing output. Its potential lies in guiding us to a more thorough analysis of social deficiency and a better formulation of training strategies.

In the perceptual and cognitive areas, social psychology directs attention to problems of perceptual and attributional error and problem-solving deficiency. In the behavioral area it provides some of the knowledge required to know how, in practical terms, we should be assessing and training patients. At a microlevel it unlocks mechanisms of interaction which are below awareness, such as the rules of conversation. If patients have conversation deficiencies, we may know it is deficient but not how, nor how we might modify the defect. The answer may lie in faults in the pitch contour, lexical choices, gaze pattern, and so on, which negatively influence the interaction. We need a fairly sophisticated knowledge of such microprocesses to identify such faulty elements in a sequence. At a higher level, we need knowledge of social norms before being in a position to train more effective skills. To give one example, the form of a greeting will differ greatly according to local conventions, social class and sex of interactors, and a therapist (e.g., middle class, female, northerner) may go badly wrong by modeling a greeting (e.g., for a working-class male southerner). To give another example, knowledge of first-, second-, and third-stage behaviors in friendship formation are clearly important for training social isolates. Knowledge of such stages gives precision to the content of training which has often been lacking.

It is nonetheless wise to temper the enthusiasm of theorizing with the caution of empiricism. We started with the suggestion that social psychology was a comparatively neglected source, as Kopel and Arkowitz (1975) argued. However, we should point out that, despite great enthusiasm, this same research group disconfirmed their own attribution hypothesis in two carefully conducted studies (Miller & Arkowitz, 1977). We should perhaps end by reiterating and extending their comment that, despite great promise and potential, we should be cautious about "premature acceptance of social psychological formulations of psychopathology and behavior change" (p. 667).

SUMMARY

The chapter reviews some aspects of social psychology which may help develop the technique of social skills training. It begins with a description of theoretical models and their component processes, including the social skills model and the processes of perception, cognition, and performance. It continues with a description of the elements of verbal and nonverbal behavior which are known to mediate communication, proceeding to analyze some of the ways these are organized to form higher order units, such as situation-related sequences (e.g., rules of conversation) and person-related sequences (e.g., expression of interpersonal attitudes). Some attention is given to the way individuals generate appropriate behavior according to cultural, situational, and interpersonal rules, and the way personal tendencies interact with these situational constraints. Finally, the stages of longer term interactions, particularly friendship formation, are examined.

REFERENCES

Altman, I. The communication of interpersonal attitudes: An ecological approach. In E. L. Hudson (Ed.), *Foundations of interpersonal attraction.* New York: Academic, 1974.

Argyle, M. *Social interaction.* London: Methuen, 1969.

Argyle, M. *Bodily communication.* London: Methuen, 1975.

Argyle, M. Personality and social behavior. In R. Harré (Ed.), *Personality.* Oxford: Blackwell, 1976.

Argyle, M. Predictive and generative rules models of P × S interaction. In D. Magnusson & N. S. Endler (Eds.), *Personality at the crossroads: Current issues in interactional psychology.* Hillside, N.J.: Erlbaum, 1977.

Argyle, M., & Cook, M. *Gaze and mutual gaze.* Cambridge: Cambridge University Press, 1976.

Argyle, M., & Dean, J. Eye contact, distance and affiliation. *Sociometry,* 1965, *28,* 289–304.

Argyle, M., & Kendon, A. The experimental analysis of social performance. *Advances in Experimental Social Psychology,* 1967, *3,* 55–98.

Argyle, M., Salter, V., Nicholson, H., Williams, M., & Burgess, P. The communication of inferior and superior attitudes by verbal and nonverbal means. *British Journal of Social and Clinical Psychology,* 1970, *9,* 22–231.

Argyle, M., Alkema, F., & Gilmour, R. The communication of friendly and hostile attitudes by verbal and nonverbal signals. *European Journal of Social Psychology,* 1972, *1,* 385–402.

Aronovitch, C. D. The voice of personality: Stereotyped judgements and their relation to voice quality and sex of speaker. *Journal of Social Psychology,* 1976, *99,* 207–220.

Asch, S. *Social psychology.* Englewood Cliffs, N.J.: Prentice-Hall, 1952.

Bandura, A. *Social learning theory.* Englewood Cliffs, N.J.: Prenctice-Hall, 1977.

Barker, R., & Wright, H. T. *Midwest and its children.* New York: Harper & Row, 1955.

Barnlund, D. C. Communicative styles in two cultures: Japan and the United States. In A. Kendon, R. M. Harris, & M. R. Key (Eds.), *Organization of behavior in face to face interaction.* The Hague: Mouton, 1975.

Baron, R., & Lawton, S. Environmental influences on aggression: The facilitation of modeling effects by high ambient temperatures. *Psychonomic Science,* 1972, *26,* 80–82.

Beattie, G. W. The dynamics of interruption and the filled pause. *British Journal of Social and Clinical Psychology,* 1977, *16,* 283–284.

Beattie, G. W. Floor apportionment and gaze in conversational dyads. *British Journal of Social and Clinical Psychology,* 1978, *17,* 7–15.

Berger, C. R., Gardner, R. R., Parks, M. R., Schulman, L., & Miller, G. R. Interpersonal epistomology and interpersonal communication. In G. R. Miller (Ed.), *Explorations in interpersonal communication.* Beverly Hills: Sage, 1976.

Birdwhistell, R. L. Some relations between American kinesics and spoken American English. In A. G. Smith (Ed.), *Communication and culture.* New York: Holt, Rinehart and Winston, 1966.

Brazil, D. C. Discourse intonation. *Discourse analysis monographs 1.* Birmingham: English Language Research, 1975.

Brown, B. L., Strong, W. J., & Rencher, A. C. Perceptions of personality from speech: Effects of manipulations of acoustical parameters. *Journal of the Acoustical Society of America,* 1973, *54,* 29–35.

Buck, R., Miller, R. E., & Caul, W. F. Sex, personality, and physiological variables in the communication of effect via facial expression. *Journal of Personality and Social Psychology,* 1974, *30,* 587–596.

Byrne, D., & Clore, G. L. Effectance arousal and attraction. *Journal of Personality and Social Psychology Monographs,* 1967, *6* (Whole No. 638).

Byrne, D., McDonald, R. D., & Mikawa, J. Approach and avoidance affiliation motives. *Journal of Personality,* 1963, *31,* 21–37.

Campbell, A., & Rushton, J. P. Bodily communication and personality. *British Journal of Social and Clinical Psychology,* 1978, *17,* 31–36.

Cautela, J. R., & Baron, M. G. Covert conditioning: A theoretical analysis. *Behavior Modification,* 1977, *1,* 351–368.

Chapple, E. Experimental production of transients in human interaction. *Nature,* 1970, *228,* 630–633.

Charny, E. J. Psychosomatic manifestations of rapport in psychotherapy. *Psychosomatic Medicine,* 1966, *28,* 305–315.

Cherulnik, P. D., Neely, W. T., Flanagan, M., & Zachan, M. Social skill and visual interaction. *Journal of Social Psychology,* 1978, 104, 263–270.

Clark, H. H., & Clark, E. V. *Psychology and language.* New York: Harcourt, Brace, Jovanovitch, 1977.

Clarke, A. C. An examination of the operation of residential propinquity as a factor in mate selection. *American Sociological Review,* 1952, *17,* 17–22.

Clore, G. L., Wiggins, N. H., & Itkin, S. Judging attraction from nonverbal behavior: The gain phenomenon. *Journal of Consulting and Clinical Psychology,* 1975, *43,* 491–497.

Collett, P. The rules of conduct. In P. Collett (Ed.), *Social rules and social behaviour.* Oxford: Basil Blackwell, 1977.

Cook, M. Experiments on orientation and proxemics. *Human Relations,* 1970, *23,* 261–276.

Cook, M. *Interpersonal perception.* Harmondsworth: Penguin, 1971.

Condon, W. S., & Ogston, W. D. A segmentation of sound film analysis. *Journal of Psychiatric Research,* 1967, *5,* 221–235.

Coulthard, M. *An introduction to discourse analysis.* London: Longman, 1977.

Cozby, P. C. Self-disclosure: A literature review. *Psychological Bulletin,* 1973, *79,* 73–91.

Cunningham, M. R. Personality and the structure of the nonverbal communication of emotion. *Journal of Personality,* 1977, *45,* 564–584.

Davis, J. D. When boy meets girl: Sex roles and the negotiation of intimacy in an acquaintance exercise. *Journal of Personality and Social Psychology,* 1978, *36,* 684–692.

Derlega, V. J., & Chaikin, A. L. Norms affecting self-disclosure in men and women. *Journal of Consulting and Clinical Psychology,* 1976, *44,* 376–380.

Dion, K., Berscheid, E., & Walster, E. What is beautiful is good. *Journal of Personality and Social Psychology,* 1972, *24,* 285–290.

Duck, S. *The study of acquaintance.* Farnborough, Hants.: Saxon House, 1977.

Duncan, S., & Fiske, D. W. *Face-to-face interaction: Research, method and theory.* Hillsdale, N.J.: Erlbaum, 1977.

Duval, S., & Hensley, V. Extensions of objective self-awareness theory: The focus of attention—causal attribution hypothesis. In J. H. Harvey, W. J. Gickes, & R. F. Kidd (Eds.), *New directions in attribution research* (Vol. 1). Hillsdale, N.J.: Erlbaum, 1976.

Eisler, R. M., Miller, P. M., & Hersen, M. Components of assertive behavior. *Journal of Clinical Psychology,* 1973, *29,* 295–299.

Ekman, P. Universals and cultural differences in facial expressions of emotion. In J. Cole (Ed.), *Nebraska symposium on motivation, 1971.* Lincoln: University of Nebraska Press, 1972.

Ekman, P. Facial signs: Facts, fantasies, and possibilities. In T. Sebeok (Ed.), *Sight, sound and sense.* Bloomington, Ind: Indiana University Press, 1978.

Ekman, P., & Friesen, W. V. The repertoire of nonverbal behavior: Categories, origins, usage and coding. *Semiotica,* 1969, *1,* 49–98.

Ekman, P., & Friesen, W. V. Nonverbal leakage and clues to deception. In S. Weitz (Ed.), *Nonverbal communication: Readings with commentary.* New York: Oxford University Press, 1974.

Ekman, P., & Friesen, W. V. *Unmasking the face.* Englewood Cliffs, N.J.: Prentice-Hall, 1975.

Ekman, P., Friesen, W. V., & Scherer, K. R. Body movement and voice pitch in deceptive interaction. *Semiotica,* 1976, *16,* 23–27.

Endler, N., & Magnusson, D. Toward an interactional psychology of personality. *Psychological Buelletin,* 1976, *33,* 956–974.

Exline, R. V. Visual interaction: The glances of power and preference. In S. Weitz (Ed.), *Nonverbal communication.* New York: Oxford University Press, 1974.

Ferguson, J. *Interruptions in spontaneous dialogue.* Paper presented at the British Psychological Society conference, Stirling, 1975.

Feshback, S., & Singer, R. P. The effects of personal and shared threat upon social prejudice. *Journal of Abnormal and Social Psychology,* 1957, *54,* 411–416.

Festinger, L., Schacter, S., & Back, K. *Social pressures in informal groups: A study of human factors in housing.* New York: Harper and Row, 1950.

Freedman, J. L. *Crowding and behavior: The psychology of high-density living.* New York: Viking, 1975.

Gibbins, K. Communication aspects of women's clothes and their relation to fashionability. *British Journal of Social and Clinical Psychology,* 1969, *8,* 301–312.

Goffman, E. On face work: An analysis of ritual elements in social interaction. *Psychiatry,* 1955, *18,* 213–231.

Goffman, E. *Presentation of self in everyday life.* Edinburgh: Edinburgh University Press, 1956.

Goffman, E. *Encounters.* Harmondsworth: Penguin, 1961.

Goffman, E. *Stigma.* Englewood Cliffs, N.J.: Prentice-Hall, 1963.

Goffman, E. *Relations in public: Microstudies of the public order.* Harmondsworth: Penguin, 1972.

Goldfried, M. R., & Goldfried, A. P. Cognitive change methods. In F. H. Kanfer & A. P. Goldstein (Eds.), *Helping people change.* New York: Pergamon Press, 1975.

Gove, W. R. Societal reaction as an explanation of mental illness: An evaluation. *American Sociological Review,* 1970, *35,* 873–884.

Griffitt, W., & Veitch, R. Hot and crowded: Influences of population density and temperature on interpersonal affective behavior. *Journal of Personality and Social Psychology,* 1971, *7,* 92–98.

Gump, P. V. The behavior setting: A promising unit for environmental designs. *Landscape Architecture,* 1971, *61,* 130–134.

Gumperz, J. J. Sociolinguistics and communication in small groups. In J. B. Pride & J. Holmes (Eds.), *Sociolinguistics.* Harmondsworth: Penguin, 1972.

Hall, E. T. *The hidden dimension.* Garden City, N.Y.: Doubleday, 1966.

Harré, R., & Secord, P. F. *The explanation of social behaviour.* Oxford: Blackwell, 1972.

Harvey, O. J., Hunt, D. E., & Schroder, H. M. *Conceptual systems and personality development.* New York: Wiley, 1961.

Hayes-Roth, B. Evolution of cognitive structures and processes. *Psychological Review,* 1977, *84,* 260–278.

Henley, N. M. *Body politics: Power, sex and nonverbal communication.* Englewood Cliffs, N.J.: Prentice-Hall, 1977.

Heslin, R. *Steps toward a taxonomy of touching.* Paper presented to the Midwestern Psychological Association, Chicago, May 1974.

Hoffman, M. L. Sex differences in empathy and related behaviors. *Psychological Bulletin,* 1977, *84,* 712–722.

James, W. T. A study of the expression of bodily posture. *Journal of General Psychology,* 1932, *7,* 405–437.

Jefferson, G. A case of precision timing in ordinary conversation: Overlapped tag-positioned address terms in closing sequences. *Semiotica,* 1973, *9,* 47–96.

Jones, E. E., & Gordon, E. M. Timing of self-disclosure and its effects on personal attraction. *Journal of Personality and Social Psychology,* 1972, *24,* 358–365.

Jones, E. E., & Nisbett, R. E. *The actor and the observer: Divergent perceptions of the causes of behavior.* Morristown, N.J.: General Learning, 1971.

Jones, S. C., & Panitch, D. The self-fulfilling prophecy and interpersonal attraction. *Journal of Experimental Social Psychology,* 1971, *7,* 356–366.

Kanfer, F. H., & Karoly, P. Self-control: A behavioristic excursion into the lion's den. *Behavior Therapy,* 1972, *3,* 398–416.

Kelley, H. H. The warm-cold variable in first impressions of persons. *Journal of Personality,* 1950, *18,* 431–439.

Kendon, A. Some functions of gaze-direction in social interaction. *Acta Psychologica,* 1967, *26,* 22–63.

Kendon, A. The role of visible behavior in the organization of social interaction. In M. von Cranach & I. Vine (Eds.), *Social communication and movement.* London: Academic, 1973.

Kendon, A., & Cook, M. The consistency of gaze patterns in social interaction. *British Journal of Psychology,* 1969, *60,* 481–494.

Kendon, A., & Ferber, A. A description of some human greetings. In R. P. Michael & J. H. Crook (Eds.), *Comparative ecology and behaviour of primates.* London: Academic, 1973.

Kent, G. G., Davis, J. D., & Shapiro, D. A. Resources required in the construction and reconstruction of conversation. *Journal of Personality and Social Psychology,* 1978, *36,* 13–22.

Knapp, M. L. *Nonverbal communication in human interaction.* New York: Holt, Rinehart and Winston, 1978.

Knapp, M. L., Hart, R. P., Friedrich, G. W., & Shulman, G. M. The rhetoric of goodbye: Verbal and nonverbal correlates of human leave-taking. *Speech Monographs,* 1973, *40,* 182–198.

Knapp, M. L., Hart, R. P., & Dennis, H. S. An exploration of deception as a communication construct. *Human Communication Research,* 1974, *1,* 115–129.

Kopel, S., & Arkowitz, H. The role of attribution and self-perception in behavior change: Implications for behavior therapy. *Genetic Psychology Monographs,* 1975, *92,* 175–212.

La France, M., & Mayo, C. Racial differences in gaze behavior during conversations: Two systematic observational studies. *Journal of Personality and Social Psychology*, 1976, *33*, 547–552.

Lanzetta, J. T., & Kleck, R. E. Encoding and decoding of nonverbal affect in humans. *Journal of Personality and Social Psychology*, 1970, *16*, 12–19.

Laver, J. Communicative functions of phatic communion. In A. Kendon, R. M. Harris, & M. R. Key (Eds.), *Organization of behavior in face-to-face interaction*. The Hague: Mouton, 1975.

Lennard, H. L., & Bernstein, A. *The anatomy of psychotherapy*. New York: Columbia University Press, 1960.

Leonard, R. L. Cognitive complexity and the similarity attraction paradigm. *Journal of Research in Personality*, 1976, *10*, 83–88.

Levinger, G. A three level approach to attraction: Towards an understanding of pair relatedness. In E. L. Huston (Ed.), *Foundations of interpersonal attraction*. New York: Academic, 1974.

Levy, P. K. The ability to express and perceive vocal communication of feelings. In J. R. Davitz (Ed.), *The communication of emotional meaning*. New York: McGraw-Hill, 1964.

Liebman, K. The effects of sex and race norms on personal space. *Environment and Behavior*, 1970, *2*, 208–246.

Lomax, A. Culture-style factors in face-to-face interaction. In A. Kendon, R. M. Harris, & M. R. Key (Eds.), *Organization of behavior in face-to-face interaction*. The Hague: Mouton, 1975.

McGovern, K. B., Arkowitz, H., & Gilmore, S. K. Evaluation of social skill training programs for college dating inhibitions. *Journal of Counseling Psychology*, 1975, *22*, 505–512.

Mackey, W. C. Parameters of the smile as a social signal. *Journal of Genetic Psychology*, 1976, *129*, 125–130.

Magnusson, D., & Endler, N. S. Interactional psychology: Present status and future prospects. In D. Magnusson & N. S. Endler (Eds.), *Personality at the crossroads: Current issues in interactional psychology*. Hillsdale, N.J.: Erlbaum, 1977.

Mahoney, M. *Cognition and behavior modification*. Cambridge, Mass.: Ballinger, 1974.

Matarazzo, J. D., & Wiens, A. N. Speech behavior as an objective correlate of empathy and outcome in interview and psychotherapy research. *Behavior Modification*, 1977, *1*, 453–480.

Mehrabian, A. *Nonverbal communication*. Chicago: Aldine Atherton, 1972.

Meichenbaum, D. *Cognitive behavior modification*. New York: Plenum, 1977.

Meltzer, L., Morris, W. N., & Hayes, D. P. Interruption outcomes and vocal amplitude: Explorations in social psychophysics. *Journal of Personality and Social Psychology*, 1971, *18*, 392–402.

Milgram, S. The experience of living in cities. *Science*, 1970, *167*, 1461–1468.

Milgram, S. *Obedience to authority*. New York: Harper and Row, 1974.

Miller, G. A., Galanter, E., & Pribram, K. H. *Plans and the structure of behavior*. New York: Holt, 1960.

Miller, W. R., & Arkowitz, H. Anxiety and perceived causation in social success and failure experiences: Disconfirmation of an attribution hypothesis in two experiments. *Journal of Abnormal Psychology*, 1977, *86*, 665–668.

Mischel, W. *Personality and assessment*. New York: Wiley, 1968.

Mischel, W., Ebbeson, E., & Zeiss, A. Selective attention to the self: Situational and dispositional determinants. *Journal of Personality and Social Psychology*, 1973, *27*, 129–142.

Moos, R. H. Situational analysis of a therapeutic community milieu. *Journal of Abnormal Psychology*, 1968, *73*, 49–61.

Morris, D. *Manwatching*. London: Cape, 1978.

Morton, T. L. Intimacy and reciprocity of exchange: A comparison of spouses and strangers. *Journal of Personality and Social Psychology*, 1978, *36*, 72–81.

Murstein, B. I. The stimulus–value–role (SVR) theory of dyadic relationships. In S. Duck (Ed.), *Theory and practice in interpersonal attraction*. London: Academic, 1977.

Natale, M. Convergence of mean vocal intensity in dyadic communication as a function of social desirability. *Journal of Personality and Social Psychology*, 1975, *32*, 790–804.

Nisbett, R. E., Borgida, E., Crandall, R., & Reed, H. Popular induction: Information is not always informative. In J. Carroll & J. Payne (Eds.), *Cognitive and social behavior*. Potomac, Md.: Erlbaum, 1976.

Ogden, C. K., & Richards, I. A. *The meaning of meaning*. London: Routledge and Kegan Paul, 1923.

Osmond, H. Function as the basis of psychiatric ward design. *Mental Hospital*, 1956, *8*, 23–29.

Robbins, G. E. Dogmatism and information gathering in personality impression formation. *Journal of Research in Personality*, 1975, *9*, 74–84.

Rogers, W. T., & Jones, S. E. Effects of dominance tendencies on floor holding and interruption behavior in dyadic interaction. *Human Communication Research*, 1975, *1*, 113–122.

Rosenthal, R. On the social psychology of the self-fulfilling prophecy: Further evidence for pygmalion effects and their mediating mechanisms. *MSS Modular Publications*, 1973, *53*, 1–28.

Ross, L. The intuitive psychologist and his shortcomings: Distortions in the attribution process. *Advances in Experimental Social Psychology*, 1977, *10*, 173–220.

Ross, L., Lepper, M., & Hubbard, M. Perseverance in self-perception and social perception: Biased attributional process in the debriefing paradigm. *Journal of Personality and Social Psychology*, 1975, *32*, 880–892.

Rutter, D. R., Morley, J. E., & Graham, J. C. Visual interaction in a group of introverts and extraverts. *European Journal of Social Psychology*, 1972, *2*, 371–384.

Sacks, H., Schegloff, E. A., & Jefferson, G. A simplest systematics for the organization of turn-taking in conversation. *Language*, 1974, *50*, 696–735.

Saegart, S., Swap, W., & Zajonc, R. B. Exposure, context, and interpersonal attraction. *Journal of Personality and Social Psychology*, 1973, *25*, 234–242.

Scheflen, A. E. The significance of posture in communication systems. *Psychiatry*, 1964, *27*, 316–331.

Scheflen, A. E. Quasi-courtship behavior in psychotherapy. *Psychiatry*, 1965, *28*, 245–257.

Scheflen, A. E., & Scheflen, A. *Body language and the social order*. Englewood Cliffs, N.J.: Prentice-Hall, 1972.

Schegloff, E. A., & Sacks, H. Opening up closings. *Semiotica*, 1973, *8*, 289–327.

Scherer, K. R. Acoustic concomitants of emotional dimensions: Judging affect from synthesized tone sequences. In S. Weitz (Ed.), *Nonverbal communication: Readings with commentary*. New York: Oxford University Press, 1974.

Scherer, K. R. Social markers in speech. *ECSP*, Paris, October 1977.

Scott, M. B., & Lyman, S. M. Accounts. *American Sociological Review*, 1968, *33*, 46–62.

Searle, J. R. A taxonomy of illocutionary acts. In K. Gunderson (Ed.), *Minnesota studies in the philosophy of language*. Minneapolis: University of Minnesota Press, 1975. (a)

Searle, J. R. Indirect speech acts. In P. Cole & J. L. Morgan (Eds.), *Syntax and semantics* (Vol. 3), *Speech acts*. New York: Seminar Press, 1975. (b)

Secord, P. F. The role of facial features in interpersonal perception. In R. Tagiuri & L. Petrullo (Eds.), *Person perception and interpersonal behavior*. Stanford, Calif.: Stanford University Press, 1958.

Sinclair, J. McH., Forsyth, I. J. Coulthard, R. M., & Ashby, M. C. *The English used by teachers and pupils*. Unpublished final report to SSRC, Birmingham University, 1972.

Snyder, M. Self-monitoring of expressive behavior. *Journal of Personality and Social Psychology*. 1974, *30*, 526–537.

Snyder, M., & Monson, T. C. Persons, situations, and the control of social behavior. *Journal of Personality and Social Psychology*, 1975, *32*, 637–644.

Snyder, M., Tanke, E. D., & Berscheid, E. Social perception and interpersonal behavior: On the self-fulfilling nature of social stereotypes. *Journal of Personality and Social Psychology*, 1977, *35*, 656–666.

Spivack, G., Platt, J. J., & Shure, M. B. *The problem-solving approach to adjustment*. San Francisco: Jossey Bass, 1976.

Thorngate, W. Must we always think before we act? *Personality and Social Psychology Bulletin*, 1976, *2*, 31–35.

Triandis, H. C. *Interpersonal behavior*. Monterey, Calif.: Brooks/Cole, 1977.

Warr, P. B., & Knapper, C. *The perception of people and events*. New York: Wiley, 1968.

Wells, W., & Siegel, B. Stereotyped somatotypes. *Psychological Reports*, 1961, *8*, 77–78.

Wener, A. E., & Rehm, A. P. Depressive affect: A test of behavioral hypotheses. *Journal of Abnormal Psychology*, 1975, *84*, 221–227.

Witkin, H., & Goodenough, D. Field dependence and interpersonal behavior. *Psychological Bulletin*, 1977, *84*, 661–689.

Young, G. C. D. *Social skills and superordinate constructs*. Paper presented at the Second International Congress on Personal Construct Psychology, Christ Church, Oxford, 1977.

CHAPTER 2

Sociopsychological Factors in Psychopathology

Alan E. Kazdin

INTRODUCTION

Throughout history, the nature and basis of abnormal behavior has been conceptualized in many different ways encompassing demonology and biology (Zilboorg & Henry, 1941). In contemporary psychiatry, emphasis has been placed on biological causes of disordered behavior. The biological basis of psychiatric disorders has been advanced through many discoveries. Prominent among these was the finding that general paresis was in fact a neurological disorder caused by the syphilitic spirochete. The many psychological symptoms that often accompany general paresis include intellectual impairment, grandiose ideation, and distorted perception. Demonstration of organic etiology of these symptoms advanced the general notion that many other psychological disorders might be associated with underlying organic pathology.

For several psychiatric disorders, evidence of organic etiology has not emerged. Thus, the search for causes within psychiatry expanded to include disordered psychological processes that might serve as the basis for the psychiatric symptoms. An intrapsychic disease model was adopted that suggested that psychiatric disorders, as physical disorders, represent diseased psychological processes that are reflected in a specific pattern of symptomology. Abnormal behavior can be considered to reflect illness, "mental illness," and the model of diagnosis and treatment

Alan E. Kazdin • Department of Psychology, Pennsylvania State University, University Park, Pennsylvania 16802. Completion of this chapter was facilitated by a grant (MH 31047) from the National Institute of Mental Health.

characterizing physical disorders was extended to psychological disorders as well.

To be sure, many disorders within the domain of psychiatry have been shown to have a clear organic basis (e.g., various brain syndromes and specific forms of retardation). Yet, considerable debate exists about whether the general model applied to organic disorders extends to those forms of aberrant behavior that have been called functional or psychogenic, i.e., disorders not known to be associated with organic pathology. Indeed, many writers have objected to the general approach that is implied by looking for psychological "disease" or disordered psychological processes.

Among the many critics of an intrapsychic disease approach, Szasz (1960) has referred to mental *illness* as a "myth." Szasz has stressed that abnormal behavior reflects problems that people have in their everyday living and is not a disease in any sense. Abnormality appears to be defined in the context of social behavior. Unlike physical illness that can be defined by the characteristics of the individual person, so-called mental illness is defined on the basis of an observer who plays a crucial diagnostic role in judging and defining behavior as abnormal.

Many other authors have stressed the social context in which psychopathology is identified. For example, Adams (1964) has noted that the label "mental illness" has been applied to arbitrarily designated forms of abnormal behaviors. The behaviors are often associated with discomfort, social rejection, and problems in interpersonal relationships, but these are more psychosocial than medical in nature. The social nature of aberrant behavior is suggested by the fact that the behaviors tend to be disturbing rather than disturbed (Ferster, 1965; Ullmann & Krasner, 1975). Behavior is bothersome to others and is identified as a manifestation of mental illness.

Sociologists have attempted to elaborate the nature of abnormal behavior to define and account for the kind of behavior that is delineated as mental illness. For example, Scheff (1966) has suggested that symptoms of mental illness consist of violations of norms. Behavior that violates normative standards may result from organic, psychological, or environmental events. Yet, the behavior itself is defined as abnormal by society. It is the societal reaction rather than the behavior itself that designates the important ingredient in initiating and sustaining the career of a psychiatric patient (Becker, 1963).

The labeling process, according to the sociopsychological view of deviance, plays a crucial role in sustaining the career of someone whose behavior has been deviant. The label often is viewed as the factor that crystallizes deviant behavior. The label is considered to influence both the person labeled, in terms of the role expectations that he or she begins to perform, and the expectations that others have. Labeling the person as a psychiatric patient or as mentally ill thus is thought to develop a career of behavior that otherwise may have been transient.

The sociopsychological position emphasizes deviance in behavior rather than "illness" and the importance of social forces that may develop and sustain that behavior. Independent lines of evidence have suggested the importance of select tenets of the sociopsychological view. For example, research has pointed to the influence of labeling others as deviant or "mentally ill" both on the person who is labeled and others who know that someone has been identified as deviant (e.g., Farina, Gliha, Boudreau, Allen, & Sherman, 1971; Farina, Holland, & Ring, 1966). Yet, few adherents maintain the view that deviant behavior is generated and sustained solely from sociopsychological influences. Research from different areas of psychopathology has demonstrated the role of all sorts of factors including genetic, biochemical, prenatal, familial, sociocultural, and others. Needless to say, any singular position cannot easily account for the incidence of psychogenic disorders or even a particular type of disorder.

The sociopsychological model, as any other, does not by itself completely explain the onset, amelioration, and prognosis of various disorders. Yet, it is instructive to examine the role of social factors in psychological disorders. Social factors and interpersonal behavior are involved in each stage of the career of someone who eventually enters psychiatric treatment. Identifying deviant behavior, selecting individuals for treatment, and the ultimate success of treatment all depend heavily on social factors and the interpersonal skills of the person who enters the process.

The purpose of the present chapter is to discuss the role of social factors and interpersonal behavior in psychopathology. Different stages of disordered behavior are highlighted, including identifying individuals who are deviant, providing treatment, and ensuring their adjustment when they are placed back in the community. In highlighting social factors in psychopathology, the present chapter will focus on psychoses. The bulk of research on social factors has been completed with psychotic patients and, more specifically, schizophrenics. Different lines of research revealing the importance of social behavior have been pursued and readily permit integration of the stages of identifying and hospitalizing individuals with severe psychiatric impairment. The exclusive focus of the chapter on the social behavior of psychotic patients is not intended to imply a narrow theoretical framework about the etiology of psychoses. Rather, the treatment of social factors is designed to point to the multiple role of social processes and to draw implications that existing data may bear for treatment.

SOCIAL VARIABLES IN IDENTIFYING PSYCHOPATHOLOGY

Deviance In and Out of the Hospital

The role of social factors and social behavior begins and may even be most significant at the point where psychopathological behavior is identified. Apparently, deviant behavior is readily evident in everyday experience and it is not merely the psychiatric patient whose behavior may reflect severe impairment. Several studies have shown a significant percentage of individuals functioning in everyday life whose behavior is impaired.

In the now-classic study conducted in Midtown Manhattan, Srole, Langner, Michael, Opler, and Rennie (1962) demonstrated that almost one-quarter of the subjects interviewed suffered psychiatric symptoms and some impairment as a result of these. In addition, the approximately one-fifth of the people sampled were rated as incapacitated in terms of the extent to which their problems interfered with their adjustment. Several other studies encompassing different countries and urban and rural populations have consistently shown that a large percentage of individuals who are not hospitalized evince psychiatric symptoms to a significant degree (e.g., Essen-Möller, 1956; Leighton, Lambo, Hughes, Leighton, Murphy, & Macklin, 1963; Phillips, 1966). In general, many people outside of hospitals appear to have sufficient impairment to qualify them for treatment. The fact that they are not identified for and placed in treatment has suggested to some authors that social contingencies rather than "symptoms of illness" might account for hospitalization.

Additional research suggesting the influence of social variables has looked at the nature of the deviance of those who are hospitalized for treatment. Presumably, the deviance of hospitalized patients is more severe than that of those people in the community who suffer select symptoms but successfully remain out of the hospital. Yet, several studies have suggested that hospitalized patients do not always warrant retention in the hospital in light of their psychiatric status. For example, Cooper and Early (1961) found that a large percentage (84%) of psychiatric patients who were hospitalized did not have any serious disturbance that warranted hospitalization. Similarly, Scheff (1966) found that many of the patients (43%) did not meet requirements for hospitalization but were still retained in the hospital. In an earlier study, Scheff (1964) found that the majority of the patients (61%) sampled did not clearly meet the requirements for hospitalization to begin with.

The specific percentages cited in these studies are not important since they would be expected to fluctuate considerably as a function of the population sampled, the criteria for retention, and the specific measures used to implement these criteria. Yet, the pattern of data is sufficiently clear to suggest that individuals on both sides of the hospital walls may have impairment. Moreover, those who are in the hospital may have more severe impairment but in many cases are not seriously disturbed. Many writers have stressed that the behavior of those who are designated as mentally ill is not distinct from many of those who are not so designated (Kitsuse, 1962; Laing, 1967). One of the distinguishing characteristics of those identified for treatment is the problems they evince in interpersonal relationships (Lemert, 1962). Persons whose behaviors are socially problematic are eventually brought to treatment and begin the career of a psychiatric patient (Lemert, 1946).

Research showing the occasionally disturbed behavior of those on the outside and the occasionally undisturbed behavior of some patients on the inside of the hospital has suggested looking at factors other than psychiatric status that dictate hospitalization. Major social variables that relate to identifying psychiatric disorders and hospitalizing individuals whose behaviors are deviant have been studied in epidemiological research.

Epidemiological Factors in Psychopathology

The epidemiological study of psychopathology attempts to assess the distribution of disorders according to various social and cultural variables. The information that results reflects factors that relate to the onset and diagnosis of psychopathology. The purpose in noting epidemiological factors here is to stress the relevance of social and cultural factors in the incidence of severe psychopathology.

There are various problems in interpreting the correlations of social factors with psychopathology. Perhaps the most significant problem worth noting at the outset is the criterion used to identify psychopathology. Usually, hospitalization rather than psychiatric impairment *per se* is used to identify psychopathology, although some studies have examined impairment in the population at large. When hospitalization is used as the criterion for psychopathology, incidence of diagnosed disorders rather than prevalence of deviance is studied. In many cases, whether the empirically demonstrated relationships would hold when measuring impairment among hospitalized and nonhospitalized populations can be argued. Nevertheless, the epidemiology of psychiatric disorders is instructive in pointing to social factors and their relevance for considering both the identification and possible treatment of psychiatric patients.

Social Class. One of the better demonstrated findings pertains to the relationship between socioeconomic standing and the incidence of psychiatric disorders. Socioeconomic standing usually has been defined by income, occupation, education, and area of residence within a city. In general, the highest rates of diagnosed psychoses are associated with lower socioeconomic status (Derogatis, Yevzeroff, & Wittelsberger, 1975; Hollingshead & Redlich, 1958; Kaplin, Ree, & Richardson, 1956; Malzberg, 1969). The relationship between social class and serious psychiatric impairment is not linear. The preponderence of diagnosed psychopathology occurs with the lowest socioeconomic class. In any case, the relationship is well demonstrated, indicating that one's socioeconomic standing is a correlate of diagnosed impairment (cf. Mishler & Scotch, 1963).

Social Disorganization and Deterioration. Social disorganization and deterioration refer to the types of changes and the level of development of a given community. Signs of disorganization or deterioration in a community include high levels of crime and delinquency, poverty, broken homes, and conflict of values (e.g., Leighton, 1959). The level of community disorganization, when measured by the above characteristics, is related to the incidence of psychoses with more disorganized communities having higher levels of psychoses (Jaco, 1954). Within a given city, areas of the greatest disorganization or community deterioration show the highest level of schizophrenia, a finding revealed in the classic study of Faris and Dunham (1939).

The finding that community disorganization and deterioration are related to psychiatric impairment has been suggested to result from the movement of impaired individuals toward impoverished and unstable areas of the community in light of their inability to function adequately in other areas of the community. Yet, little evidence for geographic movement among psychotics exists that would explain the relationship between community disorganization and incidence of diagnosed psychoses (Kleiner & Parker, 1963, 1969). Thus, the relationship suggests that the existing living conditions rather than those conditions into which individuals move correlate with psychopathology.

Migration. Migration refers to changes in residence which include moving across different cities or across state or county boundaries (Kantor, 1969; Malzberg, 1969). The relevance of migration and psychiatric disorders perhaps was first suggested from the influx of foreign immigrants to the United States (Malzberg, 1969). In general, foreign-born persons have a greater incidence of psychiatric hospitalization than those who are native to the host country or than second-generation individuals who are of the same ethnic origin. Of course, these data do not show specifically that migration *per se* leads to hospitalization since individuals entering a country have all sorts of adjustment problems to contend with once they arrive.

The importance of migration has been clearer in research examining admissions to psychiatric hospitals from those who move within the United States. Persons who migrate across state lines have higher hospital admission rates than those who are native to the states (e.g., Lazarus, Locke, & Thomas, 1963; Locke, Kramer, & Pasamanick, 1960; Malzberg & Lee, 1956). These relationships tend to hold up across gender, race, and states in which the studies were conducted although the magnitude of the differences in hospital admissions among migrating and native individuals have varied as a function of each of these variables (Kantor, 1969). Additional research has suggested that movements within smaller geographical boundaries than states are correlated with the incidence of psychopathology. People who move more across county lines or within a city have a higher incidence of psychoses than those who are less mobile (Freedman, 1950; Tietze, Lemkau, & Cooper, 1942).

The relationship between migration and psychiatric impairment has been interpreted to reflect the problems that impaired individuals have in adjusting to their residence. Such adjustment problems result in increased movement and change of residence. Since other factors such as social class may covary with migration, many other interpretations might be advanced.

Marital Status. The incidence of diagnosed psychosis has been known for many years to be related to marital status. Severe psychiatric impairment is much greater for those who are single, separated, divorced, or widowed than for those who are married (Langner & Michael, 1963; Rose & Stub, 1955; Wanklin, Fleming, Buck, & Hobbs, 1956). Although this relationship has been found for both males and females, the difference in incidence of serious impairment among unmarried and married males is particularly great. Among individuals who are diagnosed as psychotic, marital status continues to be an important variable. Patients who are married have a better prognosis than those who are single (Buss, 1966).

The relationship of marital status to psychosis occasionally has been attributed to a selection factor. Those individuals who are predisposed toward psychotic disorders are less likely to become married or to stay married. However, selection becomes implausible as an adequate explanation when the data are examined closely. For example, individuals who are widowed show greater incidence of psychopathology than those who are single, suggesting that failing to become married to begin with is not the crucial factor. Similarly, individuals who are married and later become divorced have a higher rate of psychopathology than those who never married (Fried, 1964). This finding too suggests that stresses associated with loss of a partner may be related to psychopathology.

Loss of Parents. The loss of one or both parents, by their divorce, hospitalization, or death, has been related to psychopathology. Persons who become psychotic have a significantly higher incidence of losing a parent prior to adolescence or early adulthood than individuals not diagnosed as having severe impairment (e.g., Brown, 1961; Lidz & Lidz, 1949). Actually, the loss of a parent is a correlate of diverse forms of deviance including psychoses, neuroses, sociopathic disorders, alcoholism (e.g., Gregory, 1958). There is some evidence that serious disorders such as psychoses are associated with greater rates of parental loss than are less severe disorders such as neuroses (e.g., Berg & Cohen, 1959), but this has not been demonstrated widely within or across various disorders.

Social Deprivation and Stress

The above overview enumerates many of the relationships between social factors and psychopathology. The list is by no means exhaustive and other relationships might be elaborated (Dohrenwend & Dohrenwend, 1974). For example, unemployment, religious affiliation, religious mobility (changing religious affiliation), ethnic group affiliation, gender, and race are related to psychiatric impairment and hospitalization (e.g., Fried, 1964; Roberts & Myers, 1954; Srole & Langner, 1969). These factors, along with the other ones described earlier, only represent correlates of psychopathology.

Although many interpretations might be placed on the separate factors and their role in precipitating psychological disorders, they can be viewed in a unified fashion by considering the social deprivation and stress they provide for individuals who are impaired. The greater the sources of deprivation (e.g., due to socioeconomic status or race) or stress (e.g., due to loss of a parent or migration), the greater the likelihood of psychiatric impairment (Fried, 1964).

Zubin and Spring (1977) have advanced the notion of vulnerability that may help explain the role of social factors that contribute to psychopathology. These authors have proposed that individuals differ in vulnerability to psychiatric disorders (although they specifically are concerned with schizophrenia). Individuals are endowed with a degree of vulnerability that under suitable circumstances may express itself in psychopathology. Vulnerability may originate from many different sources (e.g., genetic, biochemical, family processes, social deprivation). Independently of the source of vulnerability, psychopathology may not be precipitated unless sufficient stress occurs.

Sources of stress may include social factors, physical disease, perinatal complications, family experiences, and others. When stresses surpass the individual's vulnerability threshold, the individual can no longer cope and psychopathology may be precipitated. A breakdown in coping ability does not necessarily cause the disorder but presents an opportunity for psychopathology to develop. A person's degree of vulnerability may determine the severity and longevity of impairment in light of a particular set of stressful circumstances.

A vulnerability interpretation of psychopathology has important implications for treatment. Different treatment approaches might be supported. First, psychopharmacological interventions might be used to inhibit or reduce vulnerability. Indeed, drug therapy is often explicitly directed at reducing the stress value or impact of events that may precipitate episodes of psychopathology. Second, larger scale social interventions might be used to alter conditions (e.g., unemployment, living conditions) that are implicated in stress. Third, treatment can focus on the individual's coping skills so that sources of stress can be more readily managed. Improvements in social competence and skills to handle sources of stress would be important. Of course, increasing a person's ability to cope with sources of stress would not alter the initial level of vulnerability but would reduce the influence of events that might make vulnerability manifest in psychopathology.

SOCIAL BEHAVIORS AND COMPETENCE OF PSYCHIATRIC PATIENTS

The epidemiological evidence pertains to the social factors that are related to psychiatric impairment. Social factors have received considerable attention in conjunction with other facets of social behavior of patients. Typically, research has focused on schizophrenia since that diagnosis accounts for the majority of hospitalized psychiatric patients, and hence will serve as the basis for the present section. Interrelated lines of research have implicated the social behavior of psychotic patients in different ways.

Process-Reactive Schizophrenia

A major area of research has evaluated the differences between *reactive* and *process* schizophrenics. Reactive schizophrenia is characterized by the sudden onset of symptoms with little or no history of disordered behavior. Process schizophrenia, on the other hand, is characterized by a long history of psychiatric impairment and a gradual onset of symptomatic behaviors. Reactive schizophrenics have a much more favorable prognosis for recovery than do process schizophrenics. The different categorizations of schizophrenia, of course, are viewed as end points on a continuum rather than discrete categories (Becker, 1956).

The distinction between reactive and process schizophrenia is based on several variables including physical health, family difficulties, types of delusions, and many others (e.g., Kantor, Wallner, & Winder, 1953). Interestingly, many of the differences between process and reactive schizophrenics refer to social behaviors. Reactive schizophrenics tend to be introverted, lack heterosexual behaviors and interests, often are physically aggressive, and have had difficulties at school. In contrast, process schizophrenics tend to be extroverted, evince heterosexual behaviors and interests, are more likely to engage in verbal rather than physical aggression, and were well adjusted in their early school years. While these indicators do not exhaust the bases for distinguishing process and reactive schizophrenics, it is significant to note that social behavior plays an important role in overall prognosis and responsiveness to treatment.

Premorbid Status

The importance of social behavior in schizophrenia can be examined by looking at the premorbid (preillness) adjustment of schizophrenics. Phillips (1953) proposed a scale to assess the variables included in a patient's case history that were associated with prognosis. The scale evaluates several aspects of adjustment including sexual adjustment, marital status, types of friends, and social behavior. Many of the items in these categories refer to concrete aspects of social interaction. For example, premorbid adjustment depends on the extent to which a person has dated during adolescence, mixed closely with boys and girls as a child, maintains a heterosexual relationship, has close friends, and is married and has children. Individuals with poor premorbid adjustment tend to be isolated, aloof, and seclusive, have few or no close friends, have little contact with members of the opposite sex, and little or no sexual interest in men or women.

Standing on each of the above variables is weighted in varying degrees to yield a numerical score that reflects overall premorbid adjustment. The social adjustment prior to the onset of schizophrenia, as measured by the above variables, is related to important dimensions of psychiatric impairment, such as the patient's prognosis (Phillips, 1953) and responsiveness to treatment (Farina, Garmezy, & Barry, 1963; Farina & Webb, 1956).

Social Competence

Social functioning in relation to psychopathology has been measured in other ways. The measurement of an individual's level of social competence has received considerable attention in relation to psychopathology (Phillips & Zigler, 1961; Zigler & Phillips, 1960). Social competence refers to a multidimensional variable that includes age, intelligence (IQ), education, occupation, employment history, and marital status. Although social competence subsumes individual variables related to social standing, it has been conceived of as a broader dimension that reflects the adequacy of one's social functioning. Individuals high in social competence are considered to be able to meet the demands of everyday functioning and to be equipped to handle participation and responsibility for their own welfare and the welfare of others.

Although standing on the variables comprising social competence is considered to reflect social adequacy and overall maturity, the variables do not immediately reflect social skills. Such variables as age, occupation, and intelligence, for example, do not necessarily convey how well an individual can cope in social situations. Yet, the overall level of competence is considered to reflect adaptiveness to societal demands in general (Zigler & Phillips, 1961).

As with measures of premorbid adjustment mentioned above, social competence predicts prognosis and responsiveness to treatment. Patients, including both schizophrenic and nonschizophrenic diagnoses, with higher social competence have a better prognosis than those with lower scores (Zigler & Phillips, 1962). Interestingly, social competence also has been found to be related to the type of symptom pattern. High-social-competence patients tend to show symptoms related to self-deprivation and turning against the self (e.g., self-depreciation, suicidal ideas, bodily complaints), whereas low-social-competence patients tend to show more symptoms related to the avoidance of others (e.g., withdrawal, suspiciousness, apathy). Of course, these low-social-competence patients have a relatively poor prognosis. From the findings pertaining to premorbid adjustment, social competence, and types of symptoms, the role of social factors in psychopathology is quite clear. Individuals with more favorable social behavior and status prior to the onset of their disorder have a much better prognosis. Moreover, those whose symptoms do not show high levels of avoidance of others have a better prognosis than those who do.

Social Isolation and Withdrawal

Research on social competence addresses a person's standing on variables of social significance but does not directly evaluate actual social performance. Additional research has examined how individuals later diagnosed as schizophrenic perform socially.

Different indicators of social performance have been used encompassing both personality measures, self-report, and overt behavior.

Bower, Shellhamer, and Daily (1960) found that preschizophrenics were characterized by a withdrawn personality. In another study, comparisons of preschizophrenic and normal control subjects revealed that the former were characterized by social withdrawal and lack of social adeptness (Schofield & Balian, 1959). Kohn and Clausen (1955) reported that approximately one-third of the schizophrenic patients they sampled reported a lack of activities and few friendships. In contrast, a very small portion of a normal control sample (4%) reported a similar pattern of isolation.

In a particularly interesting study of social isolation, Barthell and Holmes (1968) examined high school yearbooks as an archival measure of social activity (e.g., participation in clubs, student government, special-interest groups). High school graduates later diagnosed as schizophrenics had participated in significantly fewer social activities in high school than did normal control subjects. Neurotic subjects fell between normals and schizophrenic subjects in terms of level of social activity.

Social behavior is also important in distinguishing schizophrenic patients with different prognoses. Indeed, research on social competence suggests differential prognosis among patients diagnosed as schizophrenic. Social behavior has a similar prognostic role. A patient's social behavior on the ward appears to predict the extent of chronicity (Nuttall & Solomon, 1965). Social behavior more than severity of symptoms has been identified as a crucial determinant of duration of hospitalization (Jenkins & Gurel, 1959).

The importance of social behavior of schizophrenic patients was demonstrated by Depue and Dubicki (1974), who classified patients as withdrawn or active a few days after they were admitted. Patients classified as withdrawn had different careers and prognoses than those who were classified as active. Withdrawn patients had first been hospitalized at an earlier age, had spent more time in the hospital on each admission, were lower in their premorbid adjustment, had a higher incidence of delusions and hallucinations, and reported themselves as being quieter in a group situation than did active patients.

Aside from the social behavior of psychotic patients, some research has suggested that patients may serve as cues for minimal and negative social interaction from others. Farina and his colleagues demonstrated that physical attractiveness and psychopathology were correlated in hospitalized psychiatric patients (Farina, Fischer, Sherman, Smith, Groh, & Mermin, 1977). Psychiatric patients were found to be less attractive than normal controls. More important, within the hospital population, physical attractiveness was related to several other variables. Less attractive patients were less socially responsive in an interview, had more severe diagnoses, were hospitalized for longer periods, and received fewer visitors from the community. Physical attractiveness was related to duration of hospitalization independently of severity of pathology. Interestingly, unattractive patients had relatively poor interpersonal relationships even during childhood and adolescence. These findings suggest that patients in general, but particularly those who are less attractive, may suffer poor interpersonal relationships throughout their development which continues through their lengthy hospitalization. The paucity of visitors and the lengthy duration of hospitalization suggest that important social concomitants are related to overall attractiveness.

Problems in diverse aspects of interpersonal interaction have been demonstrated in a large number of studies extending beyond social withdrawal. For example, research has demonstrated that schizophrenic patients have difficulty in interpreting nonverbal cues of others in interpersonal situations (Argyle, Alkema, & Gilmour, 1971; Newman, 1977). Apparently, schizophrenics are less sensitive to nonverbal cues than "normals" in social interaction. Reduced sensitivity to cues presented to them might account for inappropriate interpersonal interactions even if the patients had the appropriate responses in their repertoires.

In addition, psychiatric patients, relative to normals, evince deficits in generating solutions to interpersonal problems (Platt, Siegel, & Spivack, 1975; Platt & Spivack, 1972). Although patients can recognize appropriate resolutions to interpersonal problems, they apparently have difficulty in generating the means of solving problems on their own. In addition, psychiatric patients have been found to be less able to generate the rationale for choosing a particular course of action in interpersonal problem situations (Platt *et al.*, 1975). Other deficits in interpersonal interactions have been noted. Schizophrenic patients less readily disclose personal feelings in interpersonal situations and indeed engage in avoidance responses in these situations (Levy, 1976; Shimkunas, 1972).

INSTITUTIONAL TREATMENT

Throughout the history of psychiatry, different treatments have been proposed that stress the importance of the social behavior of the institutionalized patient. These treatments differ in conception and design but are based on the notion that a patient's inappropriate behaviors can be traced to problems in social interaction. Although conceptually independent, the positions implied by these different treatments maintain that developing appropriate social interaction is an important end in itself and essential for community placement independently of the origin of the patient's problems.

Social behaviors occupy an important role in the definition of behavior of psychotic patients as evident by withdrawal, irrational statements, blunted affect, and difficulty in communication skills. These behaviors do not begin to exhaust the symptoms of psychiatric patients but illustrate the role of social behavior in identifying psychopathology. The importance of social skills in psychopathology can also be conveyed by highlighting three different treatments—moral treatment, milieu therapy, and the token economy.

Moral Treatment

Moral treatment was the result of a humanitarian reform in the 18th and 19th centuries (see Bockoven, 1963). Through the efforts of Pinel and Esquirol in France and Tuke in England during this period, hospitals were established which eliminated the cruel and punitive treatments that had characterized existing asylums. Under the humanitarian reform, psychiatric patients received improved care and attempts were made to equate treatment of mental illness with that of physical illness.

Moral treatment was based on the view that a patient's social functioning had deteriorated because of the stresses to which he or she had been subjected. Psychological and emotional stresses were considered to be primarily responsible for abnormal behavior. The term "moral" treatment was used to stress the psychological nature of the disorder. Also, the term "moral" conveyed the general view of this form of treatment, namely that the individual patient had a moral responsibility and treatment placed considerable responsibility for improvement on the patient (Bockoven, 1963).

Moral treatment consisted of providing comfortable living conditions for the patient in a family-like atmosphere. Facilities were small to help sustain this family-like organization and so that staff and patients could know each other. During the day, patients could engage in a variety of purposeful activities, including games, sports, social activities, and outdoor hobbies.

Many aspects of moral treatment stressed the role of social interaction. Indeed, social behavior constituted a crucial part of treatment. The home-like atmosphere of the setting was enhanced by frequent meetings of patients and staff. Staff and patients would usually eat meals together in a congenial atmosphere and have group meetings. The attending physician usually knew each patient personally and frequently interacted with each one. Some of the treatment consisted of dyadic interaction where a patient and the physician conversed. The conversations occasionally took the form of distracting the patient from unpleasant feelings and emotional states (Carlson & Dain, 1960). In general, attempts were made in the setting to reduce the social distance between staff and patients and to foster appropriate social interaction and personal responsibility.

Additional social factors pertain to the types of behaviors that were modeled by the patients and the expectations of the staff for appropriate social deportment. Staff modeled appropriate interpersonal behavior by providing concern, understanding, and respect with the view that these would be reciprocated by the patients. In addition, the staff expected patients to engage in socially appropriate behavior. Finally, the freedom of the setting and encouragment of activity allowed patients to pursue and develop various interests (e.g., in manual and intellectual activities) that would provide alternatives to inappropriate behavior. The congenial physical and social environments were considered sufficient to provide the conditions for patient improvement.

As is well known, the historical verdict on the success of moral treatment is generally quite favorable. Moral treatment was associated with high discharge rates. Moreover, the decline of moral treatment and the increase of custodial care in larger institutional facilities were associated with marked decreases in discharge and increases in the rate of chronic patients (Bockoven, 1963; Dain, 1964).

Milieu Therapy

After moral treatment declined, a medical and custodial approach characterized treatment. Patients were considered to be passive agents and subject to treatment or merely housed in a large custodial facility. Eventually, long after the decline of moral treatment, a technique emerged with somewhat similar principles and procedures. This procedure, referred to as "milieu therapy," represented a refinement of moral treatment in the sense that the social conditions considered to produce change were incorporated more systematically. As with moral treatment, milieu therapy is committed to developing the social responsibility of the patients.

Milieu therapy, also sometimes referred to as the "therapeutic community," can be traced to the period after War World War II (Jones, 1953; Main, 1946), although earlier procedures can be identified in the history preceding the war.[1] The concept of a community stresses the

[1] Occasionally, the "therapeutic community" and "milieu therapy" are distinguished (see Rossi & Filstead, 1973). Yet, the distinction is not widely ascribed to. In general, milieu therapy has not been concretely defined in a consistent fashion, making it difficult to draw even finer distinctions among possibly related terms (Magaro, Gripp, & McDowell, 1978).

importance of the social structure of the individual ward or total hospital as a central feature of treatment.

Milieu therapy is based on the view that psychiatric disorders are manifestations of inappropriate interpersonal behaviors. Inappropriate behaviors can be alleviated by providing the appropriate social atmosphere and expectations in the hospital that foster prosocial behavior. The hospital is viewed as a community in which sociopsychological forces can be brought to bear to influence patient behavior. A hospital ward based on milieu treatment usually houses only a small number of patients (e.g., up to 40) (Almond, 1971). Patients are given responsibility for their own behavior in a number of different ways. The central therapeutic ingredient is considered to be group discussion meetings (Lewis, Beck, King, & Stephen, 1971). Discussions are considered to facilitate the communication of feelings, attitudes, and values, and to build up group cohesiveness. Yet, the goal extends beyond expression of feelings. The group is utilized to enhance the limited and inadequate communication skills thought to characterize psychotic patients, to improve social functioning, and to make the individual patient sensitive to social influences.

The group meetings usually include patients and staff. A number of interpersonally relevant topics are addressed such as handling rejection, dealing with dependence and independence, making decisions, and resolving interpersonal tensions likely to arise as a part of group living. Patients are given an equal or near-equal role in reaching important decisions. Patients are encouraged and expected to make suggestions for management of ward functioning, their own treatment, and occasionally even to develop plans for their discharge.

Patients are also encouraged to participate in work, recreational, and educational activities. Participation in activities is assumed to enhance development of an individual's responsibility for his or her own behavior and to instill personal competence. At the same time many activities are social in nature (e.g., games, recreation) and thought to provide important interpersonal conditions for developing social behaviors within the group (Visher & O'Sullivan, 1970).

In general, milieu therapy is characterized by democratization of the hospital ward. Staff and patients work cooperatively and an attempt is made to break down traditional barriers about status differences among staff and patients and the medical orientation where one party treats another. Patients are expected to take an active and major role in their own rehabilitation and the rehabilitation of others.

In general, much of the evidence evaluating the effects of milieu therapy has suggested that improvements are achieved in discharging patients from hospitals (e.g., Dodd & Petrovich, 1967; Ellsworth, 1964;

Ellsworth & Stokes, 1963; Sanders, Smith, & Weinman, 1967). Yet, the results are not clear for several reasons. Some of the comparisons have been retrospective, comparing current discharge rates of milieu therapy with rates during a previous period. Differences in discharge philosophy might account for the results. Second, some demonstrations showing significant improvements in discharge rates associated with milieu therapy have reported that these effects are lost when relapse rates are examined (e.g., Galioni, Adams, & Tallman, 1953; Sanders *et al.*, 1967). Finally, improvements can rarely be attributed to any specific aspect of milieu therapy as opposed to some form of active therapy superimposed over custodial care. Indeed, occasionally milieu therapy has not shown that such ingredients as group meetings and activities even enhance patient performance in the hospital (Visher & O'Sullivan, 1970). Yet, these restrictions on the results do not dismiss all of the generally favorable outcome data. Indeed, recent investigations have suggested that milieu therapy can lead to reduction of inappropriate behavior on the ward and increased discharge rates and community adjustment (Greenberg, Scott, Pisa, & Friesen, 1975; Paul & Lentz, 1977).

The Token Economy

Recently, reinforcement programs have been applied with institutionalized psychiatric patients. The programs, referred to as "token economies," are designed to restructure the entire ward environment so that concrete incentives are provided for adaptive and prosocial behavior (Kazdin, 1977). Incentives in the form of tokens (marks, points, coins, checkmarks) are provided for adaptive behaviors that are considered to be incompatible with inappropriate behavior. The tokens can be exchanged for a wide variety of backup rewards such as privileges, activities, and special events that the patients regard as rewarding. The general philosophy underlying the token economy is that the usual institutional environment ordinarily helps foster maladaptive behaviors and apathy. Behaviors need to be encouraged and directly rewarded which compete with apathy and help foster skills that will be adaptive in the community.

Many token economies have focused on behaviors such as grooming, dressing, bathing, and attending ward activities. Programs focusing on adaptive behaviors have shown improvements in many social and interpersonal skills such as increased patient cooperativeness, social interaction, and communication skills (e.g., Maley, Feldman, & Ruskin, 1973; Shean & Zeidberg, 1971). Other programs have focused on specific bizarre behaviors such as hallucinations, irrational speech, depression, and multiple symptoms characteristic of psychoses (e.g., Gripp & Magaro, 1971; Hersen, Eisler, Alford, & Agras, 1973; Schwartz & Bellack, 1975). Bizarre behaviors have been effectively altered either by focusing directly on the performance or by developing other adaptive behaviors on the ward.

Several programs have focused more specifically on social interaction. Usually, very specific social responses have been focused on, such as greeting others, being in close proximity to and conversing with staff or other patients, participating in group meetings, and making suggestions for improvements in patient care or treatment (e.g., Doty, 1975; Wallace, Davis, Liberman, & Baker, 1973). In other programs, patients are encouraged to become involved in group meetings and to interact in a larger social matrix (e.g., Greenberg *et al.*, 1975; Pomerleau, Bobrove, & Harris, 1972).

An interesting side effect of token programs pertains to the social behavior of the staff who conduct these programs. Apparently, staff who administer token economies engage in more social contact with the patients than occurs with traditional treatment (e.g., Paul & Lentz, 1977). Also, the nature of the social interaction changes. Staff become much more positive in the attention they provide and increase in praise, smiles, and approval for appropriate patient behavior (Trudel, Boisvert, Maruca, & Leroux, 1974). Indeed, staff even come to view patients more favorably and believe that patients can profit more from treatment as a function of administering the program (McReynolds & Coleman, 1972). Aside from the potential benefits of increased social contact with the patients, improvements in attitudes may lead staff to expect more from the patients in terms of normative levels of behavior.

The outcome of token-reinforcement programs has been quite favorable. Patients whose adaptive behaviors are increased through reinforcement techniques have shown higher rates of discharge and lower rates of recidivism (see Kazdin, 1977). Yet, many programs suggest that patients may not perform the behaviors rewarded during treatment as soon as the program is withdrawn. Thus, long-term maintenance of the gains has not been widely demonstrated. Moreover, many of the skills that might be needed for community adjustment are not always explicitly developed in the program. Hence, these programs may need to be supplemented by aftercare programs that sustain treatment gains and ensure that social behaviors adaptive to the community setting continue to be performed.

Recently, a major comparative outcome study was completed over several years that contrasted milieu therapy and social learning therapy (Paul & Lentz, 1977). The social learning therapy included primarily a token economy supplemented with several other behavioral techniques. A range of outcome measures was used while the programs were in effect and up to 1 1/2 years of follow-up when patients had been placed in the community. In general, both programs made improvements over the course of treatment relative to changes achieved by another hospital giving routine care. However, the social learning program led to markedly superior gains in reducing and eliminating bizarre behaviors, improving appropriate social interaction and communication skills, and developing activity and overall adequate functioning across a range of measures. The gains made during treatment were reflected in successful placement in the community.

Another study has suggested that improvements associated with token reinforcement procedures may be enhanced by adding milieu therapy to treatment (Greenberg *et al.*, 1975). In this study, patients participated in a token economy or a token economy supplemented with a milieu approach. Both programs included incentives for self-care, social, and work behaviors, but one group received incentives for group decision making for such issues as devising treatment programs for members of the groups. During the year of the program, patients in the token economy plus milieu therapy spent more days out of the hospital than did patients without the milieu condition.

AFTERCARE AND COMMUNITY ADJUSTMENT

Posthospital Performance

Research from different lines of inquiry has pointed to the importance of social behaviors and interpersonal functioning of patients after they leave treatment. To begin with, research encompassing several somatic, pharmacological, and psychotherapeutic treatments has shown that treatment provided in the hospital has little relationship to community adjustment (Fairweather, Sanders, Maynard, & Cressler, 1969; Freeman & Simmons, 1963; Staudt & Zubin, 1957). Recent evidence has suggested more favorable results for specific treatments such as social learning programs and foster home care (e.g., Greenberg *et al.*, 1975; Linn, Caffey, Klett, & Hogarty, 1977; Paul & Lentz, 1977).

Several studies have suggested that institutional programs may not provide adequate social skills to facilitate successful community adjustment. The successes of institutional programs, when evident, usually are reflected in positive changes in such areas as self-care, antisocial behavior, and symptomatology (Berger, Rice, Sewell, & Lemkau, 1963; Gurel, 1965; McPartland & Richart, 1966). Fewer gains have been noted in areas related to community adjustment including adequacy in family and social relationships. In a comprehensive evaluation of patient change, Ellsworth, Foster, Childers, Arthur, and Kroeker (1968) found that hospital treatment did not markedly improve such instrumental role behaviors of patients as social contact and interpersonal relations. In contrast, symptomatic improvements were demonstrated.

The demands of institutional life may even compete with developing social skills that later will foster successful adjustment after discharge. For example, Ellsworth *et al.* (1968) found that hospital treatment was associated subsequently with negative effects on patient productivity and motivation and only negligible increases in social activity. Also, patients were found to be cooperative and compliant in the hospital; these characteristics were considered to reflect the demands of hospital treatment that might be maladaptive after discharge.

A central feature in evaluating institutional treatment is the extent to which behaviors in the institution reflect or are associated with subsequent adjustment. Performance along many dimensions in the hospital has often been found to be uncorrelated with or even negatively correlated with later adjustment. For example, severity of anxiety in the hospital has been found to be negatively correlated with acceptability of behavior and social skills after the patient is discharged (Ellsworth *et al.*, 1968). The lack of relationship between improvements in symptoms and social behavior, mentioned above, also reflects the possible limited relevance of improvements in the hospital and social behavior. Indeed, it may not be the symptomatic behaviors in the hospital that predict how well patients will do when released. Rather, social behavior in the hospital tends to be correlated with community adjustment (Fairweather, 1964; Fairweather *et al.*, 1969).

Successful stay in the community is strongly associated with several signs of adequate social adjustment. A major area of adjustment is employment. Ex-patients who are functioning marginally or who return to the hospital have a higher rate of unemployment than those who remain in the community (Brown, 1959; Freeman & Simmons, 1963; Lamb, 1968). Another area of adjustment pertains to social activities. Ex-patients who remain in the community have greater participation and membership in social groups than those who return to the hospital (Davis, Freeman, & Simmons, 1957; Freeman & Simmons, 1963).

The pattern of social interaction among ex-patients vis-à-vis successful community stay is particularly interesting. Research has suggested that individuals who are successful in staying in the community live in less isolation from their friends than those who are rehospitalized. On the other hand, successful-stay patients have been found to have fewer contacts with their relatives than those who are rehospitalized (Freeman & Simmons, 1963).

Lack of contact with one's relatives may depend on the type of interactions in which relatives engage. Brown, Birley, and Wing (1972) found that the quality of the relationship between a schizophrenic patient and the relative with whom he lives predicts relapse within the 9 months after discharge. Through interviews, relatives were classified on the basis of *expressed emotion,* which included such characteristics as hostility and the number of critical comments made by the relative about the patient and his disorder. Patients returning to a relative high in expressed emotion showed a significantly higher relapse rate than those returning to a relative low in the level of expressed emotion. Relapse was less likely among patients from high-expressed-emotion homes if the patients either were on medication or avoided close contact with the family. These findings were replicated by Vaughn and Leff (1976), who also demonstrated the lack of patient contact with their relatives who are high in emotion expressed is associated with a better prognosis than patients in more frequent contact with such relatives. The findings that avoidance of relatives with certain types of social interaction patterns is associated with improved relapse rates strongly suggests the importance of interpersonal variables in rehabilitation of schizophrenic patients (Brown *et al.*, 1972; Vaughn & Leff, 1976).

Additional aspects of family relationships are associated with the community adjustment of psychiatric patients. Freeman and Simmons (1963) found that ex-patients living in conjugal families were more likely to perform at higher levels in occupational and social behavior than those patients living with their parents or siblings. Also, family expectations and role requirements of the ex-patient relate to performance. In families with fewer full-time workers, ex-patients were more likely to perform at higher occupational levels than were ex-patients in families with more full-time workers. For males, work and social performance were likely to be better if they were regarded as the chief breadwinner than if they were not.

Evidence suggests that patients who remain in the community have relatives who place more demands on them for instrumental role performance (Freeman & Simmons, 1963). For example, patients whose relatives expect high levels of accomplishment perform better on a range of occupational, social, and household activities than do patients with relatives of low expectations. And the expectations are related to ultimate success in staying in the community, with higher expectations being associated with greater success in community stays.

Community-Based Treatments

The above discussion suggests the role of several variables that influence community adjustment. The variables encompass the role expectations placed on the patient and the extent to which the individual functions interpersonally with his or her relatives. A few of the variables shown to influence community stay have been actively incorporated into programs that have attempted to improve community adjustment.

A number of different programs have been designed to improve community stay. These programs are directed primarily at the findings showing the lack of relationship between institutional treatment and relapse rate and at the debilitating effects that institutions are likely to have on social behavior. Since social behaviors are strongly implicated in community adjustment, programs have attempted to enmesh treatment into community resources to facilitate the transition from treatment to the community. Two such efforts illustrate the general thrust of these efforts at enhancing successful community placement.

Lodge Placement. One of the most dramatic demonstrations of treatment designed to facilitate community adjustment and placement of chronic psychiatric patients grew out of a series of studies by Fairweather and his colleagues (Fairweather, 1964; Fairweather *et al.*, 1969). Fairweather's initial research began by evaluating different treatments used in inpatient psychiatric care. The treatments (individual or group psychotherapy, living and working together in small groups) were found not to differ in terms of community adjustment after release relative to patients who received traditional care. Improvements in the hospital setting simply were unrelated to community adjustment. Rather, community adjustment was related to premorbid status of the patients (e.g., whether patients were chronic or acute). After release from the hospital, those who had shown improvements in hospital adjustment did not perform better in remaining in the community than did those who had not improved.

To help the transition from hospital to community life, Fairweather (1964) developed a socially based program in the hospital. Patients were placed in problem-solving groups. The group members were given responsibility for taking care of each other and solving problems that developed as a part of hospital and group living. Individuals given problem-solving tasks presumably might be able to cope better with community adjustment. Although the groups functioned well within the hospital, their relapse rate proved to be similar to patients receiving traditional care.

Treatment was extended in a subsequent study to continue the group living conditions into the community (Fairweather *et al.*, 1969). This extension was not only based on the previous demonstration that in-hospital improvement was unrelated to community stay, but also on the finding that a patient's posthospital living conditions were strongly related to relapse and readmission. In this extension, patients participated in the small groups within the hospital but differed in the manner in which they were discharged. One group of patients was discharged to a lodge, a dormitory situation where the group could be kept intact. In addition, work was provided along with extensive supervision by professional and lay consultants who helped organize the living conditions to facilitate the transition from hospital to community life. Eventually, group members were able to manage by themselves and became completely autonomous from the hospital.

The treatment program within the hospital required individual responsibility for various tasks including self-care, punctuality on job assignments, and acceptable job completion. Money and passes away from the hospital were used as incentives. Patients met on their own in small groups to evaluate each other's progress and to make decisions about such things as job assignments and dispensation of money and privileges. In addition, the group had the task of solving individual patient problems that arose. After successful progress in individual tasks, patients were discharged to a lodge.

The lodge initially raised many problems related to features of everyday life such as developing rules, providing work, and continuing adherence to medication. Eventually, patient group meetings and various forms of patient governing bodies were developed to take over the supervision and organization that initially had been handled by professionals and lay consultants.

A major feature of the lodge was to provide employment for the patients. As mentioned earlier, employment status has been related to success in staying in the community. Employment at the lodge consisted of janitorial, maintenance, and gardening tasks, which provided wages and a means of support for the residents. The business was patient owned and operated and provided services to the local urban areas.

Follow-up evaluation conducted up to 40 months after treatment revealed that patients discharged to the lodge remained in the community longer and had higher rates of successful employment than did patients who had been discharged in the usual fashion. Since both the lodge and the usual discharge group received the small-group treatment program, the posthospital program appeared to account for change. Interestingly, on a number of outcome criteria patients who had been discharged to the lodge or to the community did not differ. Both groups were generally satisfied with community living and evinced similar social behaviors. Hence, individual psychosocial adjustment did not appear to be the crucial factors. However, the major differences in duration of community stay and employment were marked.

Halfway House. Halfway houses consist of transitional living facilities that are intermediate between the psychiatric hospital and community living (Raush & Raush, 1968). The rationale for halfway houses is based on concern over the debilitating conditions of custodial institutions. Traditional custodial institutions have not proven particularly effective in returning patients to community life. To facilitate reentry into the community, patients are often discharged to halfway houses. These houses are structured similarly to boarding houses. A relatively small number of patients (e.g., usually 6–15 residents) live in the house. A small number is utilized to facilitate home-like living conditions. The facility usually consists of old residences (e.g., convalescent homes, motels, apartments, town houses) where a small number of patients can reside. Extensive support is given to encourage normalized patterns of living.

The residents may share in the management of the facility and complete chores for their own self-care or for the successful running of the house. Nonprofessional staff may direct the house but do not usually have responsibility for direct treatment. Treatment in the form of medication or psychotherapy may be provided to the residents but usually is conducted at other locations (e.g., hospital, outpatient clinic). Group interaction may be encouraged by small meetings to solve problems that arise in the house, but this is not viewed as treatment *per se*. Rather, the goal of the house is to help ex-patients to adjust to the demands of community living.

The halfway house is usually located in an urban area where direct contact with community facilities is available. To facilitate community adjustment and self-sufficiency, residents are encouraged to obtain employment either on their own or through a rehabilitation counselor. Also, residents are encouraged to make plans for leaving the facility and to become completely self-sufficient. Many leave the facility eventually and return to their families or to foster home care (Raush & Raush, 1968).

The halfway house and lodge-placement procedures, mentioned earlier, sample the efforts to facilitate community adjustment of ex-patients. These programs, and others like them, emphasize the importance of providing ex-patients with a means of reentering community life. The success of reentry appears to depend on the patients' social skills, as reflected in such areas as developing friendships or peer relations and in securing employment. Abrupt reentry into the community after institutional treatment is less successful than gradually entering into the many demands of community functioning.

IMPLICATIONS FOR TREATMENT

Earlier sections of this chapter have highlighted the pervasive role that social factors and interpersonal variables play in psychopathology.

Identification of individuals who are diagnosed as "mentally ill" has been viewed by many authors as indicative of problems of interpersonal behavior rather than specific symptoms of psychopathology. The role of social behaviors rather than specific symptoms alone in initiating the career of a psychiatric patient is suggested in part by the pervasiveness of psychiatric impairment among individuals who are not hospitalized and who successfully function in the community. This consistent finding has raised questions about the reasons why some people are hospitalized and others are not. The answers do not lie simply in severity of impairment but rather encompass the ability of the individual to function in everyday life.

The problems in social behaviors that psychiatric patients evince apparently do not simply emerge immediately preceding their entrance into the career of a patient. Persons later seen in treatment have often shown a history of poor social relations and withdrawal from interpersonal relationships. Possibly, there would be value in identifying individuals with such interpersonal deficits prior to the point at which this is compounded with additional problems of pathology in order to intervene early. Indeed, improvements in coping skills in general, and particularly in the area of social interaction, have been suggested as possible factors that might decrease a susceptibility to schizophrenia (Zubin & Spring, 1977).

After a person has been diagnosed as psychiatrically impaired and is hospitalized, social variables and interpersonal behaviors continue to play a major role. Patients may show many different signs of social behavior in the kind of information they possess, how they spend their time in the hospital, the number of contacts with visitors, and so on (e.g., Braginsky, Braginsky, & Ring, 1969; Brown, 1959; Depue & Dubicki, 1974). Patients who evince greater social interaction and less withdrawal spend less time in the hospital and stay in the community longer than those who show less adequate social skills.

Of course, the evidence is correlational and shows only that social skills are related to such criteria as responsiveness to treatment and community stay. It does not necessarily follow that altering social skills will improve hospital and community adjustment. It may well be that some other variable that accounts for both severe pathology and poor social skills is the critical one to alter. However, social skills are related to adjustment after treatment. The many different ways in which social problems become manifest in community adjustment argue strongly for its importance in the ultimate success in rehabilitation.

After a patient leaves a hospital, the probability for staying in the community is related to many different interpersonal skills. Ex-patients who have more frequent contact with friends, as opposed to those who remain relatively isolated, adjust better and remain outside of the hospital (Freeman & Simmons, 1963). An important focus of treatment might be in developing skills that would help foster friendships. This might be accomplished in different ways. First, ex-patients can be placed in a group situation such as the community-based lodge discussed earlier (Fairweather *et al.*, 1969) where they interact with individuals they knew in the hospital. Indeed, transplanting several patients already functioning as a cohesive group from the hospital to the community would seem to be a viable treatment approach. Second, patients may receive training in interpersonal skills that directly focuses on developing friendships. As part of this, individuals could be encouraged to affiliate with social groups since participation in formal groups is also related to community stay.

The importance of social skills is suggested in findings pertaining to interaction with one's family. Evidence reviewed earlier suggests that the kind of family to which a patient returns is related to successful community stay. Social skills training might reduce the dependency of success on family determinants and provide greater independence for social adjustment on the coping skills that the patient has when coming into contact with his or her family.

Patients who return to families that are high on the expressed emotion variable tend to have greater relapse rates. Patients who avoid contact with these relatives have a better chance of remaining in the community. While it may be useful to train individuals to avoid certain interactions that precipitate or exacerbate their condition, this is unlikely to be wise as a general strategy. Avoidance behavior in social situations is not something that most patients need to learn; it apparently has been acquired all too well long before hospitalization. On the other hand, it is likely to be worthwhile to train individuals to handle interpersonal relations such as those that arise in problematic family situations. Such coping skills are more likely to permit development of positive relationships not only with one's family but with others as well. Of course, the training in social skills should not be restricted to interactions with friends and families. Research has demonstrated that psychotic patients lack skills in solving interpersonal problems more generally and tend to be insensitive to many interaction cues to which nonpsychotic patients normally react.

Other areas of functioning related to successful community tenure no doubt would profit from social skills training. For example, ex-patients who are not successfully employed have a higher relapse rate than those who are. Employment is no doubt multiply determined and cannot be traced to a single area such as social deficits. Yet, it is clear as well that social skills are directly related to procuring jobs independently of the performance skills for the specific job task. Indeed, job-training programs focus on several areas of interpersonal functioning including preparation for meeting an employer, interaction with a potential employer, and problem-solving skills related to the job (cf. Azrin, Flores, & Kaplan, 1975). Similarly, marital status is strongly related to community adjustment, as well as to initial hospitalization. Those individuals returning to conjugal roles perform better in the community than those who do not. Skills in handling problems related to sustaining a meaningful interpersonal relationship might enhance community tenure.

CONCLUSION

Social factors and interpersonal behaviors play a systematic role in psychiatric diagnosis, treatment, and adjustment of patients to community life. Of course, patients suffer more problems than difficulties in social skills. Yet, problems in social behavior appear as a consistent theme throughout the career of a psychiatric patient. The presence of such interpersonal problems as isolation and avoidance of relationships suggests intriguing treatment possibilities. Specifically, developing social skills would seem to be of potential value at different developmental stages in the career of someone who eventually becomes a psychiatric patient.

Evidence reviewed earlier suggests that persons diagnosed as psychotic have a long history of interpersonal problems, particularly those patients with a poor prognosis. This suggests the value of training in social skills early in life as an attempt to decrease the risk of identified psychiatric impairment later in life. Social skills training may not be sufficient to eliminate the many symptomatic behaviors that psychotic patients manifest. Yet, it would be worth examining whether treating individuals with severe interpersonal problems early in life influences subsequent psychiatric impairment, responsiveness to later treatment, and prognosis in general.

After persons are identified as candidates for psychiatric treatment, social behaviors continue to be relevant as a treatment focus. Social behavior in the hospital is associated with responsiveness to treatment and successful community stay after release. Social skills training programs can focus directly on returning patients to the specific situations in which interpersonal problems exist. Integrating social skills training with functioning in the community would help patients test the progress made in treatment. Specific areas of social skills training may warrant attention, such as developing contacts with friends, participating in social groups and activities, procuring and sustaining employment, developing heterosexual relationships, and interacting adequately with family members. Programs in social skills have yet to examine the different areas of interpersonal functioning that present problems for psychiatric patients and the relative impact that training in separate areas has for successful community adjustment.

Many different treatments for psychiatric patients have recognized the importance of social behavior in severe psychopathology. Programs such as moral treatment, milieu therapy, token economies, community-lodge placement, and halfway houses, highlighted earlier, either provide conditions where social behavior may develop or include experiences designed to foster appropriate interaction, Yet, few programs have concretely identified the specific social skills deficits that patients evince and then systematically develop behavior to rectify these deficits. The present chapter has identified some of the areas where social behaviors are problematic for psychiatric patients and warrant further examination as a focus of treatment.

REFERENCES

Adams, H. B. "Mental illness" or interpersonal behavior? *American Psychologist*, 1964, *19*, 191–197.

Almond, R. The therapeutic community. *Scientific American*, 1971, *224*, 34–42.

Argyle, M., Alkema, F., & Gilmour, R. The communication of friendly and hostile attitudes. *British Journal of Social and Clinical Psychology*, 1971, *10*, 386–401.

Azrin, N. H., Flores, T., & Kaplan, S. J. Job-finding club: A group-assisted program for obtaining employment. *Behaviour Research and Therapy*, 1975, *13*, 17–27.

Barthell, C. N., & Holmes, D. S. High school yearbooks: A nonreactive measure of social isolation in graduates who later became schizophrenic. *Journal of Abnormal Psychology*, 1968, *73*, 313–316.

Becker, H. S. *Outsiders: Studies in the sociology of deviance*. New York: The Free Press, 1963.

Becker, W. C. A genetic approach to the interpretation and evaluation of the process-reactive distinction in schizophrenia. *Journal of Abnormal and Social Psychology*, 1956, *53*, 229–336.

Berg, M., & Cohen, B. B. Early separation from the mother in schizophrenia. *Journal of Nervous and Mental Disease*, 1959, *128*, 365–369.

Berger, D. G., Rice, C. E., Sewall, L. G., & Lemkau, P. V. Factors affecting the adequacy of patient community adjustment information obtained from the community. *Mental Hygiene*, 1963, *47*, 452–460.

Bockoven, J. S. *Moral treatment in American psychiatry*. New York: Springer, 1963.

Bower, E. M., Shellhamer, T. A., & Daily, J. M. School characteristics of male adolescents who later became schizophrenic. *American Journal of Orthopsychiatry*, 1960, *30*, 712–729.

Braginsky, B. M., Braginsky, D. D., & Ring, K. *Methods of madness: The mental hospital as a last resort*. New York: Holt, Rinehart and Winston, 1969.

Brown, F. Depression and childhood bereavement. *Journal of Mental Science*, 1961, *107*, 754–777.

Brown, G. W. Experiences of discharged chronic schizophrenic patients in various types of living groups. *Milbank Memorial Fund Quarterly*, 1959, *37*, 105–131.

Brown, G. W., Birley, J. L. T., & Wing, J. K. Influence of family life on the course of schizophrenic disorders: A replication. *British Journal of Psychiatry*, 1972, *121*, 241–258.

Buss, A. H. *Psychopathology*. New York: Wiley, 1966.

Carlson, E. T., & Dain, N. The psychotherapy that was moral treatment. *American Journal of Psychiatry*, 1960, *117*, 519–524.

Cooper, A. B., & Early, D. F. Evolution in the mental hospital: Review of a hospital population. *British Medical Journal*, 1961, *1*, 1600–1603.

Dain, N. *Concepts of insanity in the United States, 1789–1865*. New Brunswick, N.J.: Rutgers University Press, 1964.

Davis, J. A., Freeman, H. E., & Simmons, O. G. Rehospitalization and performance level among former mental patients. *Social Problems*, 1957, *5*, 37–44.

Depue, R. A., & Dubicki, M. D. Hospitalization and premorbid characteristics of withdrawn and active schizophrenics. *Journal of Consulting and Clinical Psychology*, 1974, *42*, 628–632.

Derogatis, L. R., Yevzeroff, H., & Wittelsberger, B. Social class, psychological disorder, and the nature of the psychopathologic indicator. *Journal of Consulting and Clinical Psychology*, 1975, *43*, 183–191.

Dodd, B. B., & Petrovich, D. V. Out of the back ward. *American Journal of Nursing*, 1967, *67*, 2124–2128.

Dohrenwend, B. S., & Dohrenwend, B. P. (Eds.), *Stressful life events: Their nature and effects*. New York: Wiley, 1974.

Doty, D. W. Role playing and incentives in the modification of the social interaction of chronic psychiatric patients. *Journal of Consulting and Clinical Psychology*, 1975, *43*, 676–682.

Ellsworth, R. B. The psychiatric aide as rehabilitation therapist. *Rehabilitation Counseling Bulletin*, 1964, *7*, 81–86.

Ellsworth, R. B., & Stokes, H. A. Staff attitudes and patient release. *Psychiatric Studies and Projects*, 1963, *7*, 1–6.

Ellsworth, R. B., Foster, L., Childers, B., Arthur, G., & Kroeker, D. Hospital and community adjustment as perceived by psychiatric patients, their families, and staff. *Journal of Consulting and Clinical Psychology*, 1968, *32* (5, P. 2), Monograph Supplement.

Essen-Möller, E. Individual traits and morbidity in a Swedish rural population. *Acta Psychiatrica et Neurologica Scandinavica*, 1956, suppl. 100.

Fairweather, G. W. (Ed.). *Social psychology in treating mental illness: An experimental approach*. New York: Wiley, 1964.

Fairweather, G. W., Sanders, D. H., Maynard, A., & Cressler, D. L., *Community life for the mentally ill*. Chicago: Aldine, 1969.

Farina, A., & Webb, W. W. Premorbid adjustment and subsequent discharge. *Journal of Nervous and Mental Disease*, 1956, *124*, 612–613.

Farina, A., Garmezy, N., & Barry, H. Relationship of marital status to incidence and prognosis in female schizophrenic patients. *Journal of Abnormal and Social Psychology*, 1963, *67*, 624–630.

Farina, A., Holland, C. H., & Ring, K. The role of stigma and set in interpersonal interaction. *Journal of Abnormal Psychology*, 1966, *71*, 421–428.

Farina, A., Gliha, D., Boudreau, L. A., Allen, J. G., & Sherman, M. Mental illness and the impact of believing others know about it. *Journal of Abnormal Psychology*, 1971, *77*, 1–5.

Farina, A., Fischer, E. H., Sherman, S., Smith, W. T., Groh, T., & Mermin, P. Physical attractiveness and mental illness. *Journal of Abnormal Psychology*, 1977, *86*, 510–517.

Faris, R. E. L., & Dunham, H. W. *Mental disorders in urban areas.* Chicago: University of Chicago Press, 1939.

Ferster, C. B. Classification of behavioral pathology. In L. Krasner & L. P. Ullmann (Eds.), *Research in behavior modification.* New York: Holt, Rinehart and Winston, 1965.

Freedman, R. *Recent migration to Chicago.* Chicago: University of Chicago Press, 1950.

Freeman, H. E., & Simmons, O. G. *The mental patient comes home.* New York: Wiley, 1963.

Fried, M. Social problems in psychopathology. In L. J. Duhl (Ed.), *Urban America and the planning of mental health services.* New York: Group for the Advancement of Psychiatry, 1964.

Galioni, E. G., Adams, F. H., & Tallman, F. F. Intensive treatment of backward patients—a controlled pilot study. *American Journal of Psychiatry,* 1953, *109,* 576–583.

Greenberg, D. J., Scott, S. B., Pisa, A., & Friesen, D. D. Beyond the token economy: A comparison of two contingency programs. *Journal of Consulting and Clinical Psychology,* 1975, *43,* 498–503.

Gregory, I. Studies of parental deprivation in psychiatric patients. *American Journal of Psychiatry,* 1958, *115,* 432–442.

Gripp, R. F., & Magaro, P. A. A token economy program evaluation with untreated control ward comparisons. *Behaviour Research and Therapy,* 1971, *9,* 137–149.

Gurel, L. *Patterns of mental patient posthospital adjustment.* Washington, D.C.: Veterans Administration Psychiatric Evaluation Project, 1965.

Hersen, M., Eisler, R. M., Alford, G. S., & Agras, W. S. Effects of token economy on neurotic depression: An experimental analysis. *Behavior Therapy,* 1973, *4,* 392–397.

Hollingshead, A. B., & Redlich, F. C. *Social class and mental illness.* New York: Wiley, 1958.

Jaco, E. G. The social isolation hypothesis and schizophrenia. *American Sociological Review,* 1954, *19,* 567–577.

Jenkins, R. L., & Gurel, L. Predictive factors in early release. *Mental Hospital,* 1959, *10,* 11–14.

Jones, M. *The therapeutic community.* New York: Basic Books, 1953.

Kantor, M. B. Internal migration and mental illness. In S. C. Plog and R. B. Edgerton (Eds.), *Changing perspectives in mental illness.* New York: Holt, Rinehart and Winston, 1969.

Kantor, R. E., Wallner, J. M., & Winder, C. L. Process and reactive schizophrenia. *Journal of Consulting Psychology,* 1953, *17,* 157–162.

Kaplan, E., Ree, R. B., & Richardson, W. A comparison of the incidence of hospitalized and non-hospitalized cases of psychoses in two communities. *American Sociological Review,* 1956, *21,* 472–479.

Kazdin, A. E. *The token economy: A review and evaluation.* New York: Plenum, 1977.

Kitsuse, J. I. Societal reaction to deviant behavior. Problems of theory and method. *Social Problems,* 1962, *9,* 247–257.

Kleiner, R. J., & Parker, S. Goal-striving, social status, and mental disorder: A research review. *American Sociological Review,* 1963, *28,* 189–203.

Kleiner, R. J., & Parker S. Social mobility, anomie, and mental disorder. In S. C. Plog & R. B. Edgerton (Eds.), *Changing perspectives in mental illness.* New York: Holt, Rinehart and Winston, 1969.

Kohn, M. L., & Clausen J. A. Social isolation and schizophrenia. *American Sociological Review,* 1955, *20,* 265–273.

Laing, R. D. *The politics of experience.* New York: Pantheon, 1967.

Lamb, H. R. Release of chronic psychiatric patients into the community. *Archives of Psychiatry,* 1968, *9,* 38–44.

Langner, T. S., & Michael, S. T. *Life stress and mental health: The Midtown Mahattan study.* London: Free Press, 1963.

Lazarus, J., Locke, B. Z., & Thomas, D. S. Migration differentials in mental disease. *Milbank Memorial Fund Quarterly*, 1963, *41*, 25–52.

Leighton, A. H. *My name is Legion*. New York: Basic Books, 1959.

Leighton, A. H., Lambo, T. A., Hughes, C. C., Leighton, D. C., Murphy, J. M., & Macklin, D. B. *Psychiatric disorder among the Yoruba*. Ithaca, N.Y.: Cornell University Press, 1963.

Lemert, E. M. Legal commitment and social control. *Sociological Society for Research*, 1946, *30*, 370–378.

Lemert, E. M. Paranoia and the dynamics of exclusion. *Sociometry*, 1962, *25*, 2–20.

Levy, S. M. Schizophrenic symptomatology: Reaction or strategy? A study of contextual antecedents. *Journal of Abnormal Psychology*, 1976, *85*, 435–445.

Lewis, D. J., Beck, P. R., King, H., & Stephen, L. Some approaches to the evaluation of milieu therapy. *Canadian Psychiatric Journal*, 1971, *16*, 203–208.

Lidz, R., & Lidz, T. The family environment of schizophrenic patients. *American Journal of Psychiatry*, 1949, *106*, 332–345.

Linn, M. W., Caffey, E. M., Klett, J., & Hogarty, G. Hospital vs. community (foster) care for psychiatric patients. *Archives of General Psychiatry*, 1977, *34*, 78–83.

Locke, B. Z., Kramer, M., & Pasamanick, B. Immigration and insanity. *Public Health Reports*, 1960, *75*, 301–306.

McPartland, T. S., & Richart, R. H. Social and clinical outcomes of psychiatric treatment. *Archives of General Psychiatry*, 1966, *14*, 179–184.

McReynolds, W. T., & Coleman, J. Token economy: Patient and staff changes. *Behaviour Research and Therapy*, 1972, *10*, 29–34.

Magaro, P. A., Gripp, R., & McDowell, D. J. *The mental health industry: A cultural phenomenon*. New York: Wiley, 1978.

Main, T. F. The hospital as a therapeutic institution. *Bulletin of the Menninger Clinic*, 1946, *10*, 66–70.

Maley, R. F., Feldman, G. L., & Ruskin, R. S. Evaluation of patient improvement in a token economy treatment program. *Journal of Abnormal Psychology*, 1973, *82*, 141–144.

Malzberg, B. Are immigrants psychologically disturbed? In S. C. Plog and R. B. Edgerton (Eds.), *Changing perspectives in mental illness*. New York: Holt, Rinehart and Winston, 1969.

Malzberg, B., & Lee, E. *Migration and mental disease*. New York: Social Science Research Council, 1956.

Mishler, E. G., & Scotch, N. A. Sociocultural factors in the epidemiology of schizophrenia. *Psychiatry*, 1963, *26*, 315–343.

Newman, E. H. Resolution of inconsistent attitude communications in normal and schizophrenic subjects. *Journal of Abnormal Psychology*, 1977, *86*, 41–46.

Nuttall, R., & Solomon, L. Factorial structure and prognostic significance of premorbid adjustment in schizophrenia. *Journal of Consulting Psychology*, 1965, *29*, 362–372.

Paul, G. L., & Lentz, R. J. *Psychosocial treatment of chronic mental patients: Milieu versus social-learning programs*. Cambridge, Mass.: Harvard University Press, 1977.

Phillips, D. L. The "true prevalence" of mental illness in a New England state. *Community Mental Health Journal*, 1966, *2*, 35–40.

Phillips, L. Case history data and prognosis in schizophrenia. *Journal of Nervous and Mental Disease*, 1953, *117*, 515–525.

Phillips, L., & Zigler, E. The action–thought parameter and vicariousness in normal and pathological behaviors. *Journal of Abnormal and Social Psychology*, 1961, *63*, 137–146.

Platt, J. J., Siegel, J. M., & Spivack, G. Do psychiatric patients and normals see the same solutions as effective in solving interpersonal problems? *Journal of Consulting and Clinical Psychology*, 1975, *43*, 279.

Platt, J. J., & Spivack, G. Problem-solving thinking of psychiatric patients. *Journal of Consulting and Clinical Psychology,* 1972, *39,* 148–151.

Pomerleau, O. F., Bobrove, P. H., & Harris, L. C. Some observations on a controlled social environment for psychiatric patients. *Journal of Behavior Therapy and Experimental Psychiatry,* 1972, *3,* 15–21.

Raush, H. L., & Rausch, C. L. *The halfway-house movement: A search for sanity.* New York: Appleton–Century–Crofts, 1968.

Roberts, B. H., & Myers, J. K. Religion, national origin, immigration, and mental illness. *American Journal of Psychiatry,* 1954, *110,* 759–764.

Rose, A. M., & Stub, H. R. Summary of studies on the incidence of mental disorders. In A. M. Rose (Ed.), *Mental health and mental disorder.* New York: W. W. Norton, 1955.

Rossi, J. J., & Filstead, W. J. (Eds.). *The therapeutic community.* New York: Behavioral Publications, 1973.

Sanders, R., Smith, R. S., & Weinman, B. S. *Chronic psychoses and recovery.* San Francisco: Jossey-Bass, 1967.

Scheff, T. J. The societal reaction to deviance: Ascriptive elements in the psychiatric screening of mental patients in a midwestern state. *Social Problems,* 1964, *11,* 401–413.

Scheff, T. J. *Being mentally ill: A sociological theory.* Chicago: Aldine, 1966.

Schofield, W., & Balian, L. A comparative study of the personal histories of schizophrenic and nonpsychiatric patients. *Journal of Abnormal and Social Psychology,* 1959, *59,* 216–225.

Schwartz, J., & Bellack, A. S. A comparison of a token economy with standard inpatient treatment. *Journal of Consulting and Clinical Psychology,* 1975, *43,* 107–108.

Shean, J. D., & Zeidberg, Z. Token reinforcement therapy: A comparison of matched groups. *Journal of Behavior Therapy and Experimental Psychiatry,* 1971, *2,* 95–105.

Shimkunas, A. M. Demand for intimate self-disclosure and pathological verbalizations in schizophrenics. *Journal of Abnormal Psychology,* 1972, *80,* 197–205.

Srole, L., & Langner, T. S. Protestant, Catholic, and Jew: Comparative psychopathology. In S. C. Plog & R. B. Edgerton (Eds.), *Changing perspectives in mental illness.* New York: Holt, Rinehart and Winston, 1969.

Srole, L., Langner, T. S., Michael, S. T., Opler, M. K., & Rennie, T. C. A. *Mental health in the metropolis: The midtown Manhattan study.* New York: McGraw-Hill, 1962.

Staudt, V. M., & Zubin, J. A biometric evaluation of the somatotherapies in schizophrenia. *Psychological Bulletin,* 1957, *54,* 171–196.

Szasz, T. S. The myth of mental illness. *American Psychologist,* 1960, *15,* 113–118.

Tietze, C., Lemkau, P., & Cooper, M. Personality disorder and spatial mobility. *American Journal of Sociology,* 1942, *48,* 29–39.

Trudel, G., Boisvert, J., Maruca, F., & Leroux, P. Unprogrammed reinforcement of patients' behaviors in wards with and without token economy. *Journal of Behavior Therapy and Experimental Psychiatry,* 1974, *5,* 147–149.

Ullmann, L. P., & Krasner, L. *A psychological approach to abnormal behavior* (2nd ed.). Englewood Cliffs, N.J.: Prentice-Hall, 1975.

Vaughn, C. E., & Leff, J. P. The influence of family and social factors on the course of psychiatric illness. *British Journal of Psychiatry,* 1976, *129,* 125–137.

Visher, J. S., & O'Sullivan, M. Nurse and patient responses to a study of milieu therapy. *American Journal of Psychiatry,* 1970, *127,* 451–456.

Wallace, C. J., Davis, J. R., Liberman, R. P., & Baker, V. Modeling and staff behavior. *Journal of Consulting and Clinical Psychology,* 1973, *41,* 422–425.

Wanklin, J. M., Fleming, D. F., Buck C., & Hobbs, G. E. Discharge and readmission among mental hospital patients. *Archives of Neurology and Psychiatry,* 1956, *76,* 660–669.

Zigler, E., & Phillips, L. Social effectiveness and symptomatic behaviors. *Journal of Abnormal and Social Psychology*, 1960, *62*, 231–238.

Zigler, E., & Phillips, L. Social competence and outcome in mental disorder. *Journal of Abnormal Psychology*, 1961, *63*, 264–271.

Zigler, E., & Phillips, L. Social competence and process-reactive distinction in psychopathology. *Journal of Abnormal and Social Psychology*, 1962, *65*, 215–222.

Zilboorg, G., & Henry, G. *A history of medical psychology*. New York: W. W. Norton, 1941.

Zubin, J., & Spring, B. Vulnerability—a new view of schizophrenia. *Journal of Abnormal Psychology*, 1977, *86*, 103–126.

CHAPTER 3

Behavioral Assessment of Social Skills

Alan S. Bellack

INTRODUCTION

There has been considerable controversy over the definition of "behavior therapy" (cf. Bellack & Hersen, 1977a, chap. 1). While there is no consensus on precisely what behavior therapy is, there is general agreement that behavioral assessment is a hallmark of the field. Behavior therapists uniformly stress the need for detailed, objective assessment, in which the idiosyncratic behavior pattern of the individual patient (or subject) is systematically and comprehensively analyzed. However, such an assessment is rarely, if ever, achieved. In contrast to the sound methodological basis of much of behavior therapy, a large portion of our assessment methodology has "feet of clay." With a few notable exceptions (e.g., Goldfried & D'Zurilla, 1969; Patterson, Ray, Shaw, & Cobb, 1969), behavioral assessment strategies have been developed on an *ad hoc* rather than on an empirical basis. Generally, insufficient attention has been paid to fundamental psychometric requirements (Cone, 1977; Hersen & Bellack, 1977; Jones, 1977). Many assessment procedures are supported more by face validity and consensual agreement than by sound empirical analysis (Bellack, Hersen, & Turner, 1978; Lick & Unger, 1977). In addition, situational factors in the assessment process (such as demand characteristics), and the importance of tailoring an assessment to each individual, typically have received much less attention in practice than they have in the abstract.

Nowhere is the contrast between conceptual rigor and applied convenience more apparent than in the assessment of social skills. Scores of self-report devices and behavior-observation strategies for assessing so-

Alan S. Bellack • Department of Psychology, University of Pittsburgh, Pittsburgh, Pennsylvania 15260.

cial skill have been developed over the past 10 years. Some have been infrequently used aside from the research group responsible for the procedure, whereas others have received relatively widespread acceptance (cf. Rehm & Marston's, 1968, Situation Questionnaire; Wolpe–Lazarus Assertiveness Inventory, Wolpe & Lazarus, 1966). Nevertheless, no instrument or procedure has yet been shown to have sufficient reliability and validity to warrant general adoption (Arkowitz, 1977; Eisler, 1976; Hersen & Bellack, 1977). Of course, several of the procedures currently in vogue seem to be quite useful, and some may prove to be psychometrically sound. However, current assessment strategies in the area are in the unenviable position of relying on consensual validation and anecdotal clinical reports, an unacceptable status for use by an empirical discipline.

My approach to this chapter is based on the uncertain state of the field. Most of the specific individual assessment devices currently available have been evaluated in several recent reviews (Arkowitz, 1977; Eisler, 1976; Hersen & Bellack, 1977). In addition, the chapters in Part Two of this volume also discuss strategies for use with specialized populations. Rather than reprise this material, I will present an overview of the general assessment strategies which are employed, including both clinical and empirical issues. Major limitations of these approaches will then be highlighted. The final section of the chapter will focus on a variety of issues which have received insufficient attention in the literature.

THE FOCUS OF ASSESSMENT

The ultimate goal or focus of any behavioral assessment is a functional analysis of the relevant target behavior. Before this goal can be accomplished, the target behavior must be identified and defined, an assessment task in its own right. Consequently, before discussing the specific techniques which have been employed to assess social skills, it is appropriate to specify the nature of the target behavior and the relevant issues on which a functional analysis must be based.

No one definition of social skills has been universally accepted and no current definition is sufficiently comprehensive to represent adequately the diverse work in the area. Most of the numerous definitions which have been offered seem designed to reflect the particular interests of the persons offering the definition (cf. Argyris, 1965; Hersen & Bellack, 1977; Libet & Lewinsohn, 1973; Trower, Bryant, & Argyle, 1978; Wolpe, 1969). Thus, the Libet and Lewinsohn definition is primarily applicable to the behavior of depressives, and Wolpe's deals mainly with assertion. However, while the specifics of a definition are at issue, there is probably a fair amount of agreement about the general conception of social skills. There seem to be four elements which are common (albeit by implication) to most definitions. *First,* performance in interpersonal situations depends on a set of discrete verbal and nonverbal response components (Trower *et al.,* 1978). Thus, specific parameters can be identified which determine the adequacy of social behavior. *Second,* the particular parameters which comprise adequate behavior, and their configuration, vary according to the situation (Eisler, Hersen, Miller, & Blanchard, 1975). Social skills are situationally specific; what is appropriate in one situation or context is not necessarily appropriate in another. *Third,* the various component elements which comprise adequate social behavior are learned-response capabilities, i.e., skills. The socially adept individual has learned how, when, and where to employ the various response components. Consequently, individuals may vary in their overall level of social skill, as well as their skillfulness in different specific situations. *Fourth,* when specific social skill deficits can be identified, they can be targeted and remediated by training.

As will be discussed further below, there are other aspects of interpersonal behavior which should probably become part of the basic skills model and be a focus of assessment (e.g., social perception, information-processing capability). However, current assessment strategies are primarily based on these four conceptions. In this context, Bellack and Hersen (1978, p. 178) have identified four questions which guide social skills assessments:

1. "Does the individual manifest some dysfunctional interpersonal behavior?"

This is the entry-level question which determines whether or not any subsequent assessment is necessary. In most empirical studies this question is answered by an arbitrarily defined cutoff score on a self-report inventory. When the research involves treatment, self-referral is also employed as definitional of dysfunction. In clinical contexts, identification of dysfunction is much more dependent on the clinician who conducts the intake or diagnostic interview and/or other clinical staff who have contact with the patient. Interpersonal history, behavior during interviews, and interactive behavior on a hospital ward or in other group settings can all serve as sources of data to suggest the existence of interpersonal difficulties. When the presence of some social dysfunction is established, the assessment proceeds to the next question.

2. "What are the specific circumstances (i.e., situations) in which the dysfunction is manifested?"

The situational specificity of behavior is now well-recognized (cf. Mischel, 1968) and applies to social skills as well as most other behaviors. In fact, social skills are situationally specific in two ways. First, the particular combination of response components which comprise a skillful response varies across situations (Trower *et al.*, 1978). For example, much behavior which is appropriate with a spouse is not appropriate with a co-worker of opposite sex, and some behavior which is appropriate with a spouse in the home is not appropriate when performed in public. Second, individuals vary in their ability to behave skillfully in diverse situations. Thus, Eisler *et al.* (1975) and Hersen, Bellack, and Turner (1978) showed that psychiatric patients performed differentially in positive and negative assertion situations, with males and females, and with familiar people as opposed to strangers.

These two types of situational specificity have important implications for assessment. The source and nature of any particular deficit cannot be identified unless and until the situational context is identified. On a preliminary level this might entail such gross categorization as: assertiveness (i.e., deficit in situations requiring assertion), dating skills, marital communication. However, the degree of specification ultimately must be substantially increased, for example: unassertive with what type of people (e.g., boss? spouse?), in what context (e.g., resisting unfair treatment? expressing emotion?), under what circumstances (e.g., when others are present? when alone?)? Of course, some severely dysfunctional individuals will manifest deficits across a wide variety of situations. But it is not reasonable to presume such situational generality.

3. "What is the (probable) source of the dysfunction?"

The skill model ascribes *some* interpersonal difficulties to specific response deficits. These deficits are, conceptually, responsive to skills-training programs. However, there are a variety of other factors which could cause the same pattern of social failures. Anxiety has been shown to interfere with many types of behavior patterns and to serve an inhibitory function over others (cf. Bellack, in press; Martin, 1971). Interpersonal anxiety can interfere with effective social performance in both of these ways, even in the absence of any specific skill deficits (Arkowitz, Lichtenstein, McGovern, & Hines, 1975; Wolpe & Lazarus, 1966). Cognitive disturbances, such as are common in schizophrenia, can interfere with or distort interpersonal communication as well as produce bizarre behavior which results in interpersonal failure (Bellack & Hersen, 1978). Faulty attributions or expectations about the consequences of certain behaviors have also been shown to affect the course of interpersonal behavior independent of skill level (Eisler, Frederiksen, & Peterson, 1978; Warren & Gilner, 1978). Yet another pattern exists for individuals who have response capabilities in their repertoires but fail to emit the response because they have not been reinforced (or have been punished).

Given the numerous potential reasons for any particular dysfunction, the assessment must determine the relevant factors in each individual case. While it has not yet been clearly demonstrated, it is likely that different treatment procedures would be applicable for different types of impairment. It should also be pointed out that these various factors are not necessarily orthogonal. For example, a skill-deficient individual might well be highly anxious in the pertinent situation, or a highly anxious individual may have lost certain skills through disuse or failure to learn them properly. The required assessment task is neither differential diagnosis (i.e., A *or* B) nor historical analysis of "first" cause. Rather, what is required is a detailed analysis of the current maintaining factors, those problems which currently prevent adequate performance.

4. "What specific social skills deficits does the patient have?"

The term "social skills" is frequently applied to rather broad categories of behavior, such as assertiveness, heterosexual or dating skill, and ability to develop and maintain friendships. However, these terms are generic labels for complex behavior patterns consisting of a host of subtle response components. Identifying a deficit in "assertiveness" is little more than a preliminary diagnostic labeling. Effective treatment requires the specification of each response component which appears at inappropriate frequency, duration, intensity, or form. For example, eye contact or gaze might occur at too high or low a duration, or at the wrong times. The individual might provide too much or too little social reinforcement to maintain a casual conversation, or may provide reinforcement at the wrong times (Fischetti, Curran, & Wessberg, 1977).

Of the four assessment questions, this final one is the most difficult to answer. Each of the first three questions can often be answered by self-report and informal clinical observation. In contrast, the degree of specification required for this question characteristically requires some systematic, intensive observation. A long list of behaviors must be examined in *each* situation identified (in Question 2) as problematic in order to determine the precise nature of the dysfunction. Unfortunately, the content of that list can only be roughly estimated at the present time. As will be discussed further below, few of the generic categories of interpersonal behavior have been sufficiently well analyzed so as to identify the critical combination of components necessary for effective performance (cf. Mehrabian, 1972; Trower *et al.*, 1978). It appears to be much easier to (subjectively) determine when interpersonal performance is somehow "not quite right" than it is to specify exactly what was wrong (e.g., Arkowitz, 1977; Curran, 1977).

ASSESSMENT STRATEGIES

A wide variety of assessment procedures have been employed to appraise social skills. The empirical literature on social skills training and assessment has emphasized two assessment strategies: self-report inventories and behavioral observation of simulated interpersonal encounters. These two strategies appear to have been widely adopted primarily because of their economy and convenience in administration, and their apparent validity for gross screening or differentiation of high- and low-skill groups. Other procedures, such as self-monitoring, peer ratings, and physiological assessment, have also been utilized, but much less frequently. Interviewing is almost universally employed for clinical assessments, but has not been systematically incorporated as a research method.

In the following sections the use of each of these approaches will be considered in the context of the four assessment questions identified above. Despite the absence of a sound empirical base, interviewing is perhaps the most sensitive and critical assessment tool; hence, it will be discussed first. Subsequent sections will then consider self-report inventories, self-monitoring, behavioral observation, and physiological assessment.

Interviewing

Interviewing has traditionally been the backbone of assessment in psychology and psychiatry (Bellack & Hersen, in press, chap. 5; Matarazzo & Wiens, 1972). Yet in spite of its preeminence (or perhaps because of it), behavior therapists have eschewed the interview because of its reliance on self-report and subjective clinical impression. There is little question that data collected in interviews are especially susceptible to bias. As with any self-report, interview responses are limited by the patient's potentially selective and distorted self-observation and recall, as well as by a host of demand characteristics (cf. Bellack & Hersen, 1977b). Likewise, the interviewer is a less than perfect data-collection device, listening selectively and being subject to a variety of biases (Bellack & Hersen, 1979, Chap. 5; Matarazzo & Wiens, 1972). Thus, the interview can produce a distorted account of the patient's distorted report of his/her experience. Nevertheless, the interview may often be the best and the only way to gather critical information. The patient comes to the interview with more information about him/herself than anyone else has, and with some types of information which are not privy to anyone else or any assessment procedure other than the interview (e.g., subjective valuations, self-verbalizations).

The interview can provide two general types of information about social skills: the interpersonal history and informal observational data (Bellack & Hersen, 1978). The interpersonal history comprises both a current and retrospective account of the patient's interactions with others. It is a blueprint of the quality, quantity, and form of the individual's interpersonal behavior from adolescence to the present. Some of these data can be secured by self-report inventories and demographic data forms. However, the interview format generally allows for informative (often critical) elaboration of the out-of-context responses to questionnaires. In fact, self-report inventories often are more useful as interview guides than as primary sources of data. The interactive format of the interview has the unique advantage of providing context and qualitative elaboration to objective reports of discrete events and response frequencies. For example, self-report of expectations of punishment for assertive behavior can substantially alter the diagnostic formulation and treatment plan for an "unassertive" patient (Eisler *et al.*, 1978).

One major product of the interpersonal history is a semiobjective account of the nature and frequency of various types of social interactions. The patient can be probed for details about each relevant generic category of social skill, as well as a general picture of the history of interpersonal activity. Examples include: dating history, friendship patterns, ability to deal with diverse interpersonal problems, participation in social activities and social organizations, response to anger and frustration, etc. Interpersonal behavior patterns of family members can also be informative. These data can be useful in several ways. Most directly, the patient can report specific interpersonal problem areas and the precise situational contexts in which they occur. The data can also be compared with population norms to identify other problem areas. Accessory resources and deficits can be identified, such as lack of sufficient social contacts to practice new skills, ignorance of social norms, and insufficiency of information needed to maintain a conversation (e.g., knowledge of sports, current events). Finally, the interpersonal history and family description can provide clues to the onset and course of the dysfunction.

The interpersonal history can also provide important data on the patient's subjective evaluation of his/her behavior in various interpersonal encounters. For example: How *difficult* does he/she find it to make various responses? How uncomfortable or *anxious* is the patient in various situations? Does the patient feel *guilty* about the expression of anger, and does he differentiate anger and assertiveness? These data are critical for determining what the patient regards as the relevant focus of treatment. They can also be of considerable value in determining the nature of the dysfunction (e.g., skills deficit versus anxiety versus both) and in formulating a treatment plan. Thus, I recently treated a patient who experienced intense heterosocial anxiety in addition to having a marked heterosocial skill deficit. Treatment included a systematic *in vivo* shaping program to reduce the anxiety (Leitenberg, Agras, Allen, Butz, & Edwards, 1975) as well as social skills training. If untreated, the anxiety would have prevented exposure to heterosocial situations regardless of skill level.

The second major pool of data in the interview comes from the interviewer's observation of the patient's behavior. The interview is, of course, an interpersonal encounter. Consequently, a wide variety of social amenities, conversational skills, and other interpersonal responses are displayed between the initial greeting and final parting comments. Personal grooming habits, response to questions, autistic gestures, eye contact, relevance of responses, posture, and reaction to social reinforcement are a few of the many response parameters which are manifest in the course of the interaction.

A major limitation of these data is the potential situational specificity of interview behavior. It is a relatively unique interaction and the patient's interview behavior might not be representative of behavior in other interactions. Consequently, these observational data must be interpreted with caution. Nevertheless, they can still be highly useful for formulating hypotheses and planning further assessment, and for testing hypotheses generated from other sources of data. For example, gross deficits in conversational ability in the absence of indications of substantial anxiety or depression might well be representative of behavior in other settings. Alternatively, a highly polished interactive style accompanied by reports of social dysfunction suggests either an anxiety problem or a highly compartmentalized skill deficit. Even if both of these (or any other) hypotheses are subsequently shown to be invalid, valuable information could be secured about the breadth and range of the patient's interpersonal response repertoire.

In the context of the four assessment questions, the interview would appear to be of tremendous value for Question 1 (preliminary identification of an interpersonal problem), Question 2 (situational context of the problem), and Question 3 (determining the reasons for the problem). It can also be an important source of secondary data about specific component deficits (Question 4). However, little empirical data currently exist to substantiate these contentions. Interviewing has traditionally been an inexact technique, varying substantially in form and substance from interviewer to interviewer, and patient to patient. Data produced by interview frequently have uncertain reliability, and rarely are in the quantifiable form required for empirical investigation. The potential value of interviewing warrants further research in an effort to rectify these limitations. One possible solution is the use of standardized interview schedules designed specifically for the assessment of social skills (cf. Kiesler, 1975). Interviewer behavior would be held relatively consistent by specifying a set of questions and the order and manner in which they are to be asked. The subjective quality of interpretation could be reduced by the use of Likert scales or other quantitative scaling for each item. In addition to structured questioning, standard prompts or subtle interpersonal cues could be built in, and each patient's response to these stimuli could be compared to derived norms. Finally, the interview could be videotaped and subsequently analyzed for appropriateness of interviewer behavior and interrater reliability of scoring. Similar devices, such as the Hamilton Rating Scale for Depression (Hamilton, 1960), have proven useful in the evaluation of other types of dysfunctions. A standardized, structured social skills interview could provide a useful compromise between a totally free-wheeling "traditional" interview and the exclusion of interviewing from the assessment process.

Self-Report Inventories

In a recent review of the literature, Hersen and Bellack (1977) described 7 representative self-report inventories used in the assessment of heterosocial skill and 11 used in the assessment of assertiveness. This listing was not exhaustive then, and several additional inventories have since been developed. In addition, various items from the Fear Survey Schedule (Wolpe & Lang, 1964) and *ad hoc* fear and skill "thermometers" (global 10- or 100-point self-ratings) are also commonly administered. Clearly, self-report inventories (and related techniques) are among the most frequently employed assessment procedures; several are included in almost every empirical study of social skills and social skills training. Many of the specific inventories are described and evaluated in subsequent chapters. Consequently, the comments below are meant to provide a critical overview of the general strategy.

While interviewing has been almost exclusively a clinical tool, the various self-report inventories have primarily been research devices. Most have been developed in the context of specific research programs for one of two purposes: (1) to categorize potential subjects as high or low on a particular skill, and (2) to serve as a dependent measure in treatment-outcome studies. In both cases the focus has been on gross labeling rather than highly specific analysis of the individual's response pattern. As with most self-report inventories, social skill tests are designed to yield single composite scores based on the sum of item scores; responses to individual items typically are ignored. This approach is generally not useful for the clinician working with a single patient. While some information can be gained by comparing a patient's score to peer group norms, these tests are not precise enough to provide rigorous cutting scores. Thus, the clinician could not safely make major decisions on the basis of scores from any of these tests. However, the scales can be employed as interview guides, in which case the patient would be queried about responses to specific test items.

Social skill inventories are subject to the same limitations attendant on any self-report scales (cf. Bellack & Hersen, 1977b). Individual items are subject to interpretation by the testee, and thus can serve as quite different questions for different individuals. Data required to answer items are often not available to the testee (e.g., he/she might not know what he/she would do in situation *X*). Alternatively, responses may be distorted by numerous factors such as faulty recall, demand characteristics (as in posttherapy assessments), and dissimulation. Various semiquantitative terms (e.g., very much, frequently) are defined in a highly variable manner by different people. Thus, the same response can refer to substantially different levels or frequencies for different subjects. These various problems can often be minimized or avoided by careful item construction, but they have received relatively little attention in the development of most social skill inventories. In fact, with few exceptions, these instruments have not been developed in accordance with sound psychometric principles. In general, they have not been subjected to adequate empirical analysis and they have uncertain reliability and validity.

This phenomenon is particularly well illustrated by the manner in which items are constructed. Scaling procedures have been a particular source of difficulty. Various inventories have used true–false formats; 3-, 4-, 5-, and 7-point Likert-type scales; and simple unanchored numerical ratings. These diverse formats vary in psychometric adequacy and cannot be presumed to yield equivalent results. Almost all scales describe a series of interpersonal situations. But subjects are variously asked to specify how much *difficulty* they would have dealing with each situation, how *anxious* they would be, how *frequently* they would perform a specified response, or how *effectively* they could perform in the situation. These various adjectives refer to different phenomenal and behavioral dimensions. A frequency scale cannot be assumed to reflect the same behavioral pattern as a difficulty scale or an anxiety scale. Simply labeling a scale a "skill" test does not guarantee that it measures skill even if items pertain to interpersonal behavior. The particular behavioral correlates (if any) of these various item types must be explored. While the importance of empirical determination of item content has been considered (e.g., Goldfried & D'Zurilla, 1969; Goldsmith & McFall, 1975), the empirical determination of item form has been neglected.

The use of summative scores on self-report inventories also warrants comment. These scores are statistical abstractions which purport to provide a general picture of the test taker's behavior. However, there are two problems with this assumption. First, it is not legitimate to sum the scores on each test item unless the items have been empirically selected and shown to have equivalent psychometric properties (cf. Anastasi, 1961). As stated above, this is not the case with most existing scales. Second, summative scores fail to consider the situational variability of social skills. Overall scores imply a general level of skill which does not adequately represent the stimulus-specific pattern of the test taker's response capability. In fact, such scores obscure any situational patterns which might be indicated by specific items. In addition, they do not reflect the specific response components which are problematic. Consequently, these devices can, *at best*, only provide a gross picture of the individual's standing in relation to his/her peers.

Self-Monitoring

Self-monitoring (SM) has become one of the most widely used behavioral assessment procedures, yet it has been only a peripheral tool for the assessment of social skills. The only empirical literature which has regularly employed SM is research on heterosocial skills (cf. Chapter 5). Subjects have frequently been asked to keep dating logs, recording information ranging from simple dating frequency to extensive self-evaluations of specific aspects of performance (Arkowitz, 1977). In addition to monitoring behavior, subjects have also been asked to rate subjective distress or anxiety and to note discriminative stimuli and environmental consequences to their behavior. Similar forms of SM would seem to be applicable for assessing other social skills as well, although anecdotal clinical reports outnumber research examples of this application. Unassertive individuals, those who have difficulty establishing friendships, and others could simply be directed to monitor their behavior in relevant situations. Recording of antecedent and consequent events could help elucidate the context for the interpersonal dysfunction. It is highly recommended that patients be instructed to evaluate the difficulty of making targeted responses and the degree of anxiety associated with the behavior. These subjective ratings could provide valuable data for isolating the nature of the dysfunction and appraising the progress of treatment. For example, continued reports of high anxiety despite increased skill might suggest the need for supplemental anxiety-reduction treatment.

Self-monitoring is a unique assessment tool in that it plays an intermediary role between self-report and behavioral observation. Many of the subjective distortions, recall errors, and biases associated with other self-report procedures can be minimized because the patient is instructed to observe ongoing events systematically. This observation is especially attractive for the assessment of social skills because so much interpersonal behavior occurs outside the purview of the researcher or clinician. However, SM remains a self-report procedure, and the various types of error associated with self-report are not entirely eliminated. The most notable and well-documented threats to the validity of SM are unreliability and reactivity (Kazdin, 1974; Nelson, 1977). Unreliability refers to inaccuracy or inconsistency of observation or record keeping, while reactivity refers to change in the target behavior as a function of self-observation. While valuable information can be secured by SM, these two sources of potential error limit the confidence one can place in the data.

Reactivity is not as serious a problem in the assessment of social skills as it is for some other target behaviors. The effect of reactivity is, generally, to change behavior in a therapeutic direction (Kazdin, 1974). If the patient or subject actually has a skill deficit, there should be little capability for improvement solely as a function of self-observation. Conversely, unreliability is a very serious problem. Reliability is a fundamental requirement for any data set; unreliable data are simply not interpretable. The assessor can employ two less-than-perfect strategies to increase reliability: (1) He/she can enlist the aid of a significant other (e.g., spouse, friend) to observe the patient or subject, in much the same way as a reliability checker is used in research. This approach has obvious limitations, not the least of which are the other person's direct involvement in some problem interactions and the ever-changing context for many target behaviors (e.g., changing dating partners). (2) The observation and record keeping procedures should be objectified and simplified as much as possible so as to reduce unnecessary confusion and inconvenience. For example, brief records on a time-sampling schedule are more desirable than narrative descriptions of all target events.

These two strategies might be effective for monitoring the occurrence and general form of molar behavior sequences, such as going on a date or asking a boss for a raise. But it is questionable whether or not the molecular response components which comprise social skill can be self-monitored reliably. Appropriate use of eye contact, adequacy of voice volume and intonation, and response timing are not easily discriminable responses. They are difficult for trained raters to appraise in isolation on videotapes, let alone for patients to recognize their fleeting occurrence in the midst of an interaction. While a well-trained patient or subject *might* be able to provide gross data on these types of responses (e.g., "I tried to speak louder and look at him while I spoke"), reliable high-fidelity self-monitored records seem an impossibility. Consequently, SM is, at best, an imperfect alternative when behavior cannot be observed by a trained, independent rater.

Behavioral Observation

Behavioral assessment is characterized by a "direct, sampling" approach to data collection (cf. Goldfried & Kent, 1972). In keeping with this approach, direct observation of the target behavior in its natural setting is the penultimate assessment strategy. Unfortunately, the special character of interpersonal behavior frequently makes such observation impractical or impossible. Intimate and private interactions generally are not privy to outside observers, and many public interactions (e.g., assertion situations, meeting a stranger) occur too infrequently and unpredictably to permit easy access. In response to these restrictions, researchers and clinicians have turned to staged interactions in the laboratory or clinic. This section will highlight the three major observational strategies which have been employed: *in vivo* observation, naturalistic interactions in the laboratory or clinic, and role-play.

In Vivo Observation. As stated above, this is the most desirable and least practical approach. It has rarely been employed in the social skills literature *per se.* King, Liberman, Roberts, and Bryan (1977) described a strategy in which chronic psychiatric patients were escorted into the community by a trained observer who directed them to engage in a series of encounters requiring assertion (e.g., "Call friends and invite them to house", "Obtain job application"). It seems unlikely that such forced and observed responses would be representative of the subjects' typical style of interaction. But response capability might be ascertained (e.g., Can the subject make an effective response if pressed?). Arkowitz and his colleagues (e.g., Christensen & Arkowitz, 1974; Royce & Arkowitz, 1976) have arranged dates for heterosocially deficient college students and secured feedback from the dating partners. This somewhat appealing approach has the double-edged disadvantage of producing unreliable data from "real" dating partners or reactive data from trained experimental confederates.

The most systematic work on *in vivo* observation of interpersonal behavior emanates from research on family interaction. Both live observers and elaborate automated audiotape recording systems have been placed in the home in an effort to "eavesdrop" on critical family interactions (Jacob, 1976; Weiss & Margolin, 1977). The most representative and empirically sound procedures have been developed by Patterson and his colleagues in the Oregon group (cf. Patterson, 1974; Patterson *et al.*, 1969). Typically, observers enter the home at dinnertime (a period of high interaction). Family members are instructed to remain in one or two rooms and to refrain from watching television during the observation period so as to increase the likelihood of (observable) interactions. Alternatively, tape recorders can be placed in various rooms and programmed to cycle on and off at random or preselected intervals. Both approaches appear to provide reliable and valid data. However, validity has primarily been established for molar categories of behavior, such as positive and negative interactions. While molecular behaviors are observed, data often are collapsed across subcategories to yield summative scores (Jacob, 1975). Consequently, it has not yet been determined whether or not these procedures are valid for assessing specific response components.

The reactive effects of being observed have been well established (Johnson & Bolstad, 1973), and threaten the validity of any observational procedure. Because of their comparatively unobtrusive nature, random-onset tape-recording systems may produce somewhat less reactivity than do live observers (Johnson & Bolstad, 1975). On the other hand, they miss potentially important nonverbal aspects of interactions, and it is unlikely that reactivity to being observed can be totally overcome with either (any ?) procedure (Weiss & Margolin, 1977). The extent of reactivity should be determined for behavior categories as well as observational methods. It might well be that reactivity is not a critical issue except in the assessment of highly private responses (e.g., wife or child abuse, sexuality, affection) which people would not emit in the presence of any observation. The biggest limitation to *in vivo* observation appears to be the restricted sampling which can be achieved. Behaviors which occur in diverse locations, which occur with a variety of individuals (who have not consented to being observed), which are infrequent, and which are highly private are not amenable to *in vivo* observations. Even home observations are restricted according to the times and rooms in which observers are present or the recording machine is turned on. As stated previously, such limitations preclude the assessment of most classes of social skills, and allow circumscribed access to others. Consequently, *in vivo* observation of interpersonal behavior can play a primary role only in a few specialized situations.

Naturalistic Interactions. In an effort to circumvent the sampling and cost problems of *in vivo* observations, many researchers have attempted to secure representative samples of behavior in the laboratory. Naturalistic interactions are structured or staged interactions which are intended to parallel various *in vivo* encounters. For example, Gutride, Goldstein, and Hunter (1973) unobtrusively observed psychiatric patients interacting with a stranger in a waiting room; unknown to the patients, the stranger was a trained confederate. Arkowitz *et al.* (1975) employed two different tasks to assess heterosocial skill. In one, they instructed male subjects to interact with a female confederate (for 10 min) as if they had just met. The second task entailed a mock 5-min telephone conversation in which the subject tried to arrange a date with a female confederate. As illustrated by these examples, naturalistic interactions have included both surreptitious observation in which the subject presumably is unaware of the deception and situations in which the subject is made aware and instructed to respond "as if" the interaction were "real." Both duration of these interactions and the specific format (or instructional set) have varied substantially across studies, apparently on an *ad hoc* basis.

A third general approach, primarily employed for assessing assertiveness, consists of "critical incidents" (Hersen & Bellack, 1977). Subjects have been deliberately short-changed, pressured by phony magazine salesmen, and urged to volunteer their time for charitable activities. In some cases (e.g., the latter two) the incidents are staged in the community or by telephone rather than in the laboratory. As with the more extended interactions described above, format and duration of these procedures vary considerably. In general, these procedures have not been shown to be valid (cf. Hersen & Bellack, 1977). In addition, their use is limited because they rely on deception.

Each of the procedures described above requires the use of trained confederates to interact with the subject. The realism of the interaction thus depends on the acting ability of the confederate. Similarly, the confederate's behavior has a considerable impact on what the subject can and will do. For example, the confederate can be directed to provide personal information or respond minimally, to ask open-ended or closed-ended questions, to make some response after either 10, 60, or 120 sec of silence, etc. Such variations undoubtedly have differential effects on subject behavior. These problems have been substantially avoided in the assessment of family interactions by requesting family members to interact with one another (Jacob, 1976; Weiss & Margolin, 1977). The content of the interaction typically is structured, but the form is left open. Among the most widely used procedures are several variants of the revealed-differences technique (cf. Jacob, 1975). Family members independently complete checklists pertaining to the family or to one another. They are then informed of the various responses and are requested to resolve the differences and jointly produce one completed checklist. The conflict-resolution process is observed or recorded for subsequent evaluation.

It is difficult to evaluate the naturalistic observation strategy on the basis of the existing literature. Subjects, confederates, target behaviors, format, and content of interactions have varied substantially from study to study. It is clear that some of the variants which have been employed are substantially invalid (e.g., most critical-incident procedures). But the best way to conduct these interactions for the various social skills has not yet been empirically determined. Hence, the maximum potential validity is unknown. Many current strategies have been shown to be valid for general ratings of interpersonal behavior, such as overall assertiveness and skill and overall anxiety. The data for specific response components have generally been unimpressive (e.g., groups expected to differ in skill have not manifested differences in specific components). Whether these findings reflect invalidity of the tasks, poor selection of target behaviors, absence of skill differences between criterion groups, or some other factor is unknown. Future research in this area should focus on the development of empirically selected tasks as well as on validity *per se*. In addition, the "respond as if" approaches should be compared to the deception approaches. If the former are shown to be valid, they would be a preferred alternative as they are easier to arrange and avoid the ethical constraints associated with deception.

Role-Play Tests. Beginning with the pioneering work of Rehm and Marston (1968), role-play tests have become a hallmark of social skills assessment. In the basic format, an interpersonal vignette is described to the subject, a role model (confederate) utters a prompt line, and then the subject responds to the role model as if the interaction were actually taking place. For example:

> NARRATOR: It is 4:45 P.M. You are in a hurry to leave work and get home when your boss walks up to you and says. . . .
>
> ROLE MODEL: I've got to have this report by tomorrow morning. You'll have to stay late and finish it.

Following the subject's response, the next situation is described and enacted. Various role-play tests have included from 4 to 32 such situations. In clinical settings the therapist often serves as role model as well as narrator. No data have been presented to date to justify the use of a separate role model.

The differentiation of role-play tests from the "as if" type of naturalistic interactions is somewhat arbitrary given that both involve simulated interactions. However, there are substantial differences between the approaches which seem to make them distinct. Role-play tests are highly structured, require facility at making quick, brief responses, and rely on the subject's ability to take on roles quickly in a series of diverse interactions (i.e., even in a homogeneous test each interaction involves a distinct role). Conversely, naturalistic interactions require only one role enactment in which the subject can gradually become enmeshed due to the extended nature of the interaction. These interactions are also less structured, and place more emphasis on ability to maintain a conversation than on ability to make one or two discrete responses. The exact effects of each of these differences are not yet known, but they undoubtedly have a cumulative, if not an individual, impact. This contention is supported by a study by Bellack, Hersen, and Lamparski (in press), which found little correspondence between behavior on a role-play test and a naturalistic interaction focused on the same social skill.

Numerous variations of the basic role-play format have been employed. Scene descriptions and role model prompts have been presented on audiotape, videotape, and by live experimenters. The interactions have been limited to a single prompt and subject response, and extended by having the role model offer one or several counterresponses (e.g., resisting the subject's attempt to be assertive). Scene descriptions have varied substantially in the amount of background elaboration provided. Similarly, role models have varied the nature of their prompts (e.g., from neutral, to assertive, to angry). As with items on self-report inventories, these format and structural variations typically have been selected on an *ad hoc* or face-validity basis, even when item content has been empirically determined. Comparatively little attention has been paid to the substantial differences such variations might produce.

A few of the alternative approaches have been empirically compared. Galassi and Galassi (1976) found substantive differences between brief and extended interactions, and showed that interactions with live role models were more anxiety provoking than interactions with taped models (i.e., the subject makes a response to a taped prompt). Eisler *et al.* (1975) showed that responses varied as a function of such variables as sex and familiarity of the role model. A recently completed study in my laboratory compared scene descriptions which varied in degree of elaboration while maintaining a single critical focus (e.g., negative assertion with a familiar female). For example, "You are leaving an *evening* class with a classmate who lives across town. The classmate, who is *carrying a bulky package,* asks, 'Could you give me a lift home?' " versus "You are leaving a class with a classmate who lives across town. The classmate asks . . ." followed by the same prompt line. Preliminary analyses suggest that the more detailed scenes elicit longer and more personal replies, as well as fewer awkward responses, nervous smiles, and questions directed to the experimenter (e.g., "Do I know the person? How far away do they live?"). Apparently, the brief descriptions require the subject to provide his/her own qualifications and critical details. The subject becomes anxious and/or responds to an idiosyncratically created scene.

Because of the variations which have been employed, it is difficult to provide an overall evaluation of the validity of role-playing tests. Several studies have reported positive relationships between role-play procedures and independent criterion variables (Borgatta, 1955; Kreitler & Kreitler, 1968; Stanton & Litwak, 1955; Warren & Gilner, 1978). However, each of these studies is marked by methodological flaws, inadequate description of methodology, and/or marginal results. For example, Warren and Gilner (1978) creatively had spouses evaluate "how typical" their partner's role-played responses were of *in vivo* behavior. But ratings were made on a 4-point scale rather than one with an odd number of points. In addition, mean ratings of "how typical" only slightly exceeded three ("probably yes"), a result which probably reflects response bias rather than a confirmation of validity. Borgatta (1955), and Stanton and Litwak (1955) rated poorly defined behavioral categories, such as "autonomy," "showing solidarity," and "asking orientation." Kreitler and Kreitler (1968) rated noninteractive motor responses such as "show us how you walk in the street," as well as interpersonal interactions.

Further support for the validity of role-play tests comes from studies in which "known groups" are differentiated on role-play tests. For example, there are numerous reports in which subjects independently categorized as high or low in social skill or social anxiety have been shown to respond differently on role-play tests (Arkowitz *et al.*, 1975; Borkovec, Stone, O'Brien, & Kaloupek, 1974; Eisler *et al.*, 1975; Hersen *et al.*, 1978). Similarly, subjects receiving social skills training have been shown to improve their role-played performance and/or perform better than subjects not receiving such training (Bellack, Hersen, & Turner, 1976; Curran & Gilbert, 1975; Goldsmith & McFall, 1975). While these findings are suggestive, they must be interpreted with caution. First, the results have not been uniformly positive; such known groups have not always been differentiated (Bander, Steinke, Allen, & Mosher, 1975; Glanz, 1978). Second, groups have frequently been differentiated *only* on subjective, overall measures of anxiety and social skill. Such differences could result from factors unrelated to role-play performance, such as physical attractiveness. Third, and most important, these findings provide convergent validational support but do not demonstrate the external validity of the procedure.

Several recent studies suggest that behavior on role-play tests does *not* correspond to behavior in other settings. Bellack, Hersen, and Turner (1978) compared the role-play performance of chronic psychiatric patients with behavior in a structured interview and in a group-therapy situation. Behavior in the latter two situations was moderately correlated, but role-play was not related to either. Bellack, Hersen, and Lamparski (in press) examined the behavior of male and female college students in two situations: (1) a role-play test of heterosocial skill, and (2) a naturalistic interaction with an opposite-sex student who, unknown to the subjects, was an experimental confederate. Role-play behavior was moderately correlated with behavior in the naturalistic interaction for females but there was little relationship between the two situations for males. In a third study, Bellack, Hersen, and Turner (in press) compared the role-play behavior of chronic psychiatric patients with actual *in vivo* performance on the exact same situations. Once again, there was not extensive correspondence between role-play and *in vivo* behavior.

This series of studies raises serious questions about the validity of role-play tests as they are currently employed. The negative findings were consistent across populations (patients and college students) and social skills (heterosocial skill, positive and negative assertion). It is possible that other variations of role-play procedures might elicit more representative behavior patterns (e.g., extended interactions rather than the brief interactions employed in each of the studies). Alternatively, role-playing might be more suited for assessment of interpersonal behaviors other than those observed in this series of investigations. However, such speculations must be empirically tested. Until such time, data and conclusions based on role-play procedures must be examined with great caution.

Several aspects of the brief role-play format seem especially suspect in restricting the external validity of existing role-play tests. First, the brief interaction format may be too restrictive, failing to tap many critical interactive behaviors (Curran, 1978). Ability to initiate and maintain a conversation, response timing, and use of social reinforcers are among the potentially important response skills which cannot be assessed if the behavior sample consists only of single, initial responses. Second, the role-play format may be highly anxiety provoking and consequently elicit anomalous responses. The task demand to make a series of responses quickly, with no preparation, while being videotaped or otherwise observed, must certainly be stressful. Hersen *et al.* (1978) and Glanz (1978) both assessed heart rate during role-play tests and found anticipatory heart rates of greater than 95 bpm throughout. Glanz's results are notable in that he tested nonanxious college students. Borkovec *et al.* (1974) similarly found high heart rates with little habituation, even for low-anxious college subjects.

The third problem pertains to the "respond as if" requirement of these tasks. There are no data to suggest that most subjects can or do respond as if they are really in the designated situation. In fact, it seems unlikely that they could. First, most scene descriptions fail to contain much of the information necessary to define the situations adequately. For example, the courses of the interactions are not provided, and potential consequences are not specified. Consequently, subjects frequently may not have enough information to respond in a realistic manner. Second, the demand for a series of rapid-fire responses may not allow the subject to conceptualize each situation and consider what a characteristic response might entail. Finally, the stress generated by the entire procedure may divert the subject's focus, creating an urgency simply to emit some response rather than allowing for actual involvement in the situations described. In an effort to combat these various problems, my students and I are currently investigating several promising alternatives including more detailed scene descriptions and previewing of each scene prior to the narration so that the subject can calmly contemplate the nature of the encounter. These and other variations should be systematically examined in an effort to improve the validity of what potentially is an invaluable assessment strategy.

Ratings by Peers and Significant Others

Several limitations of the direct-observation procedures available to clinicians and researchers have been documented above, including limited access to certain interactions and brief sampling even when access is possible. In addition, these procedures often cannot detect the environmental impact of the target behaviors: how other people react to the subject, and how they appraise his/her behavior. One possible solution to these problems is to secure data from peers and/or significant others in the subject's environment. Numerous examples of this approach appear in the literature. Spouse observation is among the most commonly employed strategies for assessing marital interaction (Weiss & Margolin, 1977). Roommates and fraternity/sorority members have been requested to rate the heterosocial skill of college students (cf. Arkowitz, 1977). Classmates have been requested to evaluate the degree to which they would choose to play with withdrawn peers who were receiving social skills training (Whitehill, Hersen, & Bellack, 1978).

Several alternative strategies are available for securing data from sources in the environment, including structured interviews and questionnaires (Fiske, 1975), sociometric ratings (Kane & Lawler, 1978), and systematic observation of ongoing behavior (Weiss, Hops, & Patterson, 1973). Each of these strategies has been extensively studied, and a review is beyond the scope of this paper. In general, these procedures can provide important information with adequate reliability and validity. However, they generally cannot provide totally objective data or accurate accounts of highly specific responses. The ratings by a significant other, in particular, are susceptible to bias and are often highly reactive (Weiss & Margolin, 1977). For example a wife tracking her husband's affectionate behaviors can generally be expected to alter her own behavior in anticipation of and in response to those behaviors, thus affecting their occurrence. On the other hand, the purpose of social skills training (and assessment) is, ultimately, to have some impact on the patient's environment. Therefore, the environment's reaction to the patient is a critical factor in planning and evaluating treatment (Kazdin, 1977). The reactions and perceptions of others are important even if they do not reflect a completely accurate picture of the patient's actual behavior.

Physiological Assessment

It has been well documented that behavioral dysfunction can be manifested in any of three independent response systems: motoric, cognitive, and physiological (cf. Bellack & Hersen, 1977a, Chaps. 1, 2; Lang, 1971). Consequently, it is generally recommended that behavioral assessment encompass all three channels. Despite this general guideline, the relevance of physiological assessment for the evaluation of social skill is uncertain. Comparatively few studies on social skill have examined autonomic functioning (the potentially relevant physiological system) (Borkovec *et al.*, 1974; Hersen *et al.*, 1978; McFall & Marston, 1970; Twentyman & McFall, 1975). The results of those studies have presented an uncertain picture of the relationship of autonomic arousal to social skill. Consistent physiological differences between groups shown to differ on cognitive and motoric measures have not emerged (e.g., Schwartz & Gottman, 1976). However, as stated above, the various response channels are relatively independent. In addition, if the assessment procedures are themselves highly arousing (as suggested in the preceding section), group differences due to skill would be superseded or "washed out."

Eisler (1976) has argued that physiological assessment is not particularly useful for appraising social skill. On a pragmatic level, he suggests that the assessment procedures are highly intrusive and, thus, can reduce the realism of the observed encounter. On a more conceptual level, he questions the relationship of physiological arousal to social skill: "Some individuals may behave, to all outward appearances, in a highly socially skilled manner under conditions of high physiological arousal, while others may appear equally skilled under conditions of low or moderate arousal" (Eisler, 1976, p. 386). Bellack and Hersen (1978) have similarly suggested that arousal and skill are orthogonal. If research supports this contention, physiological assessment could be useful for two reasons: first, to determine whether or not interpersonal dysfunction is due to anxiety in cases when no specific skill deficits can be identified; second, to identify high levels of anxiety which coexist with skill deficits. While social skills training has been shown to reduce moderate levels of anxiety (e.g., Curran & Gilbert, 1975), independent anxiety-reduction treatment may well be necessary when anxiety is extreme.

NEW DIRECTIONS IN THE ASSESSMENT OF SOCIAL SKILL

Behavior therapists have only recently become interested in interpersonal behavior. Our understanding of interpersonal functioning and our strategies for assessing it are thus somewhat elemental. Many critical issues have not yet received sufficient attention and analysis. The final section of this chapter will consider three areas which require further consideration in the assessment process: (1) identification of target behaviors, (2) social perception, and (3) cognitive factors affecting interpersonal behavior.

Identification of Target Behaviors

Behavioral research on interpersonal behavior has been based on the well-founded assumption that molar skill categories, such as assertiveness and heterosocial skill, are based on identifiable sets of molecular behaviors, such as voice volume and eye contact. This assumption is generally supported by (a) the occasional correlations found between ratings of component behaviors and global ratings of the molar category (e.g., Bellack *et al.*, 1978; Eisler *et al.*, 1975), and (b) the voluminous social psychology literature documenting the importance of nonverbal (e.g., molecular) response characteristics (cf. Harper, Wiens, & Matarazzo, 1978; Mehrabian, 1972; Trower *et al.*, 1978). However, in many cases the molecular components have not shown a strong relationship to ratings of the molar categories (Arkowitz, 1977), and there is little evidence that the components typically assessed and subjected to treatment are important correlates of dysfunctional behavior as it occurs in the natural environment. That is, there is no evidence that increasing eye contact, voice volume, or the like actually affects marital interaction, dating frequency, level of depression, or any other clinically meaningful set of behaviors.

The issue is not so much the importance of molecular response components *per se* as it is the determination of exactly which behaviors are important in diverse circumstances. Apparently, the behaviors which have been studied have been selected primarily on the basis of face validity. In those instances when identified groups have not manifested different levels of the selected behaviors, it has generally been presumed that (a) the assessment procedure (e.g., role-play test) was invalid, or (b) the treatment which should have produced a difference was not effective, or (c) the molar category under study (e.g., heterosocial behavior) depends on something other than response skills (e.g., anxiety). However, it is entirely possible that the failure to find expected differences results from assessment of the wrong responses or assessing the correct responses in the wrong way.

For example, Fischetti *et al.* (1977) demonstrated that high and low heterosocially anxious males did not differ in the frequency of social reinforcers emitted, but they did differ in the *timing* of their reinforcers. If frequency rather than response placement were assessed, substantially different conclusions would be reached. Another potential error in the manner of assessment involves unidirectional scoring of behaviors which can be bidirectionally inappropriate. For example, eye contact (or gaze), speech duration, and response latency are typically timed and scored in seconds (i.e., a unidirectional scale). However, each behavior can be inappropriately short *or* long. Consequently, such scoring is often invalid for grouped data (i.e., the two errors can balance out within a group). Trower *et al.* (1978) have described a series of subjective bidirectional scales which circumvent this problem.

The most serious issue involves the selection of target behaviors. The relevant molar behavior categories must be systematically analyzed in an effort to determine precisely which response components account for the most variance in behavioral competence. The social psychology literature can provide some direction for this task by identifying potential elements. However, much of that literature is based on an analogue research model which typically examines single behaviors out of the normal interpersonal context (cf. Archer & Akert, 1977). An adequate clinical-behavioral analysis must determine the critical elements or combinations of elements as they occur in ongoing interactions. Thus, the question is not whether or not good eye contact contributes to effective assertiveness, but rather (a) which are the *most* important components?

and (b) what are the minimum requirements for an effective response? This issue is clearly reflected in a study by Waxer (1977). He found that judges placed differential emphasis on various response characteristics in appraising emotionality. Furthermore, the responses which were most informative to judges were *not* the responses on which the criterion groups actually differed the most. These data suggest the importance of social validation in the selection of target behaviors as well as the evaluation of change (Kazdin, 1977).

The identification of appropriate target behaviors is probably the most critical task facing workers in the area of social skills. Assessment procedures cannot be adequately validated until their focus is identified (e.g., which behaviors). Similarly, training procedures cannot be evaluated until the scope of training is specified. Finally, explanations of various interpersonal dysfunctions cannot be analyzed until the nature of the dysfunctions is elucidated.

Social Perception and Cognitive Factors

Current approaches to social skills assessment and training have placed almost exclusive emphasis on motoric response components. It generally has been presumed that the quality of interpersonal performance depends (almost) entirely on the response skills in the person's repertoire. However, this conception only accounts for part of the variance in social skill. While the individual cannot perform effectively without the requisite response capability, the mere presence of such skills in the repertoire does not insure effective performance. More specifically, the individual must know *when* and *where* to make various responses, as well as *how* to make them. This aspect of interpersonal skill has been referred to as social or interpersonal perception (Morrison & Bellack, 1978). It includes: (1) knowledge about the meaning of various response cues and familiarity with social mores; (2) attention to the relevant aspects of the interaction, including the context and the interpersonal response cues provided by the partner; (3) information-processing capability; and (4) the ability accurately to predict interpersonal consequences.

The importance of social perception skill is illustrated in asking an employer for a raise. The individual must know how to make the request in an appropriately assertive manner. But he/she must also determine an appropriate time, formulate a strategy based on the employer's particular response style, identify the employer's initial reaction (e.g., highly positive or negative, ambivalent), project the potential consequences of a particular response choice, etc. For example, presenting an ultimatum might be appropriate if the initial reaction was ambivalence, but it would be ill-advised if the reaction were negative and quitting was not an acceptable option. Determining whether the response is negative or ambivalent requires accurately "reading" or decoding the employer's facial expression, posture, voice tone, and other nonverbal responses, and integrating that information with speech content.

Several recent studies have demonstrated the importance of cognitive factors in interpersonal behavior sequences (Eisler *et al.*, 1978; Fiedler & Beach, 1978; Schwartz & Gottman, 1976). Fiedler and Beach (1978) showed that whether or not people act assertively is substantially affected by the consequences they anticipate for their behavior. Thus, assertive responses may well be inhibited if the individual expects negative consequences to ensue (Eisler *et al.*, 1978). These findings are consistent with anecdotal clinical data suggesting that many nonassertive patients confuse assertion with aggression, view it as inappropriate, and expect to be punished if they respond assertively. Consequently, they are often reluctant to make assertive responses even after otherwise successful training. Inappropriate or inaccurate appraisals of potential response consequences can result from faulty decoding of the cues supplied by others, or from inappropriate labeling and/or dysfunctional belief systems (e.g., "assertion is bad"). The response component model is inadequate to account for dysfunctional behavior which results from any of these difficulties.

Despite the importance of assessing the various social perception skills and associated cognitive parameters, no empirically sound assessment techniques currently exist. Guilford and his colleagues (cf. Guilford, 1968) have developed tests of "social intelligence," but these instruments appear to measure general cognitive or intellectual ability more than any specific component of interpersonal behavior. Platt and Spivack (1970) have developed the Means–Ends Problem Solving Procedure. This is a multidimensional instrument designed to assess ability to solve interpersonal problems, including both identification of the nature of problem situations and generation of solutions. Subjects are provided with the beginning and ending of an interpersonal scenario and are required to write a story which fits in between. Stories are scored on a variety of semiobjective dimensions. This instrument has promise as part of a comprehensive interpersonal behavior assessment battery.

Another promising approach is the Social Interpretation Task (SIT) developed by Archer and Akert (1977). The SIT consists of a series of videotaped interpersonal interactions. After viewing the tape, the subject is required to answer a series of questions about the nature of the interactions, relationship of the participants to one another, etc. Ability to answer these questions apparently depends on accurate decoding of verbal and nonverbal response cues provided by the actors. Morrison,

Bellack, and Hersen (in preparation) have developed a related instrument specifically for the assessment of perception skills dealing with assertiveness. Subjects are presented a series of videotapes portraying submissive, assertive, and angry-hostile responses to each of six negative assertion scenarios, and unappreciative, overly solicitous, and appropriately appreciative responses to each of six positive assertion scenarios. The subjects indicate how "effective" they think each response is and the mood or emotion displayed by each actor.

Work in this area is at an elemental stage, and well-validated instruments to assess the various perception skills and cognitive parameters of social skill have not yet been developed. This would appear to be a high-priority topic for future research. Our ability to identify the nature of interpersonal dysfunction, plan appropriate treatments, and insure generalization of treatment effects will be limited until such procedures are available.

SUMMARY

This chapter has presented a critical review of current approaches for the assessment of social skills. The goals of the assessment were first elaborated in the context of four questions which must be answered. Major assessment strategies for answering those questions were then discussed. Interviewing was described as an important strategy which has been overlooked in the research literature. Self-report inventories and self-monitoring were considered, and limitations were identified. Three strategies for behavioral observation were then evaluated—*in vivo* observation, naturalistic interactions, and role-playing—and the advantages and disadvantages of these procedures were examined. Ratings made by peers and significant others, and physiological assessment were then briefly considered. The final section of the chapter dealt with two issues which have received insufficient attention in the literature: (1) selection of target behaviors for assessment (and treatment), and (2) assessment of interpersonal perception skills and cognitive responses surrounding social behavior.

The general tone of this chapter was, at times, somewhat negative. In general, it was concluded that many current strategies are either invalid or of uncertain validity, and that new procedures are needed. This is not to say that the existing techniques, and literature based on them, should be ignored. Rather, it highlights the need for greater care and effort in the development of assessment techniques and underscores the need for further work in this area. Current techniques can tell us much about the interpersonal functioning of our patients and subjects. However, there is much that they cannot tell us. Existing work in this area must be viewed as preliminary and should serve as a basis and stimulus for future efforts.

REFERENCES

Anastasi, A. *Psychological testing* (2nd ed.). New York: Macmillan, 1961.

Archer, D., & Akert, R. M. Words and everything else: Verbal and nonverbal cues in social interpretation. *Journal of Personality and Social Psychology,* 1977, *35,* 443–449.

Argyris, C. Explorations in interpersonal competence. I. *Journal of Applied Behavioral Science,* 1965, *1,* 58–83.

Arkowitz, H. Measurement and modification of minimal dating behavior. In M. Hersen, R. M. Eisler, & P. M. Miller (Eds.), *Progress in behavior modification* (Vol. 5). New York: Academic, 1977.

Arkowitz, H., Lichtenstein, E., McGovern, K., & Hines, P. The behavioral assessment of social competence in males. *Behavior Therapy,* 1975, *6,* 3–13.

Bander, K. W., Steinke, G. V., Allen, G. J., & Mosher, D. L. Evaluation of three dating-specific treatment approaches for heterosexual dating anxiety. *Journal of Consulting and Clinical Psychology,* 1975, *43,* 259–265.

Bellack, A. S. Anxiety and neurotic disorders. In A. E. Kazdin, A. S. Bellack, & M. Hersen (Eds.), *New perspectives in abnormal psychology.* New York: Oxford University Press, in press.

Bellack, A. S., & Hersen, M. *Behavior modification: An introductory textbook.* Baltimore: Williams & Wilkins, 1977. (a)

Bellack, A. S., & Hersen, M. The use of self-report inventories in behavioral assessment. In J. D. Cone & R. P. Hawkins (Eds.), *Behavioral assessment: New directions in clinical psychology.* New York: Brunner/Mazel, 1977. (b)

Bellack, A. S., & Hersen, M. Chronic psychiatric patients: Social skills training. In M. Hersen & A. S. Bellack (Eds.), *Behavior therapy in the psychiatric setting.* Baltimore: Williams & Wilkins, 1978.

Bellack, A. S., & Hersen, M. *Introduction to clinical psychology.* New York: Oxford University Press, 1979.

Bellack, A. S., Hersen, M., & Turner, S. M. Generalization effects of social skills training in chronic schizophrenics: An experimental analysis. *Behaviour Research and Therapy,* 1976, *14,* 391–398.

Bellack, A. S., Hersen, M., & Turner, S. M. Role play tests for assessing social skills: Are they valid? *Behavior Therapy,* 1978, *9,* 448–461.

Bellack, A. S., Hersen, M., & Lamparski, D. Role play tests for assessing social skills: Are they valid? Are they useful? *Journal of Consulting and Clinical Psychology,* in press. (a)

Bellack, A. S., Hersen, M., & Turner, S. M. The relationship of role playing and knowledge of appropriate behavior to assertion in the natural environment. *Journal of Consulting and Clinical Psychology,* in press. (b)

Borgatta, E. F. Analysis of social interaction: Actual, role playing, and projective. *Journal of Abnormal and Social Psychology,* 1955, *51,* 294–405.

Borkovec, T. D., Stone, N. M., O'Brien, G. T., & Kaloupek, D. G. Evaluation of a clinically relevant target behavior for analog outcome research. *Behavior Therapy,* 1974, *5,* 503–513.

Christensen, A., & Arkowitz, H. Preliminary report on practice dating and feedback as treatment for college dating problems. *Journal of Counseling Psychology,* 1974, *21,* 92–95.

Cone, J. D. The relevance of reliability and validity for behavioral assessment. *Behavior Therapy*, 1977, *8*, 411–426.

Curran, J. P. Skills training as an approach to the treatment of heterosexual-social anxiety: A review. *Psychological Bulletin*, 1977, *84*, 140–157.

Curran, J. P. Comments on Bellack, Hersen, and Turner's paper on the validity of role-play tests. *Behavior Therapy*, 1978, *9*, 462–468.

Curran, J. P., & Gilbert, F. S. A test of the relative effectiveness of a systematic desensitization program and interpersonal skills training program with date anxious subjects. *Behavior Therapy*, 1975, *6*, 510–521.

Eisler, R. M. The behavioral assessment of social skills. In M. Hersen & A. S. Bellack (Eds.), *Behavioral assessment: A practical handbook*. New York: Pergamon, 1976.

Eisler, R. M., Hersen, M., Miller, P. M., & Blanchard, E. B. Situational determinants of assertive behaviors. *Journal of Consulting and Clinical Psychology*, 1975, *43*, 330–340.

Eisler, R. M., Frederiksen, L. W., & Peterson, G. L. The relationship of cognitive variables to the expression of assertiveness. *Behavior Therapy*, 1978, *9*, 419–427.

Fiedler, D., & Beach, L. R. On the decision to be assertive. *Journal of Consulting and Clinical Psychology*, 1978, *46*, 537–546.

Fischetti, M., Curran, J. P., & Wessberg, H. W. Sense of timing: A skill deficit in heterosexual-socially anxious males. *Behavior Modification*, 1977, *1*, 179–195.

Fiske, D. W. The use of significant others in assessing the outcome of psychotherapy. In I. E. Waskow & M. B. Parloff (Eds.), *Psychotherapy change measures*. DHEW Publication No. (ADM) 74-120, Washington, D.C.: USDHEW, 1975.

Galassi, M. D., & Galassi, J. P. The effects of role playing variations on the assessment of assertive behavior. *Behavior Therapy*, 1976, *7*, 343–347.

Glanz, L. M. *Situational determinants of assertive behaviors in a non-clinical population*. Unpublished doctoral dissertation, University of Pittsburgh, 1978.

Goldfried, M. R., & D'Zurilla, T. A behavior-analytic model for assessing competence. In C. D. Spielberger (Ed.), *Current topics in clinical and community psychology* (Vol. 1). New York: Academic, 1969.

Goldfried, M. R., & Kent, R. N. Traditional vs. behavioral personality assessment: A comparison of methodological & theoretical assumptions. *Psychological Bulletin*, 1972, *77*, 409–420.

Goldsmith, J. B., & McFall, R. M. Development and evaluation of an interpersonal skill-training program for psychiatric inpatients. *Journal of Abnormal Psychology*, 1975, *84*, 51–58.

Guilford, J. P. The structure of intelligence. In D. K. Whitla (Ed.), *Handbook of measurement and assessment in behavioral sciences*. Reading, Mass.: Addison-Wesley, 1968.

Gutride, M. E., Goldstein, A. P., & Hunter, G. F. The use of modeling and role playing to increase social interaction among asocial psychiatric patients. *Journal of Consulting and Clinical Psychology*, 1973, *40*, 408–415.

Hamilton, M. A rating scale for depression. *Journal of Neurology, Neurosurgery, and Psychiatry*, 1960, *23*, 56–61.

Harper, R. G., Wiens, A. N., & Matarazzo, J. D. *Nonverbal communication: The state of the art*. New York: Wiley, 1978.

Hersen, M., & Bellack, A. S. Assessment of social skills. In A. R. Ciminero, K. S. Calhoun, & H. E. Adams (Eds.), *Handbook for behavioral assessment*. New York: Wiley, 1977.

Hersen, M., Bellack, A. S., & Turner, S. M. Assessment of assertiveness in female psychiatric patients: Motor and autonomic measures. *Journal of Behavior Therapy and Experimental Psychiatry*, 1978, *9*, 11–16.

Jacob, T. Family interaction in disturbed and normal families: A methodological and substantive review. *Psychological Bulletin*, 1975, *82*, 33–65.

Jacob, T. Assessment of marital dysfunction. In M. Hersen & A. S. Bellack (Eds.), *Behavioral assessment: A practical handbook*. New York: Pergamon, 1976.

Johnson, S. M., & Bolstad, O. D. Methodological issues in naturalistic observation: Some problems and solutions for field research. In L. A. Hamerlynck, L. C. Handy, & E. J. Mash (Eds.), *Behavior change: Methodology, concepts, and practice*. Champaign, Ill.: Research Press, 1973.

Johnson, S. M., & Bolstad, O. D. Reactivity to home observation: A comparison of audio recorded behavior with observers present or absent. *Journal of Applied Behavior Analysis*, 1975, *8*, 181-185.

Jones, R. R. Conceputal vs. analytic use of generalizability theory in behavioral assessment. In J. D. Cone & R. P. Hawkins (Eds.), *Behavioral assessment: New directions in clinical psychology*. New York: Brunner/Mazel, 1977.

Kane, J. S., & Lawler, E. E., III. Methods of peer assessment. *Psychological Bulletin*, 1978, *85*, 555–586.

Kazdin, A. E. Self-monitoring and behavior change. In M. J. Mahoney & C. E. Thoresen (Eds.), *Self-control: Power to the person*. Monterey, Calif.: Brooks/Cole, 1974.

Kazdin, A. E. Assessing the clinical or applied importance of behavior change through social validation. *Behavior Modification*, 1977, *4*, 427–452.

Kiesler, D. J. Patient measures based on interviews and other speech samples. In I. E. Waskow & M. B. Parloff (Eds.), *Psychotherapy change measures*. DHEW Publication No. (ADM) 74–120, Washington, D.C.: USDHEW, 1975.

King, L. W., Liberman, R. P., Roberts, J., & Bryan, E. Personal effectiveness: A structured therapy for improving social and emotional skills. *European Journal of Behavioural Analysis and Modification*, 1977, *2*, 82–91.

Kreitler, H., & Kreitler, S. Validation of psychodramatic behaviour against behaviour in real life. *British Journal of Medical Psychology*, 1968, *41*, 185–192.

Lang, P. J. The application of psychological methods to the study of psychotherapy and behavior modification. In A. E. Bergin & S. L. Garfield (Eds.), *Handbook of psychotherapy and behavior change*. New York: Wiley, 1971.

Leitenberg, H., Agras, W. S., Allen, R., Butz, R., & Edwards, J. Feedback and therapist praise during treatment of phobia. *Journal of Consulting and Clinical Psychology*, 1975, 43, 396–404.

Libet, J., & Lewinsohn, P. M. Concept of social skill with special reference to the behavior of depressed persons. *Journal of Consulting and Clinical Psychology*, 1973, *40*, 304-312.

Lick, J. R., & Unger, T. E. The external validity of behavioral fear assessment: The problem of generalizing from the laboratory to the natural environment. *Behavior Modification*, 1977, *1*, 283–306.

McFall, R. M., & Marston, A. R. An experimental investigation of behavior rehearsal in assertive training. *Journal of Abnormal Psychology*, 1970, *76*, 295–303.

Martin, B. *Anxiety and neurotic disorders*. New York: Wiley, 1971.

Matarazzo, J. D., & Wiens, A. N. *The interview: Research on its anatomy and structure*. Chicago: Aldine-Atherton, 1972.

Mehrabian, A. *Nonverbal communication*. Chicago: Aldine-Atherton, 1972.

Mischel, W. *Personality and assessment*. New York: Wiley, 1968.

Morrison, R. L., & Bellack, A. S. *The role of social perception in social skill*. Unpublished manuscript, University of Pittsburgh, 1978.

Morrison, R. L., Bellack, A. S., & Hersen, M. The relationship of social perception to effective assertiveness. Manuscript in preparation.

Nelson, R. O. Methodological issues in assessment via self-monitoring. In J. D. Cone & R. P. Hawkins (Eds.), *Behavioral assessment: New directions in clinical psychology*. New York: Brunner/Mazel, 1977.

Patterson, G. R. Interventions for boys with conduct problems: Multiple settings, treatments and criteria. *Journal of Consulting and Clinical Psychology,* 1974, *42,* 471–481.

Patterson, G. R., Ray, R. S., Shaw, D. A., & Cobb, J. A. *A manual for coding of family interactions,* 1969 revision, Document #01234. (Available from ASIS/NAPS c/o Microfiche Publications, 305 East 46th St., New York, NY 10017.)

Platt, J. J., & Spivack, G. *Real-life problem-solving thinking in neuropsychiatric patients and controls.* Paper presented at the meeting of the Eastern Psychological Association, Atlantic City, April 1970.

Rehm, L. P., & Marston, A. R. Reduction of social anxiety through modification of self-reinforcement: An instigation therapy technique. *Journal of Consulting and Clinical Psychology,* 1968, *32,* 565–574.

Royce, W. S., & Arkowitz, H. *Multi-model evaluation of* in vivo *practice as treatment for social isolation.* Unpublished manuscript, University of Arizona, 1976.

Schwartz, R. M., & Gottman, J. M. Toward a task analysis of assertive behavior. *Journal of Consulting and Clinical Psychology,* 1976, *44,* 910–920.

Stanton, H. R., & Litwak, E. Toward the development of a short form test of interpersonal competence. *American Sociological Review,* 1955, *20,* 668–674.

Trower, P., Bryant, B., & Argyle, M. *Social skills and mental health.* Pittsburgh: University of Pittsburgh Press, 1978.

Twentyman, C. T., & McFall, R. M. Behavioral training of social skills in shy males. *Journal of Consulting and Clinical Psychology,* 1975, *43,* 384–395.

Warren, N. J., & Gilner, F. H. Measurement of positive assertive behaviors: The behavioral test of tenderness expression. *Behavior Therapy,* 1978, *9,* 178–184.

Waxer, P. H. Nonverbal cues for anxiety: An examination of emotional leakage. *Journal of Abnormal Psychology,* 1977, *86,* 306–314.

Weiss, R. L., & Margolin, G. Marital conflict and accord. In A. R. Ciminero, K. S. Calhoun, & H. E. Adams (Eds.), *Handbook for behavioral assessment.* New York: Wiley, 1977.

Weiss, R. L., Hops, H., & Patterson, G. P. A framework for conceptualizing marital conflict: A technology for altering it, some data for evaluating it. In L. A. Hamerlynck, L. C. Handy, & E. J. Mash (Eds.), *Behavior change: Methodology, concepts, and practice.* Champaign, Ill.: Research Press, 1973.

Whitehill, M. B., Hersen, M., & Bellack, A. S. *A conversation skills training program for socially isolated children.* Unpublished manuscript, University of Pittsburgh, 1978.

Wolpe, J. *The practice of behavior therapy.* New York: Pergamon, 1969.

Wolpe, J., & Lang, P. J. A fear survey schedule for use in behavior therapy. *Behaviour Research and Therapy,* 1964, *2,* 27–30.

Wolpe, J., & Lazarus, A. A. *Behavior therapy techniques.* New York: Pergamon, 1966.

PART TWO

TREATMENT

CHAPTER 4

Modification of Social Skill Deficits in Children

Roger C. Rinn and Allan Markle

INTRODUCTION

The importance of competent social skills to children's long-term development is well documented. For example, Roff, Sells, and Golden (1972) found that isolated children were more likely than others to be identified as delinquent several years later. Cowen, Pederson, Babijian, Izzo, and Trost (1973) demonstrated that unpopular children showed a disproportionate propensity toward mental health problems, while Roff (1961) found that a high proportion of isolated children later received bad-conduct discharges from the armed services. Ullmann (1957) and Hartup (1970) noted that unpopular, isolated children were more likely to drop out of school than popular children, and Gronland and Anderson (1963) found that isolated children had experienced multifaceted school problems. According to Strain, Shores, and Kerr (1976), withdrawn children tend to exhibit low academic performance. Cobb (1970) demonstrated that a specific social skill was associated with academic achievement: "the child who talks about academic material to his peer as well as attends to his work is more likely to succeed than the child who attends without interacting with his peers" (p. 10). Underscoring the importance of these findings is the observation by Gronland (1959) that 6% of children in grades three to six had no friends in the classroom while another 12% had only one classroom friend. Waldrop and Halver-

Roger C. Rinn • Private Practice, Huntsville, Alabama 35801. **Allan Markle** • Private Practice, Northbrook, Illinois 60062.

son (1975) found that withdrawn children demonstrated persistence in low sociability across a 5-year span.

However, children typically are not referred to mental health caregivers through concern for the long-term detrimental effects of poor social skills. Most children with such deficits are referred for treatment by parents and teachers because of the children's unhappiness, withdrawal, social incompetencies, or obnoxious interpersonal habits, among others. In our setting, a comprehensive mental health center, approximately 87% of the children admitted to services are experiencing problems with one or more components of social skills. Overall, poor competency in social skills has both long- and short-term negative consequences. The present discomfort experienced by such children and their significant others (parents, peers, etc.) may be the precursor of more severe problems in adolescence and adult life since social skills deficits appear to be chronic. Thus, social skills training, if viewed in this way, may serve as a treatment for existing problems and a preventive measure.

The term "social skills" is defined herein as a repertoire of verbal and nonverbal behaviors by which children affect the responses of other individuals (e.g., peers, parents, siblings, and teachers) in the interpersonal context. This repertoire acts as a mechanism through which children influence their environment by obtaining, removing, or avoiding desirable and undesirable outcomes in the social sphere. Further, the extent to which they are successful in obtaining desirable outcomes and avoiding or escaping undesirable ones *without inflicting pain on others* is the extent to which they are considered "socially skilled."

There are two aspects of this definition which require further explication. First, from whose perspective does the desirable-undesirable continuum evolve—the child's, parents', or society's? In actual clinical practice, the child's viewpoint is paramount as long as it does not conflict with parental and/or societal mores. In most instances, the specific goals of social skills training are reasonable to parents and socially acceptable (e.g., to make more friends or to handle a bully more effectively). Thus, the objectives of social skills training are usually agreeable to all parties (except, perhaps, the bully!). The second part of the definition involves the proviso that the socially competent child is interpersonally effective without inflicting pain on others. Pain can be administered verbally (e.g., sarcasm, name calling, threats of violence) or nonverbally (e.g., hitting, scowling, gesturing). Needless to say, some goals are more easily and rapidly reached through brute force (e.g., acquiring a desired desk in the classroom by pushing the occupant out or coercing a peer into giving up a place in line). Many parents actively encourage their children to be aggressive and cajole them to "quit acting like a sissy!" Unfortunately, excessive control through pain seems to result in peer rejection or retaliation. Hopefully, changes in societal norms will result in more acceptance of non-pain-inducing interpersonal styles.

In this chapter, we will detail our conceptualization of social skills and describe several components of the repertoire which show promise as important contributors to effective interpersonal functioning in children. Then an assessment strategy will be discussed, followed by a description of our training program for children with social skills deficits. Finally, problems and strategies for instigating and maintaining these newly acquired responses in the natural environment will be mentioned.

Conceptualization of Social Skills

Asher (1977) has delineated two types of children who are socially isolated. Using a combination of positive and negative peer nominations and peer ratings, Asher was able to discriminate between children who were actively disliked and those who were not disliked but merely ignored. It is assumed herein that isolated children who are actively disliked are interpersonally obnoxious, i.e., they emit a high rate of aversive stimuli (commands, criticisms, hits, etc.), and relatively "nonreinforcing," i.e., they emit a low rate of positive stimuli (praise, sharing, helping, etc.). Further, isolated children who are simply ignored are assumed not to be obnoxious but are nonreinforcing interpersonally. Hymel and Asher (1977) found that approximately 60% of these unpopular children were actively rejected with 40% being ignored.

It is probable that a large proportion of the aversive (obnoxious) stimuli are employed contingently to influence the behaviors of peers through punishment and negative reinforcement. This "control by pain" has been noted by Patterson and Reid (1970), who referred to it as *coercion*. Coercion occurs at a high rate in families of deviant children (e.g., Patterson, 1976) and troubled marriages (e.g., Birchler, Weiss, & Vincent, 1975). Undoubtedly, this phenomenon occurs in the peer interactions between children.

The concept of coercion was presented by Patterson and Reid in tandem with that of *reciprocity*. Accordingly, aversive (and positive) stimuli are hypothesized as being exchanged in a dyadic relationship at an equal rate ("tit for tat"). Patterson (1976) noted that

> as one member of a system applies pain control techniques, the victims will eventually learn, via modeling . . . and/or reinforcing contingencies . . . to initiate coercive interchanges. As the victims acquire coercive "skills," they will also be more likely to counter the coercive initiations of others, by coercive measures of their own. (p. 269)

Unlike the parent–child relationship where the child is clearly dependent on the parent and cannot avoid coming into contact with the source of pain (i.e., the parent), the classroom and playground afford some degree of distance among peers. In this conceptualization, actively disliked children initially "turn off" other children through coercion, receive reciprocated pain from peers, and eventually avoid peer contacts as a result. The obnoxious isolate, therefore, seems to be rejected by his peers and, in turn, rejects them.

The emphasis on positive stimuli in the interpersonal dyad is as important as the absence of aversive stimuli in the development of friendships. In a study by Hartup, Glazer, and Charlesworth (1967), it was shown that popular children were more positive than unpopular children. Charlesworth and Hartup (1967) presented data which suggested that the rate of positives given in the interpersonal setting was positively correlated with the rate of positives received. This supports Patterson and Reid's (1970) concept of reciprocity. The ignored, isolated children receive minimal positive stimulation from peers since they do not give such reinforcement. Sadly, these children are completely out of the mainstream of social functioning—they do not even exchange aversive stimuli with peers. Their socially skilled peers tend to exhibit high rates of positive stimuli and low rates of aversive stimuli, thus facilitating mutually pleasurable contacts and subsequent friendships. The foregoing discussion highlights two requirements for social skills assessment and training: (1) the children's responses should be as positive as possible, and (2) these responses should contain a minimum of aversive stimuli.

A Taxonomy of Social Skills

In our conduct of assessment and training, we have found it useful to categorize social skills into four repertoires as follows:

1. Self-Expressive Skills
   a. Expression of feeling (sadness and happiness)
   b. Expression of opinion
   c. Accepting compliments
   d. Stating positives about oneself
2. Other-Enhancing Skills
   a. Stating positives about a best friend
   b. Stating genuine argreement with another's opinion
   c. Praising others
3. Assertive Skills
   a. Making simple requests

   b. Disagreeing with another's opinion
   c. Denying unreasonable requests
4. Communications Skills
   a. Conversing
   b. Interpersonal problem solving

ASSESSMENT OF SOCIAL SKILLS DEFICITS

There have been several techniques by which researchers and practitioners have determined the levels of social skills in children. The most popular strategies include: (1) clinical interviews with significant others (e.g., parents, teachers); (2) rating scales completed by significant others; (3) self-report; (4) sociometric evaluation by peers; (5) naturalistic observation; and (6) analogue measurement. In this section, a brief discussion of each strategy will be followed by our approach to the assessment of skills deficits in children.

Clinical Interviews

The gathering of relevant clinical information about children traditionally has been aimed at the targets' significant others such as parents and teachers. Most often, the parents of socially unskilled children present therapists with details such as "She doesn't have any friends and never leaves the house after school," or "He brings other children home but they never come back again," or "She is so timid and quiet—she cries whenever another child even so much as gives her a dirty look." Teachers note analogous problems including "He will not mix with the other children at recess," or "She gets in fights in the lunchroom whenever anyone bumps into her," or "He seems to want to make friends but the other students are indifferent toward him—it's like he isn't even there!"

The clinical interview is an extremely important and useful component of social skills assessment. During the interview, the therapist can identify some of the broad problem areas (e.g., social skills deficits, noncompliance, fire setting) and develop some speculations about controlling stimuli. Additionally, the severity of the problems can often be gauged by determining the approximate frequencies of targeted responses, the environments in which they occur, etc.

Unfortunately, parents and teachers are notoriously poor informants, especially if they are unaccustomed to the behavioral approach (particularly the parenting skills of clearly specifying target behaviors and measuring their frequencies). More important, most clinical interviews are not standardized and tend to be relatively subjective due to the questionable reliability of the information and lack of standardized, quantifiable data.

Including the target children in the interview may allow for more specificity of skills deficits, but often they are somewhat inhibited or unusually well behaved during the interview ("Am I going to get a shot?"). How often have we all heard incredible stories about a child's horrible, obnoxious interpersonal antics, only to find ourselves talking to an extremely polite youngster? This points to another problem with assessing social skills deficits exclusively with an interview—such behaviors are typically situation specific.

In summary, clinical interviews are useful for determining the broad characteristics of problematic behaviors. However, such interviews do not constitute acceptable assessment devices.

Rating Scales

Under the heading of rating scales fall a wide variety of standardized and nonstandardized measures. For example, the Walker Problem Behavior Identification Checklist (Walker, 1970) has been used widely as a broad standardized measure of social functioning. The WPBIC is completed by parents and/or teachers in a few minutes. It consists of a total score and five subtests (Acting Out, Withdrawal, Distractability, Disturbed Peer Relations, and Immaturity). Despite the obvious utility of the WPBIC, definitive data have not been available concerning its applicability to diagnosing specific social skills deficits. Some preliminary findings from our setting are presented later in this section.

In a study by Rinn, Mahla, Markle, Barnhart, Owen, and Supnick, (1979a) a nonstandardized, Likert-type scale was given to teachers who were asked to rate students on a 5-point scale as to "How much do you think other children would like to play with ______?" and "How much do you think other children would like to do schoolwork with ______?" Results of a pilot research project using such data are presented later in this section. Nonetheless, data available on the efficacy of rating scales by significant others (excluding peers) in assessing social skills are scant and the area has not been well researched.

Reardon, Hersen, Bellack, and Foley (1978) employed another nonstandardized rating system, based on McFall and Lillesand's (1971) earlier "Conflict Resolution Inventory." Children were rated by teachers on a 100-point scale in terms of positive and negative assertion. Reardon *et al.* found that correlations between teacher ratings of assertiveness and analogue measures were "negligible." In conclusion, the usefulness of parent and teacher ratings of children's social skills has not been demonstrated.

Self-Report

Self-report of social skills has not been a popular assessment technique for evaluating children. However, Reardon *et al.* (1978) developed the Self-Report Assertiveness Test for Boys (SRAT-B) in which children were asked to check response alternatives to several situations requiring assertion (positive and negative) which they might use in real life. According to the investigators, their self-report measure was not discriminating and seemed more a measure of different behaviors than an analogue test of assertion. The validity of self-report measures for evaluating social skills in children has not been established.

Sociometric Evaluations

Asher (1977) has detailed the development of his sociometric evaluation strategy. Essentially, he and his colleagues have devised an ingenious system of assessment which takes advantage of (1) positive and negative peer nominations (e.g., "Name three children you don't especially like") and (2) peer ratings (e.g., "How much would you like to play with each child?" rated on a 5-point scale). As mentioned earlier in this chapter, Asher was able to discriminate between those children who are actively rejected and those who are simply ignored. Asher noted that sociometric scores have high predictive validity (e.g., Cowen *et al.*, 1973) and reliability (e.g., Roff *et al.*, 1972). However, this system of assessment is rather cumbersome for day-to-day determination of skills (as applied in a reversal design, etc.) and does not clearly specify behavioral deficits, excesses, and competencies.

Naturalistic Observation

In terms of face validity, the use of naturalistic observation is a most compelling assessment strategy for determining social skills in children. Typically, such observations have measured the rate or duration of peer interaction (e.g., O'Connor, 1969). Although Cobb (1970) noted that interacting children tended to perform academically at a level higher than their isolated (but equally attentive) peers, Jennings (1975) found that the rate of interaction was not correlated with sociometric data or cognitions about socially acceptable responses. However, Jennings demonstrated that the sociometric and cognitive (knowledge) measures were significantly related.

The absence of a consistent relationship between the rate of interactions and sociometric measures seems to be the rule rather than the exception (e.g., Deutsch, 1974). Perhaps, it is the *quality* of children's interactions rather than their quantity which determines the potency of friendship relations. Increasing the interaction rates of unskilled, "obnoxious" children may actually result in painful interactions with peers. Kirby and Toler (1970), in a single-subject reversal design, reported that "with his increased rate of interaction, Kenneth [their 5-year old subject] seemed to become much more aggressive in his play with others" (p. 313).

What is needed is a detailed observation system that is designed to measure the social skills repertoires of children in their natural environments, particularly classrooms and playgrounds. Zahavi and Asher (1978) developed a three-category system of naturalistic observation (during "play groups") which included aggressive, positive active, and inactive behaviors. Gottman, Gonso, and Rasmussen (1975) used a 16-component observation system which included such behaviors as "dispensing negative reinforcers verbally" and "alone and off task." A more discriminating system is needed to provide information concerning specific responses within the social skills repertoire and possible controlling stimuli.

Analogue Measurement

Analogue measurement of social skills has been pioneered by Hersen and his colleagues (e.g., Eisler, Miller, & Hersen, 1973). More recently, Reardon *et al.* (1978) conducted one of the first comprehensive studies of analogue assessment with grade-school boys. In their study, Reardon *et al.* rated the subjects' role-played responses to several analogue situations which were designed to elicit "positive and negative assertion." In addition to scoring the boys' responses (in terms of components such as smiles, latency of response, compliance) they rated the boys as to overall assertiveness. Those above the median were labeled "high" and those below "low." Teachers' ratings and self-ratings were also employed.

Results of the Reardon *et al.* study showed that boys judged "high" in assertion differed on several behavioral components from those "low" in assertion. Teacher and self-report ratings were not found to be predictive of the analogue measures.

Analogue measures of social skills have become popular recently, in part because of their efficiency. As noted by Reardon *et al.*, validity data are needed to demonstrate the efficacy of this approach. In attempting to provide such data, an analogue system for assessing social skills in children was developed, the results of which are presented below.

Validation of an Analogue System

In a recent investigation, we and several colleagues (Rinn *et al.*, 1979a) have developed an analogue system of assessing social skills in children. The system has been designed to measure the self-expressive, other-enhancing, and assertive repertoires mentioned earlier in this chapter.

Briefly, the subjects were 28 children (2 black females, 1 black male, 14 white females, and 11 white males) from grades three to six in a relatively small elementary school in rural Alabama. Due to the small size of the school, the children knew one another well. Videotaped situations requiring particular responses were presented to the children, whose responses were likewise videotaped. Included were 17 situations which tapped the following: self-expressive skills (expression of feeling, expression of opinion, and accepting compliments); other-enhancing skills (agreeing with another's opinions, praising others); and assertive skills (making simple requests, disagreeing with another's opinion, and denying unreasonable requests). Each response was scored as to the presence (or absence) of appropriate affect, eye contact, audibility, and latency of response (less than 1 sec). Moreover, each response was scored for the presence or absence of specific verbal content appropriate to the situation's theme (e.g., "accepting compliments" was scored as to "modest acceptance" and "specific reinforcement of compliments" while "denying unreasonable requests" was scored as to "noncompliance" and "request for alternative behavior," and one point was subtracted if the *S* included "punishment or pain" in the response). Each videotaped record was scored by an independent rater. The range of possible scores was 0–100, with actual scores ranging from 19 to 86 (mean = 65.4).

Additionally, the children were also asked to list as many positive attributes as possible about (a) themselves, and (b) their best friend. Children were asked to rate one another (within their own classroom) as to "How much do you like to play with ______" and "How much do you like to work with ______?" each on a 5-point scale. Teachers were asked to complete a similar 5-point rating scale in response to the questions "How much do you think other children would like to play with ______?" and "How much do you think other children would like to do schoolwork with ______?" Finally, teachers completed a Walker Problem Behavior Identification Checklist (Walker, 1970) on each of their students.

Results of this research were encouraging. First, the peer ratings of play were correlated significantly with the total scores on the analogue measure ($r = .34$, $p < .05$). Although peer ratings of work were not significantly correlated with the analogue measure, the relationship did approach significance ($r = .30$, $p < .10$). These findings tend to support the efficacy of the analogue system as an assessment device for peer friendships and popularity.

These correlation coefficients are relatively small. The mean peer ratings of play ranged from 2.9 to 4.8 (possible range = 1.0–5.0) while the means for work ranged from 2.3 to 4.8. These scores are more concentrated toward the desirable end of the continuum than those reported by Hymel and Asher (1977). For instance, in their study, Hymel and Asher found that 22% of their sample obtained peer ratings in the range of 1.00–3.00. In our sample, only two children (7%) fell within this range. Despite our coarctated distribution, a relationship was found. It is important to note that these children were selected from regular classrooms, and it is possible that the analogue system will be shown more strongly related to peer ratings when a sample of clinically withdrawn and/or aggressive children is compared to this normative group.

The numbers of positive attributes the children were able to list about themselves and their best friends were significantly correlated with peer ratings of play ($r = .42$, $p < .05$, and $r = .34$, $p < .05$, respectively). Moreover, best-friend positives were significantly correlated with peer ratings of work ($r = .34$, $p < .05$) while self-positives were not ($r = .02$, N.S.). These findings are intriguing. As a brief screening device, the listing of self-positives and best-friend positives by children seems to be an acceptable technique.

The teacher ratings of work and play were not correlated with the total scores on the analogue measure. This is similar to the findings of Reardon *et al.* (1978). Additionally, the total scores of the Walker Problem Behavior Identification Checklist were not correlated with the analogue measures. However, the Withdrawal subtest was correlated significantly with the total score on the analogue measure ($r = .46, p < .05$). This certainly coincides with our expectation that withdrawn children tend to have poorer social skills than others.

An additional study (Rinn, Priest, Barnhart, & Markle, 1979b) was conducted with several modifications to the analogue system in order to make the system usable by classroom teachers, outpatient therapists, and others without videotaping equipment. First, 5″ × 7″ cards were prepared with simple circular heads (similar to "Smiley Faces"), depicting various facial expressions on one side and a matching situation's narrative and prompt on the other side. The examiner was seated across a small table from the child and displayed the drawings of the facial expressions to the child's view while reading from the narrative and the prompt on the back of the card. The drawings enhanced the child's interest and attention to the task and allowed the examiner to rate eye contact (which was scored if the child looked at the card or the examiner). After four hours of training, a classroom teacher was able to score response components reliably (that is, the presence or absence of appropriate affect, eye contact, audibility, latency of response less than 1 sec, and specific verbal content) for 17 situations encompassing the skills that are listed earlier in this chapter (pp. 110–111). The range of possible scores on the analogue assessment was 0–100, with higher scores denoting greater competency.

The 23 third graders (1 black female, 1 black male, 7 white females, and 14 white males) were classmates in a small, urban elementary school from a low-income neighborhood. All children were rated on the analogue system with scores ranging from 11 to 79 ($\bar{X} = 65.3$). On the same day, a sociometric evaluation (Asher, 1977) was performed in which students (a) rated one another (as in the preceding study by Rinn *et al.*, 1979b) in terms of play and work and (b) nominated those fellow students whom they liked and whom they did not like.

Scores on the analogue system were correlated significantly with all four sociometric measures: (a) peer ratings of work, $r = .66$, $p < .001$; (b) peer ratings of play, $r = .58$, $p < .002$; (c) positive nominations by peers, $r = .48$, $p < .01$; and (d) negative nominations by peers, $r = -.49$, $p < .01$. These findings lend support for the validity of the modified analogue system. Furthermore, the relationships between the analogue and sociometric measures were of greater magnitude in the second study as compared to the first study. These differences may have been a function of intimacy. Students in the first study were merely schoolmates whereas those from the second study were classmates (thus having a greater opportunity to establish clear-cut likes and dislikes).

Overall, the preliminary results of this research support the efficacy of the analogue system as a predictor of peer-group popularity. The data also suggested that teacher ratings on the Walker Problem Behavior Identification Checklist's Withdrawal subtest were negatively correlated with the analogue measure while nonstandardized teacher ratings were not. Finally, the listing of self-positives and best-friend positives by children was shown to be a potentially useful screening device for social skills deficits. Essentially, the fewer self-positives and best-friend positives, the greater the social skills deficits. This would be particularly useful in clinical interviews.

These data along with those of Reardon *et al.* (1978) tend to support the efficacy of analogue measures of social skills deficits in children.

Present research underway is aimed at discriminating actively rejected from ignored children (identified by Asher's, 1977, system of sociometric ratings and peer nomination) in terms of the analogue system. Additional analogue research is in progress to assess the competencies of children in terms of their communication repertoires (conversational and interpersonal problem-solving skills). Once these studies are completed, it should be possible to determine a multiplicity of social skills deficits in children within a clinic setting.

TREATMENT OF SOCIAL SKILLS DEFICITS

Training programs for the establishment of social skills in children have become increasingly popular over the past 10 years. The training components commonly found in such programs include instructions, behavior rehearsal, and modeling. Instructions are usually presented initially in a teaching sequence, and may involve broad conceptual information about the repertoire (e.g., "one should be positive to his/her friends because . . . ") and specific components of the required skills (e.g., "look the other person in the eye" and "tell the other person you don't want to steal the candy"). During behavior rehearsal, trainees practice responses to stimuli from simulated situations presented by the trainer (verbally, videotaped, etc.). Instruction given during behavioral rehearsal is called coaching (e.g., "now try to speak louder"). Feedback given during behavior rehearsal can take the form of reinforcement (e.g., "Dyn-o-mite! That was just right, Mary," or "Good Work! That time you didn't look away—you looked right at him"). In modeling procedure it is expected that the trainee will acquire a response potential while observing the desirable social behaviors of others (in videotaped vignettes, live presentations, etc.).

In actual practice it is difficult to find training procedures which consist of only one of these training components. For example, Oden and Asher (1977) taught third- and fourth-grade isolated children with instructions, behavior rehearsal (without direct coaching or feedback), and a review of the instructions. In a more comprehensive package, Bornstein, Bellack, and Hersen (1977) employed all the components outlined above (i.e., instructions, behavior rehearsal, coaching, feedback, and modeling) with withdrawn children in grades three to six. The Oden and Asher (1977) intervention was assessed via sociometric measures and behavioral observations during play sessions. Their results showed enhancement of the isolated children's friendships as rated on the "play" sociometric measure. The "work" sociometric measure and the behavioral observation of play were not altered significantly. The Bornstein *et al.* (1977) procedures were shown effective by using analogue measures of assertion. In their study, Bornstein *et al.* demonstrated (using a multiple-baseline design) that their training program resulted in immediate, desired behavioral changes and that these new skills generalized across stimulus situations. Generalization to the natural environment was not examined.

In one study purported to represent a modeling procedure for enhancing social skills, O'Connor (1969) showed withdrawn nursery school children a short (23-min) sound–color film. The film depicted models interacting appropriately with their peers and receiving positive consequences for their behaviors. Social interactions of the children in the modeling condition increased significantly after treatment as compared to interactions of children in the control condition (who saw a film about dolphins).

As alluded to earlier, there are few training programs which contain a "pure" training component. Asher (1977) has made the observation that O'Connor's "film has a heavy verbal instructional component" (p. 13). Gottman (reported in Asher, 1977) analyzed the film's narrative and found a large number of actual instructions (e.g., how to interact for positive consequences). Asher concluded that "it could be then that the promising results of the O'Connor film are largely due to the verbal coaching of social skills as much as to the modeling of them" (p. 13).

All three studies presented above (i.e., Oden & Asher, 1977; Bornstein *et al.*, 1977; O'Connor, 1969) contain the component of instructions. The relative potency of the additional components of behavior rehearsal (with or without coaching and feedback) and modeling has not been demonstrated. A study "dismantling" the treatment program of Bornstein *et al.*, similar to that of McFall and Twentyman (1973), is needed. However, a conservative approach would include all the components until such research is forthcoming.

In this brief discussion of training programs for children with social skills deficits, we have not mentioned shaping procedures which are aimed at increasing isolated children's approaches to their peers. Several studies have been conducted using this approach (e.g., Allen, Hart, Buell, Harris, & Wolf, 1964; O'Connor, 1972). Nonetheless, our previous discussion of naturalistic observational assessment related the speculation (based on several studies) that the quality of peer interactions was more important than frequency in the development of friendships. If one assumes that withdrawn, unpopular, or obnoxious children primarily have a learning deficit (as opposed to a performance deficit), then the acquisition of specific skills by these children would appear a more efficacious tactic of treatment than the mere instigation of propinquity.

A Treatment Program for Outpatient Children

A treatment program of social skills training for children from grades three to six who have been referred for outpatient therapy is presented below. Although the program has been applied primarily on a one-to-one basis, it has also been used successfully with small groups of children (four or five participants per group).

The program consists of systematic training for each social skill. Self-expressive skills are trained first, followed by other-enhancing and assertive skills. A communication skills training program is currently under development. The training can be conducted by an individual clinician but seems to flow more smoothly if two therapists are present. One therapist serves as the trainer and the other as the prompter, their respective functions to be described below.

Training is performed during six weekly 30-min sessions (1 h if group training). The student is seated facing the trainer while the prompter sits to one side of the student (about 2 ft away).

Training Process. Training begins with a short introductory explanation by the trainer:

> For the next few weeks, you are going to be able to earn "Goodies" for learning some new things. We are going to show you some new ways of acting around other kids and how to make friends easier. When you are able to do these things well, you will receive points for prizes from our "Goodie Box" [trainer points to a box in the corner, filled with toys, coloring books, etc.—see below]. You will be able to earn at least six prizes—you can even choose which ones, too! These meetings are going to help you get along better with other kids and you'll learn several new ways to make friends. Any questions? [Most children ask questions directly related to the rewards in the Goodie Box.]

With the exceptions of naming positives about oneself and a best friend (see below) training of each social skill proceeds according to the following steps:

1. *Instructions.* The trainer introduces the topic by explaining the object or function of the particular skill being taught. This is followed by listing specific components of the skill.
2. *Questioning.* The trainer asks the student to repeat the instructions in the preceding steps in his/her own words. The trainer prompts the student and coaches him/her in the correct answers.

The trainer and prompter both praise the student for each appropriate response.

3. *Modeled Response.* The trainer reads aloud a situation and prompt requiring the particular skill. The prompter then models the correct behaviors.
4. *Review.* The trainer points out the cogent aspects of the modeled response for the student.
5. *Student's Response.* The student is asked to listen to the situation (read aloud by the trainer) and prompt (presented by the prompter), and then make his/her response.
6. *Feedback.* The trainer and prompter provide copious praise for correctly reproduced components of the skill. Moreover, the trainer makes suggestions to the student which will enhance the skill (e.g., "You could make that even better if you looked her right in the eye").

Prior to each training session, students are told the number of points necessary for a "goodie." If the student's responses are appropriate (e.g., all the components are reproduced in a single response) one point is awarded to him/her along with additional excited praise from the trainer and prompter. When a response is incomplete after the first trial, the trainer returns to questioning (Step 2) and completes the sequence (Steps 3–6) again. Subsequent failures are followed by the Modeled Response (Step 3) and once again the sequence is completed (Steps 4–6). Ninety-five percent of our clients acquire appropriate responses in six trials or less.

In order to train the self-expressive skill of stating positives about oneself and the other-enhancing skill of stating positives about a best friend, a different technique is employed. First, the students are asked to "Name as many good things about yourself as you can—tell us things that you like about yourself!" The students are praised by the trainer and prompter. If students list six or more positives, they are praised and receive one point toward a prize. Those who are unable to meet this criterion are assisted by the trainer and prompter through encouraging (e.g., "Try hard"), questioning (e.g., "Are you a neat person?" "Do you try not to hurt friends?"), and as a last resort, suggestion (e.g., "You really are good at baseball," "Your hair is sure clean and combed"). After the students have compiled a sizeable list of self-positives, they are once again given the original instructions. Most students complete this task after no more than four trials. Once criterion is reached, one point and praise are presented by the trainer and prompter. Training proceeds in the same way for naming positives about a best friend.

Training Situations. Samples of situations used in the training are presented below. For each skill (e.g., denying unreasonable requests) four different situations are presented during training (available on request from the authors). We have found four practice situations to be sufficient in development of competency acquisition:

1. Self-Expressive Skills
   a. Expressions of happiness and sadness
      (1) *Situation:* You have made a good grade on a test. The teacher says,
      Prompt: "You made the highest grade in the class!"
      Components: Statement expressing happiness, facial expression of happiness (e.g., smile), appropriate affect, eye contact, short latency response, and voice loud enough to be audible. (Note: The last four components are included in all responses. They will not be repeated hereafter and will be referred to by "etc.")
      (2) *Situation:* Your father tells you he has to be away from home for a month. You will really miss him. Your mom says,
      Prompt: "How do you feel about that?"
      Components: Statement expressing sadness, facial expression of sadness, etc.
   b. Expression of opinion
      Situation: You just saw the movie *Star Wars*. A friend asks you,
      Prompt: "What did you think about that movie?"
      Components: Clear expression of opinion (e.g., "I think that . . . " or "I really liked . . . "), etc.
   c. Accepting compliments
      Situation: You have a new pair of shoes. A student in your class tells you,
      Prompt: "I really dig your shoes. They look great!"
      Components: Modest acceptance, specific reinforcement to compliments (e.g., "That was nice of you to say"), etc.
2. Other-Enchancing Skills
   a. Agreeing with another's opinion (when such agreement is sincere)
      Situation: You really like your teacher and think he's a lot of fun in class. A kid in your class says,
      Prompt: "I really like him. What do you think about him?"
      Components: Clear, specific statement of agreement, specific reinforcement of the other person (e.g., "I'm glad we agree" or "You have good taste"). etc.

b. Praising others
Situation: A friend of yours has been doing really well in arithmetic. She/he says,
Prompt: "Can you believe it? I made another good grade in math!"
Components: Reinforcement of specific item, behavior, or attribute, etc.

3. Assertive Skills

a. Making simple requests
Situation: You are erasing a mistake on a paper and your only pencil breaks. You need another pencil and your friend has four. He says,
Prompt: "Looks like you broke another pencil, right?"
Components: Request for help or assistance, etc.

b. Disagreeing with another's opinion
Situation: A kid complains about a friend of yours. You really like your friend. The kid says,
Prompt: "I really think your friend is a turkey!"
Components: Specific statement of disagreement, expression of an alternative opinion, absence of verbal pain (no "zinger"), etc.

c. Denying unreasonable requests
Situation: You are bouncing a ball. You are really having fun and don't want to stop. Another guy/girl comes up and, in a mean voice, says
Prompt: "Give me that ball: I want it now—or else!"
Components: Noncompliance, request for new behavior, absence of verbal pain, etc.

Providing Motivation during Sessions. As mentioned earlier, we use a "Goodie Box" (see also Rinn, 1978) to ensure motivated students. Prior to this procedure, we had difficulty obtaining cooperation of some children, many of whom had histories of noncompliance at home and school. However, the inclusion of the Goodie Box enhances the "fun" aspects of the training procedures.

The box includes inexpensive items such as decorative candles, coloring books and crayons, wooden gliders, rubber balls, jacks, games, puzzles, jump ropes, and assorted toys. The students are given the "goodies" contingent on their performance during the sessions. Each session has specific requirements for points (see course outline, below). If all points are earned, the student receives the goodie of his/her choice. However, if a student should not receive all the required points, a "consolation prize" (e.g., a pack of gum, a dime) is awarded if any effort to perform was made (most students earn all their points each session).

The importance of *excited* praise by the trainer and prompter cannot be overemphasized. Excitement during these training sessions is infectious. Frequently, the students smile broadly after a competent response. In groups, children have often applauded one another's response. The praise used by the clinicians should include (1) specification of the appropriate response (e.g., "You certainly looked at his eyes that time"), (2) an enhancing comment (e.g., "You are really acting more grown up! You are learning these things quickly—good work!"), (3) excited affect, (4) a louder than usual voice level, (5) a smile, (6) a short-latency response, (7) eye contact, and (8) a pat on the shoulder or arm.

For the rare student who does not perform during these training sessions with the praise and Goodie Box, a home-based reward system has been effective. An occasional noncompliant child has refused to interact until more potent reinforcers were employed. For such an event, the student is given a "report card" after each session which specifies the number of points possible and the number earned. The points may be used at home for special privileges (e.g., "Staying up late on Saturdays") or other rewards (e.g., extra allowance, new clothes, going with parents to a movie).

The sessions should be as positive and fun as possible. Criticism is usually unnecessary but, if used, should be gentle and constructive. The development of a positive atmosphere can be facilitated by a high rate of positives and few criticisms by the clinicians.

Course Outline. A brief outline of the training is presented below:

Session I

A. Analogue assessment (given prior to session)
B. Introduction of trainer and prompter
C. Introductory explanation
D. Expressions of happiness and sadness (2 situations each; points = 4)
E. Expression-of-opinion situation (points = 4)
F. Goodie Box (possible points = 8)

Session II

A. Analogue assessment (given prior to session)
B. Accepting compliments (4 situations; points = 4)
C. Stating positives about oneself (points = 1)
D. Goodie Box (total possible points = 5)

Session III

A. Analogue assessment (given prior to session)

B. Stating positives about a best friend (points = 1)
C. Agreeing with others (4 situations; points = 4)
D. Goodie Box (total possible points = 5)

Session IV
A. Analogue assessment (given prior to session)
B. Praising others (4 situations; points = 4)
C. Making simple requests (4 situations; points = 4)
D. Goodie Box (total possible points = 8)

Session V
A. Analogue assessment (given prior to session)
B. Disagreeing with another's opinion (4 situations; points = 4)
C. Denying unreasonable requests (4 situations; points = 4)
D. Goodie Box (total possible points = 8)

Session VI
A. Analogue assessment (given prior to session)
B. Discussion of each skill (1 point for participating in discussion)
C. Goodie Box (total possible points = 1 plus "Bonus" prize if student earned all possible points from all sessions)

Enhancing Generalization of Social Skills. A critical but often overlooked aspect of behavioral intervention has been that of generalization. In an informative and stimulating article describing current generalization technology, Stokes and Baer (1977) concluded that

> behavioral research and practice should act as if there were no such animal as "free" generalization—as if generalization never occurs "naturally," but always requires programming. Then, "programmed generalization" is essentially a redundant term, and should be descriptive only of the active regard of researchers and practitioners. (p. 365)

Within the training program we have used, these are two specific components which were included originally to assist in generalization of a social skills repertoire. First, instructions were inserted into the training sequence of each skill since previous research suggested that the use of instructions enhanced social skills functioning as measured by sociometric measures (e.g., Oden & Asher, 1977). Conceptually, the instruction should be useful in mediating the skills across various situations. Second, the inclusion of four differing situations during behavior rehearsal should be facilitative of generalization by subjecting the students to a variety of stimulus presentations. Research is presently underway to determine the parts played by both components in the development of generalized responding.

While programming for generalization within the training session is important, the initiation of a program in the natural environment (e.g., classroom and home) is also essential. Several specific interventions have been applied by our staff. One highly successful approach has been that of training the students' parents in our child-management program, Positive Parenting (see Rinn, 1978; Rinn & Markle, 1977; Rinn, Vernon, & Wise, 1975). Briefly, parents are trained in behavior skills (e.g., pinpointing, measurement, developing effective consequences) during a didactic group training course. Once the course has been completed, the parents receive specific consultation from the trainer and prompter on which social skills responses are being learned, thus allowing the parents to reward their children's appropriate behaviors in the natural environment. Parental praise has been the most consistent reward applied, but other material (allowances, etc.) and activity (movies, etc.) rewards have been included. Occasionally, untrained parents have inadvertently punished assertion by their children during the training program. More commonly, they have ignored newly acquired prosocial responding by their withdrawn children when parental feedback (praise, etc.) would have been invaluable. Therefore, training parents in behavioral skills is routinely accomplished within our setting.

We have found that parents of excessively withdrawn children commiserate with their children about their lack of friends and perceived rejection. We have observed that teaching the parents to attend to talk about successful interactions and to ignore (as best as possible) statements of self-pity facilitates social skills acquisition.

Another generalization technique found useful has been the construction of indiscriminable contingencies (e.g., partial reinforcement of appropriate responses). Occasionally, we will ask siblings or peers to do things such as compliment the target children and record (on paper) the targets' responses. Appropriate demonstration of a particular social skill is rewarded by the sibling or peer. Moreover, the trainer, prompter, and parents provide additional praise for competent responses.

A final technique for the enhancement of generalization was mentioned earlier, the reinforcement of peer interaction. However, it is our view that peer interaction should be instigated during and after social skills training instead of being employed as a substitute for such training. Once the target children have acquired competent repertoires of social skills, then they are ready to practice with their peers. It is the responsibility of the clinician to increase the probability of peer interactions if they are low in frequency.

CONCLUSION

The area of social skills training has become an important and popular one within the field of behavior modification. As shown in this chapter, the research has been sparse but promising. Both assessment and intervention strategies are in need of further empirical examination.

Assessment strategies using analogue measures and sociometric techniques have been shown efficacious in evaluating social skills functioning in children. Our own research indicated a significant relationship between the two approaches. Hopefully, a more comprehensive study will result in a clearer understanding of this relationship.

Social skills training programs for children are admittedly in their early stages of development. However, preliminary findings imply that several techniques (e.g., instruction, modeling) are applicable to the development of socially skilled responses. Again, further research is needed to "dismantle" the training procedure in order to discern which elements are functional and should be maintained. In addition, the establishment of generalized responding may not be possible through the outpatient training of withdrawn children. The examination of various generalization schemes for social skills is an additional research topic which has been relatively unexplored.

REFERENCES

Allen, K. E., Hart, B., Buell, J. S., Harris, F. R., & Wolf, M. M. Effects of social reinforcement of isolate behavior of a nursery school child. *Child Development*, 1964, *35*, 511–518.

Asher, S. R. *Coaching socially isolated children in social skills.* Paper presented at the annual meeting of the Association for Advancement of Behavior Therapy, Atlanta, December, 1977.

Birchler, G. R., Weiss, R. L., & Vincent, J. P. A multimethod analysis of social reinforcement exchange between maritally distressed and nondistressed spouse and stranger dyads. *Journal of Personality and Social Psychology*, 1975, *31*, 349–360.

Bornstein, M. R., Bellack, A. S., & Hersen, M. Social-skills training for unassertive children: A multiple-baseline analysis. *Journal of Applied Behavior Analysis*, 1977, *10*, 183–195.

Charlesworth, R., & Hartup, W. W. Positive social reinforcement in the nursery school peer group. *Child Development*, 1967, *38*, 993–1003.

Cobb, J. A. The relationship of discrete classroom behaviors to fourth-grade academic achievement. *Oregon Research Institute Research Bulletin*, 1970, *10*, No. 10.

Cowen, E. L., Pederson, A., Babijian, H., Izzo, L. D., & Trost, M. A. Long term follow-up of early detected vulnerable children. *Journal of Consulting and Clinical Psychology*, 1973, *41*, 438–446.

Deutsch, F. Observational and sociometric measures of peer popularity and their relationship to egocentric communication in female preschoolers. *Developmental Psychology*, 1974, *10*, 745–755.

Eisler, R. M., Miller, P. M., & Hersen, M. Components of assertive behavior. *Journal of Clinical Psychology*, 1973, *29*, 295–299.

Gottman, J., Gonzo, J., & Rasmussen, B. Social interaction, social competence, and friendship in children. *Child Development*, 1975, *46*, 709–718.

Gronland, H., & Anderson, L. Personality characteristics of socially accepted, socially neglected, and socially rejected junior high school pupils. In J. Seidman (Ed.), *Educating for mental health.* New York: Crowell, 1963.

Gronland, N. E. *Sociometry in the classroom.* New York: Harper, 1959.

Hartup, W. W. Peer interaction and social organization. In P. Mussen (Ed.), *Carmichael's manual of child psychology* (Vol. 2). New York: Wiley, 1970.

Hartup, W. W., Glazer, J. A., & Charlesworth, R. Peer reinforcement and sociometric status. *Child Development,* 1967, *38,* 1017–1024.

Hymel, S., & Asher, S. R. *Assessment and training of isolated children's social skills.* Paper presented at the biennial meeting of the Society for Research in Child Development, New Orleans, March 1977.

Jennings, K. D. People versus object orientation, social behavior, and intellectual abilities in children. *Developmental Psychology,* 1975, *11,* 511–519.

Kirby, F. D., & Toler, H. C. Modification of preschool isolate behavior: A case study. *Journal of Applied Behavior Analysis,* 1970, *3,* 309–314.

McFall, R. M., & Lillesand, D. B. Behavior rehearsal with modeling and coaching in assertion training. *Journal of Abnormal Psychology,* 1971, *77,* 313–323.

McFall, R. M., & Twentyman, C. T. Four experiments in the relative contributions of rehearsal, modeling, and coaching to assertive training. *Journal of Abnormal Psychology,* 1973, *81,* 199–218.

O'Connor, R. D. Modification of social withdrawal through symbolic modeling. *Journal of Applied Behavior Analysis,* 1969, *2,* 15–22.

O'Connor, R. D. Relative efficacy of modeling, shaping, and the combined procedures for modification of social withdrawal. *Journal of Abnormal Psychology,* 1972, *79,* 327–334.

Oden, S., & Asher, S. R. Coaching children in social skills for friendship making. *Child Development,* 1977, *48,* 495–506.

Patterson, G. R. The aggressive child: Victim and architect of a coercive system. In E. J. Mash, L. A. Hamerlynck, & L. C. Handy (Eds.), *Behavior modification and families.* New York: Brunner/Mazel, 1976.

Patterson, G. R., & Reid, J. B. Reciprocity and coercion: Two facets of social systems. In C. Neuringer & J. D. Michael (Eds.), *Behavior modification in clinical psychology,* New York: Appleton-Century-Crofts, 1970.

Reardon, R. C., Hersen, M., Bellack, A. S., & Foley, J. M. *Measuring social skill in grade school boys.* Manuscript submitted for publication, 1978.

Rinn, R. C. Children with behavior disorders. In M. Hersen & A. S. Bellack (Eds.), *Behavior therapy in the psychiatric setting.* Baltimore: Williams & Wilkins, 1978.

Rinn, R. C., & Markle, A. *Positive parenting.* Cambridge, Mass.: Research Media, 1977.

Rinn, R. C., Vernon, J. C., & Wise, M. J. Training parents of behaviorally disordered children in groups: A three years' program evaluation. *Behavior Therapy,* 1975, *6,* 378–387.

Rinn, R. C., Mahla, G. Markle, A., Barnhart, D., Owen, D. O., & Supnick, J. *Analogue assessment of social skills deficits in children.* Manuscript submitted for publication, 1979. (a)

Rinn, R. C., Priest, M., Barnhart, D. L., & Markle, A. *Validation of an analogue measure of social skills in children.* Manuscript submitted for publication, 1979. (b)

Roff, M. Childhood social interactions and young adult bad conduct. *Journal of Abnormal and Social Psychology,* 1961, *63,* 333–337.

Roff, M., Sells, S. B., & Golden, M. M. *Social adjustment and personality development in children.* Minneapolis: University of Minnesota Press, 1972.

Stokes, T. F., & Baer, D. M. An implicit technology of generalization. *Journal of Applied Behavior Analysis,* 1977, *10,* 349–367.

Strain, P. S., Shores, R. E., & Kerr, M. A. An experimental analysis of "spillover" effects on the social interactions of behaviorally handicapped preschool children. *Journal of Applied Behavior Analysis*, 1976, *9*, 31–40.

Ullmann, C. A. Teachers, peers, and tests as predictors of adjustment. *Journal of Educational Psychology*, 1957, *48*, 257–267.

Waldrop, M. G., & Halverson, C. F. Intensive and extensive peer behavior: Longitudinal and cross-sectional analysis. *Child Development*, 1975, *45*, 19–26.

Walker, H. M. *Walker problem behavior identification checklist: Manual*. Los Angeles: Western Psychological Services, 1970.

Zahavi, S., & Asher, S. R. The effect of verbal instructions on preschool children's aggressive behavior. *Journal of School Psychology*, 1978, *16*, 146–153.

CHAPTER 5

Modification of Heterosocial Skills Deficits

John P. Galassi and Merna Dee Galassi

INTRODUCTION

This chapter is concerned with heterosocial skills, which are those skills necessary for social interchange between members of the opposite sex. Although these skills are relevant across the life span, the bulk of existing research has been concerned with the skills that are important in the early stages of dating relationships between college students. Unfortunately, these skills represent somewhat a will-o'-the-wisp. When globally defined, effective skills can often be distinguished from ineffective ones. However, relatively little success has been enjoyed in identifying the specific behaviors that comprise them.

Heterosocial skills include both general social skills and those skills specific to interactions with the opposite sex. Libet and Lewinsohn (1973) define social skill as "the complex ability both to emit behaviors which are positively or negatively reinforced and not to emit behaviors which are punished or extinguished by others" (p. 304). Weiss (1968) refers to social skill as the ability to develop rapport, express interest, and understanding in a social interaction.

Heterosocial skills *per se* can be defined as skills relevant to initiating, maintaining and terminating a social and/or sexual relationship with a member of the opposite sex (Barlow, Abel, Blanchard, Bristow, & Young, 1977). The behaviors required in each stage of a relationship may

John P. Galassi • Counseling Psychology, School of Education, University of North Carolina, Chapel Hill, North Carolina 27514. **Merna Dee Galassi** • Office of Developmental Counseling, Meredith College, Raleigh, North Carolina 27611.

be quite different. Given that the skills needed for competent performance change over the course of dating, a global definition of heterosocial skills is not especially meaningful.

The assumption that heterosocial skills are distinct from general social skills has been rather universally accepted. With the exception of Twentyman, Boland, and McFall (1978), little attention has been devoted to determining whether performance skills actually differ in same- and opposite-sex dyads. However, results of factor-analytic investigations of assertion (J. P. Galassi & Galassi, 1979; M. D. Galassi & Galassi, in press), social anxiety (Richardson & Tasto, 1976), and social difficulty (Bryant & Trower, 1974) indicate that items concerned with heterosocial interactions do cluster separately from items concerned with more general social interactions.

Operationally, heterosocial problems have been defined from a variety of perspectives including skill deficits, frequency of dates, and anxiety or satisfaction associated with opposite-sex interactions. The reader will encounter a myriad of terms including high- and low-frequency daters, minimal daters, skilled daters, heterosocially competent and incompetent persons, shy and confident performers, and socially anxious individuals. Such terminological confusion has resulted both from the different ways that the phenomenon has been conceptualized and operationalized and from a tendency to focus on the individual exhibiting the behaviors rather than on the behaviors themselves.

The chapter integrates the existing literature on identifying, assessing, and treating heterosocial problems. Preceding these topics, we will discuss the importance and incidence of heterosocial problems and the major theoretical formulations advanced to explain their development and maintenance.

IMPORTANCE AND INCIDENCE OF HETEROSOCIAL PROBLEMS

Heterosocial problems have been considered important because of their role as antecedents of social problems in later life, their role in the developmental sequence of adolescence and young adults, and their potential in analogue research. For hospitalized patients, premorbid social competence as measured by gross indicators such as age, marital status, and employment history has been related to the probability of making a satisfactory posthospital adjustment (Zigler & Phillips, 1961). Further, Zigler and Phillips (1959) reported differences in social competence between hospitalized and nonhospitalized patients. Deficits in heterosocial skills have been linked to the development and treatment of a variety of sexual deviations (Abel, 1976; Barlow *et al.*, 1977; Stevenson & Wolpe, 1960) including rape (Abel, Blanchard, & Becker, 1977). Klaus, Hersen, and Bellack (1977) noted that dating skill deficits may be precursors of later heterosexual deficits in psychiatric patients, a speculation supported by Bryant, Trower, Yardley, Urbieta, and Letemendia (1976).

For college students, achieving heterosocial competence is an important developmental task (Coons, 1970; Havighurst, 1972). Heterosocial incompetence constitutes a nontrivial problem with implications for present and future life adjustments. Problems relating to dating often are accompanied by anxiety, depression, and academic failure and can lead students to seek help at counseling centers (Arkowitz, Hinton, Perl, & Himadi, 1978). Martinson and Zerface (1970) reported that students were more interested in learning to get along with the opposite sex than with help in choosing a vocation or learning about their abilities, interests, intelligence, and personalities.

As an indication of the severity of the problem, Borkovec, Stone, O'Brien, and Kaloupek (1974) found that 15.5% of the males and 11.5% of the females in introductory psychology reported at least some fear of being with the opposite sex. Substantially higher percentages were reported for fear of meeting someone for the first time. Martinez and Edelstein (1977b) reported that 13% of 876 college males felt somewhat or very anxious and inhibited with women. In a survey of 3,800 students (Arkowitz *et al.*, 1978), 31% of the sample (37% of the males and 25% of the females) reported they were somewhat or very anxious about dating. Arkowitz *et al.* (1979) speculated that the greater incidence in males is due to their role as initiators of interactions. Shmurak (1973) noted that 54% of difficult social situations for males and 42% for females were concerned with dating.

Finally, heterosocial anxiety has been promoted for its potential as a clinically relevant target behavior for analogue outcome research. Borkovec *et al.* (1974) specified five criteria for selecting an analogue behavior: (1) occurs frequently in the psychiatric population; (2) interferes with daily functioning; (3) is uninfluenced by simple demand; (4) heightens physiological arousal; and (5) does not habituate rapidly. Although both Borkovec *et al.* (1974) and Heimberg (1977) presented data supporting social anxiety as a clinically relevant analogue behavior, others (Martinez & Edelstein, 1977b; O'Brien, 1977) have reported less positive findings.

Regardless of its suitability as an analogue behavior, heterosocial incompetence represents a significant problem for college students. Given the size of this population, the problem is worthy of consideration as a target behavior in its own right.

THEORETICAL FORMULATIONS AND CONTROLLING VARIABLES

Four main classes of controlling variables—conditioned anxiety, skill deficits, cognitive distortions, and physical attractiveness—have been postulated for heterosocial difficulties (Arkowitz, 1977; Curran, 1977; Perri, 1977). Except for physical attractiveness, which has been studied infrequently by clinically oriented researchers, the influence of these variables appears in both the treatment literature and the research attempting to define the nature of heterosocial competence.

Conditioned Anxiety. Heterosocial difficulties have often been attributed to conditioned anxiety. An individual has one or more unpleasant heterosocial experiences and anxiety becomes classically conditioned to various cues in those situations. Anxiety is then aroused when anticipating or responding to related situations and results in impaired performance, avoidance, or escape. The conditioned-anxiety explanation subscribes to a response disinhibition model of therapy (Bandura, 1969). This model assumes that clinical problems result from a behavioral excess (conditioned anxiety) that blocks or inhibits the expression of more appropriate behaviors (competent heterosocial performance). The problem can be eliminated only if anxiety is deconditioned thereby permitting an opportunity for expressing more appropriate behaviors. Studies using systematic desensitization as a treatment (e.g., Hokanson, 1972) exemplify the conditioned anxiety or response disinhibition model.

Skill Deficits. Heterosocial difficulties have also been attributed to skill deficits. Due to "lack of experience or opportunity to learn, faulty learning as a result of unrepresentative or faulty experiences, obsolescence of a previously adaptive response, learning disabilities resulting from biological dysfunctions, or traumatic events such as injuries or diseases that nullify prior learning or obstruct new learning" (McFall, 1976, p. 232), an individual does not possess the skills for competent performance. This lack of skill results in impaired performance and, in some instances, reactive anxiety, avoidance, or escape. A response-acquisition model of therapy (Bandura, 1969) is invoked because the problem is viewed as the result of behavioral deficits that require remediation. Treatment research by MacDonald, Lindquist, Kramer, McGrath, and Rhyne (1975) and Twentyman and McFall (1975) exemplify this model.

Cognitive Distortions. A third approach is concerned with cognitive evaluations and distortions. Many individuals experiencing difficulties in heterosocial situations are capable of emitting competent responses.

However, because of negative self-evaluations (Bandura, 1969), excessively high performance standards, unrealistic expectations, irrational beliefs, faulty perceptions, or misinterpretation of feedback, they either fail to perform or underestimate their performance. What is evidenced is a performance deficit in cognitive as opposed to behavioral skills. Such a deficit may lead to anxiety, further underestimating performance, and avoidance or escape behavior. This conceptualization may be labeled as the cognitive response acquisition model. The research on characteristics of socially anxious individuals by Arkowitz and his colleagues (Clark & Arkowitz, 1975; Glasgow & Arkowitz, 1975; Miller & Arkowitz, 1977; O'Banion & Arkowitz, 1977) and the treatment study by Glass, Gottman, and Shmurak (1976) exemplify this approach.

Physical Attractiveness. Finally, heterosocial problems may be a function of or moderated by physical attractiveness. People experiencing heterosocial difficulties may be less physically attractive than those who are not, and attractive individuals may benefit from a "halo effect" as a result of their physical characteristics. That is, physically attractive persons may be considered more desirable dating partners than less attractive persons. They may also be rated as more skillful and comfortable in heterosocial interactions regardless of their absolute skill or comfort levels. Further, they may have more opportunities to date, to refine skills, and to reduce anxiety as compared to their less attractive counterparts.

It is probably premature to single out any one of these four models as most promising. In fact, Steffen, Greenwald, and Langemeyer (1978) obtained three factors, labeled heterosocial anxiety, social self-evaluation, and dating activity, in a factor-analytic study with college women. Assessment procedures that identify the nature and contribution of these four variables for particular clients and treatments tailored to such assessments need to be developed. In addition, the potential interactions among these variables and the difficulty involved in trying to assign a causal role to one or more of them should be recognized. There is reason to believe that conditioned anxiety, skill deficits, faulty cognitions, and physical attractiveness can play a role both singly and in combination in the development of heterosocial difficulties.

IDENTIFYING TREATMENT TARGETS

Before establishing effective treatment programs, it is essential to have knowledge about normative dating patterns and specific dating skills. Traditionally, males and females have been accorded different roles in courtship. Argyle and Williams (1969) reported that, in cross-sex social contacts, adolescent females felt observed, whereas adolescent males felt themselves to be the observers. Arkowitz *et al.* (1978) found that, in their practice dating studies, males initiated contact in over 90% of the cases. Cary (1974) as cited in Curran, Little, and Gilbert (1978) found that female gazing behavior directed at males was the single best predictor of whether conversations occurred between opposite-sex strangers in a cocktail lounge.

With respect to dating habits, Klaus *et al.* (1977) reported that male and female students date at approximately the same frequency per month ($M = 5.49$ for males and $M = 5.65$ for females). Except for kissing goodnight, males and females differ concerning when they are likely to engage in various sexual behaviors. Males report engaging in such behaviors earlier than females. Females are likely to discuss "inner feelings" earlier than males.

The five most difficult stages of dating for both sexes are finding possible dates, initiating contact with prospective dates, initiating sexual activity, avoiding or curtailing sex, and ending a date (Klaus *et al.*, 1977). Females reported more difficulty finding dates, feeling at ease on dates, making conversation, ending the date, and obtaining a second date with the same partner, whereas males have more difficulty in initiating telephone contact with prospective dates. Thomander (1975) reported a similar list of difficulties. Thus, despite women's liberation, effective heterosocial skills have continued to be sex specific with males functioning as initiators and females facilitating interaction through displays of approach cues.

In addition to normative dating patterns, the specific content of heterosocial skills training programs must be identified (McFall, 1976, 1977). Do dating skills differ for men and women, and are different skills important at different points in the relationship? For the most part, identification of these skills has been guided by the known-groups approach (McFall, 1976). Based on dating-related criteria, individuals are differentiated into successful and nonsuccessful daters. They are then placed in performance situations with the opposite sex in order to identify group differences. The studies concerned with identifying differences can be divided into three groups—specific skill deficits and anxiety indices, cognitions, and physical attractiveness.

The studies we will discuss first provide information about specific skill deficits and anxiety indices in heterosocial interactions. These studies have consistently identified differences between high- and low-skill or anxiety subjects on self-report measures, and often on self and/or judges' global ratings of anxiety and skill. However, less success has been achieved in identifying and replicating specific skill deficits and anxiety signs. It is this last topic on which our discussion will concentrate, with attention directed first to males and then to females.

Identifying Specific Skill Deficits and Anxiety Indices

Males. In two of the earliest studies, Borkovec *et al.* (1974) and Arkowitz, Lichtenstein, McGovern, and Hines (1975b) compared high and low socially anxious and high- and low-frequency male daters, respectively, in behavioral performance situations. Borkovec *et al.* (1974) found specific differences between groups only on heart rate. Despite the use of a number of behavioral measures, Arkowitz *et al.* (1975b) found few specific skill differences. High-frequency daters talked significantly more and responded quicker in a taped situation test and had significantly fewer silences in an *in vivo* conversation. Both studies, however, reported differences on global ratings of social skills and self-report measures of anxiety. Arkowitz *et al.* (1975b) concluded that problems in social competence are due more to an inhibition (anxiety) than a skill deficit. Rehm and Marston's (1968) comparison of heterosocially anxious subjects with "normals" also supports this conclusion.

Glasgow and Arkowitz (1975) speculated that the failure to find behavioral differences may be due to a lack of sensitivity of the behavioral measures (simple frequency counts) and the methodology employed. They attempted to use measures (e.g., gazing given partner talk, talk balance ratio) sensitive to the reciprocal quality of interchange between male and female subjects in a naturalistic interaction rather than an interaction with a confederate. For males, neither global skill and anxiety ratings nor specific behaviors differentiated the groups, although self-ratings of skill and anxiety, the Social Avoidance and Distress Scale (SAD), physical attractiveness, and partner's preference for dates did. Glasgow and Arkowitz (1975) concluded that low-dating males are primarily characterized by negative self-evaluations rather than skill deficits.

Except for significant differences in time spent talking in a live interaction and global rating of anxiety, Martinez and Edelstein (1977a) reported findings similar to Glasgow and Arkowitz's. Unlike Borkovec *et al.* (1974), Martinez and Edelstein did not find significant heart-rate differences for males.

Expressing concern about the inadequacy of simple frequency counts, Fischetti, Curran, and Wessberg (1977) reported that high socially anxious/low-skilled and low socially anxious/high-skilled males differed only in the timing and placement (not the frequency) of a switch-pressing response that communicated understanding to a female speaker. Fischetti *et al.* (1977) emphasize that we may continue to have difficulty isolating skill differences if we do not focus on the proper skills (response placement vs. frequency of responses).

In another search for skill differences (Twentyman & McFall, 1975), diary reports revealed that shy males interacted with fewer women in fewer situations for less time than confident males. Although the groups were differentiated on pulse rate and some global anxiety ratings in laboratory situations, there was only one difference on specific behavioral indices of anxiety. In addition, confident subjects were significantly less likely to avoid initiating interactions than shy subjects. Although confident subjects were rated as more skillful and as talking longer, a subsequent reanalysis of the data (Twentyman *et al.*, 1978) which omitted avoidance responses eliminated the differences in interactional skills.

In a series of studies, Twentyman *et al.* (1978) reported that male nondaters were primarily characterized by avoiding interactions; low self-ratings of skill, confidence, and approach behaviors; and less knowledge about social cues and norms. Because the authors found few skill differences, they concluded that daters and nondaters are not very different once they get beyond the crucial initiation stage of interaction. The unreplicated skill differences included fewer silences in a face-to-face interaction, success in obtaining a date in a phone call, identifying oneself by name during a phone call, and relative talk time in conversations with males as opposed to females. The authors speculated that low daters avoid interactions because they are uncertain about their heterosocial competence as a result of being less knowledgeable about social cues and less able to judge their effect on females. For remediation, they stressed exposure to heterosocial interactions.

Perri, Richards, and Goodrich (1978) suggested that insufficient attention to assessment may account for the failure to find skill differences. Interestingly, three studies reporting solely on the development of performance tests for males have all identified skill differences. Perri and his colleagues (Perri & Richards, in press; Perri *et al.*, 1978) found that "minimal" and "regular" daters were differentiated on effectiveness of responses (quality of verbal content) in the Heterosocial Adequacy Test. Rhyne, MacDonald, McGrath, Lindquist, and Kramer (1974) found that volunteers who had been accepted for treatment of dating skills differed from those who were not on skills rated by judges using specific behavioral criteria in the Role Play Dating Interaction Test. Anxiety as assessed on a time behavioral checklist did not differentiate the groups. Finally, Barlow *et al.* (1977) reported differences in three of four behavioral categories—form of conversation, voice, and affect, but not motor behavior—between socially attractive black and white high school and college males and patients with sexually variant behaviors.

Barlow *et al.* noted that their success in identifying behavioral skills may be related to the severity of deficits in the clinical population and/or to the objective definition of the behavioral categories.

In an analogue study, Steffen and Redden (1977) found that high and low socially competent men differed only in their reactions to a female who provided negative feedback but not in their reactions to positive feedback. Low-competent men took significantly longer to react to negative feedback. High-competent men employed a wider range of responses to negative feedback and were judged more attractive and socially skilled than low-competent men. Two conclusions emerged from this study: high-competent men are better able to adjust their performance in the face of negative feedback; and studies that withhold confederate feedback during performance tests are likely to obscure skill differences. The effects of feedback and other situational influences have also been reported by Zeichner, Wright, and Herman (1977) and Zeichner, Pihl, and Wright (1977).

In a related study, Curran *et al.* (1978) found that high-anxious males were significantly less likely than low-anxious males to respond verbally to positive female approach cues (eye contact and smiles) even though both groups sat closer to the female in the positive- than in the negative-approach condition. Although the results were interpreted as support for a skills-deficit hypothesis, they can also support an anxiety hypothesis if one assumes that the anxiety in the experimental situation was sufficient to inhibit only the more complex (verbal initiation) but not the simpler (sitting proximity) approach response. Similarly, a cognitive explanation can be invoked if it is assumed that the high-anxious males perceive themselves as worthy only of close proximity to a female but not of an actual conversation with her.

For males, the search for specific behavior and anxiety indices has not been encouraging. In any one investigation only a few of the indices reveal any differences. The specific indices that have differentiated heterosocially competent and incompetent males include: heart rate and specific anxiety signs; initiating interactions and responding verbally to approach cues; response timing and placement; voice, affect, and form of conversation; verbal content measures; obtaining a date in a phone call; talk time; number of silences; and response latency. Unfortunately, these findings are not replicated across studies, and there appears to be no consistent pattern.

Females. In comparison to the number of studies of male skill deficits and anxiety signs, few such investigations have been conducted with females. The study by Glasgow and Arkowitz (1975) found only one behavioral difference, with high-frequency daters (HFD) speaking significantly longer than low-frequency daters (LFD). Differences were also reported on the SAD and on self- and partner ratings of social skill and attractiveness. Glasgow and Arkowitz (1975) interpreted the results to support a social-skill explanation for heterosocial difficulties in women as opposed to a cognitive social evaluation explanation for men.

Greenwald (1978) reported differences on a number of self-report indices of anxiety, assertion, and self-esteem between high- and low-frequency female daters. Skill differences were found in favor of HFD females on talk time in a waiting room interaction and eye contact in a role-play (Greenwald, 1977). In general, the high female daters were rated by both male confederates and female judges as more physically attractive, socially skilled, and interpersonally attractive. Global ratings of anxiety did not differentiate the groups. Stepdown *F* procedures indicated that physical attractiveness accounted for much of the variance in the data, a fact which does not support a skill-deficit model for females.

In validating the Survey of Heterosocial Interactions for females, Williams and Ciminero (1977, 1978) reported significant differences between high and low scorers on a variety of self-report measures—dates per year, participation in heterosocial behaviors, physical attractiveness, social skill, and anxiety. On a behavioral test, the high group was rated significantly higher on overall skills and on two of three skill indicators (interest and initiation). There were no differences on observer ratings of overall anxiety, on three specific anxiety measures, on physical attractiveness, on two time measures (duration and latency), on heart rate, or on number of dates. Despite the fact that actual skill differences were obtained, the authors seem to interpret the data to support a cognitive social evaluation model for women.

The three studies on females attributed their findings to different controlling variables—social skill (Glasgow & Arkowitz, 1975), physical attractiveness (Greenwald, 1977), and cognitive social evaluation (Williams & Ciminero, 1977). Few specific behavioral indices of skill (eye contact, response latency, talk time, interest, and initiation) and no specific behavioral indices of anxiety differentiated females across the studies. Interestingly, all three studies showed differences on judges' ratings of overall skill but not on anxiety (cf. O'Banion & Arkowitz, 1977) and differences on self-report or self-ratings of skills and anxiety. Two of the three revealed differences in physical attractiveness.

Methodological Considerations. The finding of few consistent specific skill and anxiety differences for both males and females may indicate that such differences are not salient. However, before such a conclusion is accepted, limitations of existing methodology and alternative hypotheses need to be considered.

First, the known-groups (e.g., comparing highs and lows) approach has several limitations (McFall, 1976). It fails to allow for the possibility that the groups differ in ways unrelated to the bases on which they were divided originally and that an unrelated factor may account for observed differences.

Second, the approach tends to treat heterosocial competence as a trait assuming that individuals perform consistently across all situations. It is unlikely that all incompetent daters have identical characteristics or behavioral deficits (Bellack & Hersen, 1977; Hersen & Bellack, 1976; cf. Kiesler, 1966). For instance, Curran, Wallander, and Fischetti (1977) have shown that males high in heterosocial anxiety could be meaningfully divided into high-anxious/high-skill and high-anxious/low-skill groups, and Glasgow and Arkowitz (1975) demonstrated that high frequency daters who were going steady were rather heterogeneous on skill and anxiety measures. Moreover, cases of social failure exhibiting various combinations of anxiety, skill, and cognitive deficits were differentially responsive to various treatment interventions (Marzillier & Winter, 1978; Trower, Yardley, Bryant, & Shaw, 1978). Two additional limitations include the use of different selection procedures across studies resulting in the identification of noncomparable subjects (Farrell, Mariotto, Conger, Curran, & Wallander, 1978; Martinez & Edelstein, 1977a) and the failure to construct sensitive performance tests.

An alternative to the known-groups approach is provided by combining the behavior-analytic/experimental method (Goldfried & D'Zurilla, 1969; McFall, 1976) and the structural approach (Duncan, 1969; Duncan & Fiske, 1977). In the behavior-analytic approach, strategies for handling a situation-specific task (e.g., Twentyman *et al.*, 1978) are empirically derived and compared for effectiveness. With respect to dating, crucial stages such as initiation would first be identified and then possible relevant responses could be enumerated and their effectiveness compared. The structural approach assumes that dyadic interaction is an organized, rule-governed phenomenon and attempts to capture the moment-to-moment flow of that interaction using variables that reflect the ongoing reciprocity between participants. In an excellent example using the structural approach, Duncan and Fiske (1977) identified behaviorally defined verbal and nonverbal signals and rules by which speaking turns are smoothly exchanged. The methodology currently employed by most clinically oriented researchers seems inadequate to reflect such subtleties in conversational flow. Consideration of the methodology used by Argyle (1969), Argyle and Kendon (1967), Duncan and Fiske (1977), Endler and Magnusson (1976), and Goffman (1959) in personality and social psychological research is in order before concluding that skill differences do not exist.

Regardless of methodology, it is possible that few specific skill and anxiety differences actually exist. If that is the case, then differences will not be reflected by specific behavioral indices, and the influence of other variables such as cognitions and physical attractiveness needs to be considered.

Identifying Cognitive Targets

Since Arkowitz *et al.*'s (1975b) initial speculation that heterosocial deficits are a function of social evaluative anxiety, evidence about the role of cognitive variables in heterosocial difficulties has been increasing.

Clark and Arkowitz (1975) investigated the contribution of self-evaluations. Although high (HSA) and low (LSA) socially anxious males did not differ in their performance with a confederate, HSA subjects significantly underestimated (overly critical) their skills relative to judges' ratings. LSA subjects overestimated their performance. Differences were also obtained on the Fear of Negative Evaluation Scale and on a modified Coopersmith Self-Esteem Inventory. The groups did not differ in their evaluation of others, thus suggesting that HSA subjects either apply more stringent standards to their own performance or selectively recall more negative aspects of it.

Further dichotomizing high-anxious males based on skill level yielded different results (Curran *et al.*, 1977). High-anxious/high-skill males underestimated their performance, whereas high-anxious/low-skill males made accurate assessments. Low-anxious/high-skill subjects also underestimated their performance, suggesting that underestimation may be a function of a variable other than anxiety. The differences between the two studies emphasize the importance of selection procedures and clear definitions of target behavior.

O'Banion and Arkowitz (1977) used females to investigate the accuracy of the selective memory for negative information hypothesis that was formulated for males by Clark and Arkowitz (1975). Females role played with a male confederate who demonstrated either positive or negative behavior toward them. All subjects received identical feedback following the role play. Results indicated that high socially anxious women remembered negative information better than did low socially anxious women regardless of success or failure during the role play. Further, self- and observer ratings of social skills and anxiety revealed skill deficits and greater anxiety for the HSA women. The findings support both cognitive factors and skill deficits in the mediation of social anxiety in women.

Similarly, Smith and Sarason (1975) found that negative feedback was rated more negatively and as able to evoke more negative emotional response in high and moderate socially anxious than in low socially anxious males and females. High-anxious subjects also had a greater expectation of negative evaluation. There were no differences among the groups on willingness to interact further with the evaluator. Females were significantly more concerned about negative evaluation than males.

The well-documented (e.g., Byrne, 1969) positive relationship between interpersonal attraction toward a stranger and attitude similarity may be moderated by social anxiety. In an analogue study, Smith and Jeffrey (1970) found that high-anxious as compared to low-anxious females performed significantly better when correct responses were followed by agreeing attitudes (social approval) and incorrect responses were followed by disagreeing attitudes (social disapproval). High-anxious females were relatively more attracted to strangers who agreed with and less attracted to strangers who disagreed with their attitudes than low-anxious females (Smith, 1972).

In contrast to O'Banion and Arkowitz (1977), Steffen and Reckman (1978) found that high socially anxious males perceive social events similarly to low socially anxious males, but interpret them differently. The contrasting findings of the two studies may be attributed to the use of different-sex subjects given that females (Smith & Sarason, 1975) react more strongly to negative evaluation than males.

As mentioned previously, Steffen and Redden (1977) found that high and low socially competent males were differentially able to alter their verbal behavior in response to negative feedback from females. Based on the findings in this section, conceivably this performance difference is a function of differing cognitive evaluations.

Perri and Richards (1977) suggested that another important difference may be in ability to exercise self-control. Students successful in increasing their dating reported significantly greater expectations of success and made significantly greater use of self-reward, stimulus control, and problem-solving procedures than unsuccessful students.

Finally, Miller and Arkowitz (1977) failed to support the prediction that high socially anxious males would attribute failure (interaction with a cold and rejecting female) internally and success (interaction with a warm and accepting female) externally, and that low-anxious subjects would show the opposite pattern.

Studies in both this and the previous section provide ample evidence for cognitive differences between individuals who experience heterosocial difficulties and those who do not. High-anxious persons attend to and/or interpret negative experiences and feedback less favorably than low-anxious persons. They are likely to expect more negative evaluations, to reward themselves less, and to make less frequent use of problem-solving and other self-control procedures. Eisler, Frederikson, and Peterson (1978), noted that "to the extent that individuals expect unfavorable consequences from behaviors in their repertoires, they will not exhibit them" (p. 426). However, it is unclear from existing studies whether cognitive distortions cause and/or maintain heterosocial difficulties or whether they simply are by-products of them. Depending on the particular role of cognitive variables (causation, maintenance, by-products), intervention strategies may need to vary accordingly.

The Role of Physical Attractiveness

Physical attractiveness has received only scant attention in the theoretical formulations of heterosocial competence and in the development of treatment programs. The failure to integrate this variable is surprising given the social psychology literature on the importance of physical attractiveness in social interaction (Berscheid & Walster, 1973). Physical attractiveness appears particularly influential in the initial stages of social interchange because it is often the only readily available information.

A number of analogue studies as well as studies involving actual interactions have reported that physically attractive individuals date more frequently (Berscheid, Dion, Walster, & Walster, 1971; Glasgow & Arkowitz, 1975; Greenwald, 1977; Kaats & Davis, 1970), are rated as more likable (Byrne, London, & Reeves, 1968; Curran, 1973; Curran & Lippold, 1975; Perrin, 1921; Walster, Aaronson, Abraham, & Rottman, 1966), more popular and more desirable as a friend or future date (Brislin & Lewis, 1968; Glasgow & Arkowitz, 1975; Tesser & Brodie, 1971), and are stereotyped as having more pleasing personalities, marrying earlier, and having happier marriages (Dion, Berscheid, & Walster, 1972) than less attractive individuals. In the early stages of a relationship, a significant correlation between physical attractiveness and desire to date the partner again (r = .89; Brislin & Lewis, 1968) and between physical attractiveness and liking date (r = .69; Tresser & Brodie, 1971) has been found regardless of one's own physical attractiveness (Huston, 1973; Walster *et al.*, 1966) and level of self-esteem (Walster, 1970). Physical attractiveness predicted liking or desire to date a partner again better than sociability (Brislin & Lewis, 1968), intelligence or personality of the partner (Walster *et al.*, 1966), couple compatability on interests and attitudes (Brislin & Lewis, 1968; Tesser & Brodie, 1971), and the best prediction based on seven behavioral variables (Glasgow & Arkowitz, 1975).

The effects of physical attractiveness may decrease when partner attitudes become more salient (Byrne *et al.*, 1968; Byrne, Ervin, & Lamberth, 1970) and as more information becomes available later in the relationship (Stroebe, Insko, Thompson, & Layton, 1971). The Stroebe *et al.* study reported that physical attractiveness had a greater effect on dating than on either liking or marrying, whereas attitude similarity had a greater effect on liking than dating.

Finally, the relationship between physical attractiveness and dating seems stronger for females than males. Physical attractiveness and number of dates correlated .61 for females and only .25 for males (Berscheid *et al.*, 1971). Physical attractiveness of the partner appeared to be more relevant to male liking of his date ($r = .68$) than to female liking of her date ($r = .51$) (Curran, 1973). Although physically attractive females evaluate themselves as happier, less neurotic, and higher in self-esteem than less attractive females, a similar relationship was not obtained for males (Mathes & Kahn, 1975). Thus, for females, level of attractiveness may also influence self-evaluation. In addition, greater physical attractiveness of females not only results in more social responsiveness from males (Barocas & Karoly, 1972), but also results in a halo effect for males who date them as reflected by judges' favorability ratings (Sigall & Landy, 1973).

Thus, it is imperative to consider physical attractiveness in assessing and treating minimal daters. In assessment, halo effects on judges' rating of social skill and anxiety due to physical attractiveness should be monitored. Given that the initiation stage of dating is crucial and physical attractiveness is especially influential at this time, greater emphasis should be devoted to this variable in the design of training programs. This recommendation appears to be particularly important with low-frequency female daters.

Finally, based on all of the studies reviewed concerning targets for treatment, there is little doubt that skills, anxiety, cognitions, and physical attractiveness each can play a role either singly or in combinations in heterosocial problems. The contribution of each for individual clients needs to be identified if effective interventions are to be employed. In this undertaking, assessment is of utmost importance.

ASSESSMENT

The importance of assessment cannot be overemphasized. Identifying treatment targets, monitoring treatment progress, assessing post-treatment gains, and evaluating outcomes of controlled research are all limited by the adequacy of assessment.

This review of assessment instruments is selective (for a more extended review, see Arkowitz, 1977; Hersen & Bellack, 1976) and discusses those instruments which are used most frequently and appear to be most valid and/or most promising. The instruments are organized according to type of assessment approach—self-report questionnaires, self-monitoring and self-ratings, evaluations by others, and direct-performance samples. We concentrate on those instruments developed especially for assessing heterosocial difficulties rather than on those adapted for use in this area.

Instrument development has rarely been the focus of heterosocial skills research. Instruments are often developed and validated as an aspect of studies that are concerned primarily with either treatment effects or characteristics of high- and low-competent daters. Such an approach confounds instrument validity with treatment effectiveness, as noted by Hersen and Bellack (1976) or with the presence of differentiating subject characteristics. Only suggestive validity data are provided by these studies (see the respective sections of this chapter for such data). In the early stages of instrument development, validation must be kept separate from identifying treatment effects or discovering differences between high and low daters. Instrumentation has also been limited by lack of knowledge concerning specific skills and other variables related to effective heterosocial performance. For a more detailed discussion of methodological issues, the reader is referred to the chapters by Bellack and Curran in this volume.

Self-Report Questionnaires

Table 1 lists a number of the most commonly used or promising questionnaires. We have chosen to highlight four of them—the Survey of Heterosexual Interactions, the Social Avoidance and Distress Scale, and the recently developed Social Interaction Self-Statement Test and Dating and Assertion Questionnaire.

Survey of Heterosexual Interactions (SHI). The SHI (Twentyman & McFall, 1975) consists of a face sheet about past dating behavior followed by 20 items to assess how a male perceives his ability to interact with women. The item content is concerned exclusively with the important first stages of dating—initiating a conversation, calling for a date, and asking for a dance. The instrument was developed as a situation-specific measure of heterosexual avoidance through extensive pilot research, with the items representing problem situations for nondating college males. Competent responses were identified through interviews with college coeds. The SHI is intended for use in both selecting subjects and assessing treatment effects. At present, we are unaware of any published data concerning reliability of the SHI.

In an empirical investigation of selection instruments for heterosocial deficits, Wallander, Conger, Mariotto, Curran, and Farrell (1978)

Table 1. Self-Report Questionnaires

Questionnaire/source	Relevant sex	Items and format	Focus and comments
Biographical Survey III (B-III), Lanyon (1967)	Males (modified version for males; Bander *et al.*, 1975)	20 items which have either a multiple-choice or numerical format	1. A measure of social competence. 2. Items were intuitively derived and ask for reports of verifiable behavior or biographical information which reflect social participation, interpersonal competence, achievement, and environmental mastery. 3. Items emphasize extraversion, activity, and decision-making at the expense of thinking or reflection. 4. B-III is noted here for its historical importance.
Dating and Assertion Questionnaire (DAQ), Levenson & Gottman (1978)[a]	Males and females	18 items Likert-type formats	1. Questionnaire is divided into two parts. Part I consists of 9 items (4 assertion, 5 dating) in which individual specifies likelihood of engaging in general behaviors. Part II consists of specific situations (5 assertion, 4 dating). Individual indicates degree of discomfort and expected incompetence. 2. Part II was empirically derived.

(*Continued*)

Table 1 (*Continued*)

Questionnaire/source	Relevant sex	Items and format	Focus and comments
			3. DAQ was developed to discriminate among specific types of social incompetence. 4. Dating items range from initiating a conversation to expressing intimate feelings with opposite sex.
Fear of Negative Evaluation Scale (FNE), Watson & Friend (1969)	Males and females	30 true-false items	1. A measure of apprehension about others' evaluations, distress over and expectation of negative evaluations, and avoidance of evaluative situations. 2. FNE is not situationally specific.
Modified S-R Inventory of Anxiousness (SRIA), based on SRIA developed by Endler, Hunt, & Rosenstein (1962)	Males (modified versions for females: Christensen & Arkowitz, 1974; Steffen, Greenwald, & Langmeyer, 1978)	5 situations each rated for 14 response modes on 5-point scale	1. A measure of anxiety for five specific heterosexual interactions. 2. Individuals rate the degree to which they generally experience each of the 14 anxiety related responses (e.g., heart beats faster) in each situation.
Situation Questionnaire (SQ), Rehm & Marston (1968)	Males (modified form for females; Curran & Gilbert, 1975)	30 7-point Likert items	1. A measure of discomfort or anxiety experienced in hierarchically arranged specific situations.

			2. Item content ranges from initiating conversations to parking with a female.
Social Activity Questionnaire (SAQ), Arkowitz *et al.* (1975b)	Males and females	7 multiple-choice items and 1 open-ended question	1. A measure of dating frequency, comfort, skill, and satisfaction with present dating behavior. 2. Major use is for preliminary screening.
Social Anxiety Inventory (SAI), Richardson & Tasto (1976)	Males and females	166 5-point Likert items (revised version by Richardson & Tasto, 1976, has 100 items; modified version with respect to content and response format, Wallander *et al.*, 1978)	1. Inventory was designed to assess the verbal-cognitive component of fear reactions to social situations. 2. Items were drawn from hierarchies used in treatment. 3. A factor analysis revealed two factors related to heterosexual contact and fears of interpersonal loss.
Social Avoidance & Distress Scale (SAD), Watson & Friend (1969)[a]	Males and females	28 true-false items (modified response format: Steffen *et al.*, 1978; modified version with Likert-type items referring specifically to females, Bander *et al.*, 1975)	1. A measure of anxiety, distress, discomfort, or fear experienced in social situations and the deliberate avoidance of those situations. 2. SAD is a trait- rather than a situation-specific measure.
Social Interaction Self-Statement Test (SISST), Glass & Merluzzi (1978)[a]	Males and females	30 5-point Likert items	1. The test has a positive self-statement scale and a negative self-statement scale each composed of 15 items.

(*Continued*)

Table 1 (*Continued*)

Questionnaire/source	Relevant sex	Items and format	Focus and comments
			2. Individuals indicate how frequently each statement characterizes their thoughts (from "hardly ever" to "very often") during a situation to which they are exposed. 3. The self-statements for the test were empirically derived.
Social Performance Survey Schedule (SPSS), Lowe & Cautela (1978)	Males and females	100 items rated on a 5-point scale	1. A general measure of social behavior containing 50 positive (e.g., has eye contact when speaking) and 50 negative behaviors (e.g., interrupts others). 2. Each behavior is rated for frequency of occurrence. 3. SPSS can be adapted for use in heterosocial situations and to rate self or others. 4. Sex differences were found for the SPSS. 5. Scoring may be adjusted to reflect standards of optimum social performance in a given population.

Social Situations Questionnaire, Bryant & Trower (1974)	Males and females	30 items rated on a 5-point scale for difficulty and 22 of the 30 items rerated for frequency	1. A measure of frequency of behavior and amount of anxiety experienced in social situations, some of which refer to opposite-sex encounters. 2. The 30 situations were selected on the basis of difficulty for psychiatric patients. 3. Items range from walking down the street to dating the opposite sex. 4. A factor analysis yielded 2 factors which related to opposite-sex interactions. 5. Norms are available for a British university population.
Survey of Heterosexual Interactions (SHI), Twentyman & McFall (1975)[a]	Males (modified version for females, SHI-F, Williams & Ciminero, 1978)	3 frequency count and 20 7-point Likert items	1. A measure of heterosexual avoidance. 2. Cover sheet concerns subject's past dating behavior. 3. Items tap perceived ability to initiate and carry out interactions with women in specific social situations. 4. Of 20 items, 4 are concerned with calling to ask out a date, 2 with asking for a dance, and 14 with initiating a conversation.

[a]Discussed in the text.

administered the SHI as well as four other questionnaires and two simulated interactions to 67 males. They concluded that correlations between questionnaires are generally low; questionnaires generally predict ratings (judge and self) poorly; dating experience (quantity and quality) generally correlated low with questionnaires; and questionnaires generally predict ratings better than dating experience. Of the self-report measures used, the SHI fared best. It correlated moderately with other questionnaires (−.56 with the Situation Questionnaire, −.54 with SAD), dating experience information, self-ratings in simulated interactions, and judges' ratings of those interactions. In each instance, the percentage of variance accounted for was usually low (7–8% with a range of 0–41%). In addition, the SHI has been used for subject selection and to identify characteristics of known groups (Martinez & Edelstein, 1977a; Twentyman & McFall, 1975; Twentyman *et al.*, 1978; Wessberg, Mariotto, Conger, Farrell, & Conger, 1978) to detect change following treatment (Hall & Edelstein, 1977; McGovern, Arkowitz, & Gilmore, 1975; Perl, Hinton, Arkowitz, & Himadi, 1977; Twentyman & McFall, 1975), and in comparisons with other self-report instruments (Wallander *et al.*, 1978).

Williams and Ciminero (1978) developed the Survey of Heterosexual Interactions for Females (SHI-F), keeping it as similar as possible to the SHI. However, the SHI was empirically derived for males, and its content and focus on initiation may not be as relevant for females. Much of the data on the SHI-F was discussed in the previous section on identifying specific skill and anxiety indicators. In addition, Williams and Ciminero reported an internal consistency value of $\alpha = .89$, test–retest reliability (interval unknown) of only .62, and normative data for 256 females. Principal-component analysis (Williams & Ciminero, 1977) yielded a single factor which accounted for only 33.8% of the variance.

Social Avoidance and Distress Scale (SAD). Although it was not developed specifically for measuring heterosocial anxiety, the 28-item SAD (Watson & Friend, 1969) has been the most frequently used self-report questionnaire, appearing in at least 20 studies. The SAD measures general rather than situational social anxiety. K-R 20 homogeneity indices of .94 have been obtained, whereas 1-month test–retest reliabilities were only .68 and .79. The scale is nonnormally distributed, and males score significantly higher on it than females. The SAD is minimally related to social desirability. Validity studies indicated that high scorers tend to avoid social interactions, prefer to work alone, and report that they talk less, worry more, and are less confident about social relationships (Watson & Friend, 1969).

The SAD has been used for subject selection and to identify characteristics of known groups (Arkowitz *et al.*, 1975b; Clark & Arkowitz,

1975; Glasgow & Arkowitz, 1975; Martinez & Edelstein, 1977a; Miller & Arkowitz, 1977; O'Banion & Arkowitz, 1977; Steffen *et al.*, 1978; Steffen & Reckman, 1978; Steffan & Redden, 1977), to detect changes following treatment (Bander, Steinke, Allen, & Mosher, 1975; Christensen & Arkowitz, 1974; Christensen, Arkowitz, & Anderson, 1975; Curran, Gilbert, & Little, 1976; Engeman, Campbell, & Steffen, 1977; Geary & Goldman, 1978; Hall & Edelstein, 1977; McGovern *et al.*, 1975; Perl *et al.*, 1977), and in comparisons with self-report instruments of heterosocial difficulties (Wallander *et al.*, 1978).

Social Interaction Self Statement Test (SISST). The SISST (Glass & Merluzzi, 1978), which was developed based on work by Schwartz and Gottman (1976), constitutes a heterosocially specific measure of cognitive evaluations that eventually may replace the Fear of Negative Evaluation Scale in heterosocial research. The items were generated empirically from males and females who listed their thoughts after imagining interactions in 10 heterosocial situations. The test is composed of both a negative and a positive self-statement subscale. Although reliability has not been reported for the instrument, a variety of validity data have been. High socially anxious women score significantly higher on the negative self-statements and significantly lower on positive self-statements than low socially anxious women. In addition, both subscales ($r = .63$; $r = -.53$) are significantly related to scores on the SHI-F, and negative self-statements are significantly related ($r = .52$) to scores on an irrational beliefs test. Positive statements are moderately but significantly related to self-ratings of skill ($r = .33$) and comfort ($r = .33$) during an actual interaction, and negative statements are moderately but significantly related to self- ($r = .35$), judge ($r = .33$), and confederate ($r = .41$) ratings of anxiety. Females scored significantly higher than males on positive self-statements and significantly lower on negative self-statements. Normative data could be collected for the SISST and used as one basis for screening in future heterosocial skills research. It also is important in cognitive treatment studies for change to be effected on the SISST or other cognitive measures before attempting to conclude that treatment was effective due to a cognitive mechanism. Finally, as with any standardized instrument, the SISST may not be sensitive to idiosyncratic cognitions and more flexible self-report cognitive assessment procedures for heterosocial deficits may be needed. As such, the reader is referred to Glass and Merluzzi (1978) for a discussion of written thought listing and videotape-aided thought recall.

Dating and Assertion Questionnaire (DAQ). The DAQ (Levenson & Gottman, 1978) was designed to: discriminate between competent and incompetent populations, with competence independently defined; discriminate among specific types of social incompetence (assertion vs.

dating); and predict differential improvement in treatments designed to ameliorate specific problems. Items were, in part, empirically derived and consisted of self-reports of frequency of performance or of discomfort and incompetence. Two- and six-week test–retest reliabilities of only .71 and .62, and internal consistency of .92, have been reported for the dating subscale. The DAQ discriminated between client and normal populations and between clients with dating and assertion problems. Clients who received treatment for dating problems showed significant changes only on the dating subscale. Assertion clients showed changes on both subscales, with more improvement on the assertion than on the dating subscale. Demonstrations of the scale's validity with behavioral and extralaboratory criteria have yet to be reported.

Self-Monitoring and Self-Ratings

Self-monitoring and self-ratings are commonly used in heterosocial skills research. In self-monitoring, the client keeps daily records of variables such as number of dates, number and range of interactions with the opposite sex, and quality of dating behavior. Twentyman and McFall (1975) reported that measures computed from such diary entries yield low (.09 to −.16) five-week test–retest reliabilities. The extent to which diaries are reactive has not been studied.

In addition to reliability and reactivity problems, other questions have been raised about measures such as dating frequency. MacDonald *et al.* (1975) stated that number of dates is not sufficiently sensitive to warrant its use as a major index of behavior change because it reflects the effects of factors other than improvement in skills. Curran and Gilbert (1975) and Klaus *et al.* (1977) pointed to the confounding of dating frequency with opportunities to date, the definition of a date, the ability to recall correctly, the fortuitous relationships between low-skilled daters who would have difficulty finding partners if they did not date each other. Because an increase in dating frequency is one index of generalization or transfer of treatment, albeit a crude one, and because a primary goal for many low-frequency daters is to increase dating, it seems imperative not to dismiss dating frequency so readily. Procedures are needed for increasing the reliability of this measure and for determining when and for whom it is a relevant variable. For low-frequency daters, it would seem extremely relevant; on the other hand, with anxious daters who have difficulty sustaining relationships it might be more appropriate to maintain or even decrease overall dating frequency, to increase the average number of dates with a few dating partners, or to ignore frequency and concentrate on increasing satisfaction instead.

Self-ratings have been used to assess a client's behavior in performance tests. These ratings are often computed on five- or seven-point anxiety and skill "thermometers" (e.g., Borkovec *et al.*, 1974; Curran *et al.*, 1977; Miller & Arkowitz, 1975). Unfortunately, low (−.10 to −.28) reliabilities have been reported (Twentyman & McFall, 1975) for such anxiety measures.

Information about general social skills (Marzillier, Lambert, & Kellet, 1976) and specific heterosocial behavior (Twentyman *et al.*, 1978) has also been obtained through standardized interviews. Although vulnerable to the usual problems of self-report measures (falsification, demand characteristics, etc.), the clinical interview is a flexible method for gathering information and monitoring treatment progress.

Evaluation by Others

Assessments of dating frequency, skill, and anxiety by peers or friends in the natural environment have been used infrequently in heterosocial skills research. Although we are unaware of reliability data for these ratings, they have been used in studies of high and low socially competent men (Arkowitz *et al.*, 1975b; Steffen & Redden, 1977) and of treatment effects (Christensen & Arkowitz, 1974; Perl *et al.*, 1977; Royce & Arkowitz, 1978). In contrast, peer ratings of physical appearance have yielded moderate to high reliabilities (Barocas & Karoly, 1972; Cross & Cross, 1971; Curran & Lippold, 1975; Kaats & Davis, 1970) but have not been used as dependent variables in treatment studies. As was true for self-monitoring, increasing reliability and validity of peer ratings is an important assessment priority in developing measures that can reflect transfer of treatment effects.

Confederates and judges have been trained to rate global and specific indices of skill and anxiety in performance tests. These measures are typically high in social validity (Kazdin, 1977) and yield a high percentage of interrater agreement (e.g., Rhyne *et al.*, 1974). In the absence of clearly defined behavioral indices of heterosocial difficulties, changes on global ratings of skill and anxiety in conjunction with self-report and self-ratings have constituted the primary basis for claims about the effectiveness of heterosocial skill training programs.

Performance Tests

A variety of methods have been used to obtain performance samples of behavior in analogue heterosocial situations. The procedures have included taped interactions requiring a single response (e.g., Arkowitz *et al.*, 1975b; Glass *et al.*, 1976; Melnick, 1973; Rehm & Marston, 1968; Twentyman & McFall, 1975), extended live interactions (e.g.,

Christensen *et al.*, 1975; Curran, 1975; Geary & Goldman, 1978; Twentyman & McFall, 1975), waiting room interactions with a live confederate (e.g., Greenwald, 1977; Melnick, 1973; Wessberg *et al.*, 1978), interactions with another subject (e.g., Glasgow & Arkowitz, 1975; Greenwald, 1977), telephone conversations (e.g., Arkowitz *et al.*, 1975b), and *in situ* interactions with a live confederate (Twentyman *et al.*, 1978). Global or specific skill and anxiety ratings, physiological measures, and physical attractiveness ratings are computed for performance in these situations. Our discussion focuses on those performance tests which have been derived with attention to psychometric considerations. Table 2 summarizes a number of these measures.

Tests with Audiotaped Confederates. The Taped Situation Test (TST) (Rehm & Marston, 1968), which consists of two alternate forms of 10 situations for males, is presented on audiotape. The situations are described by a male narrator and followed by a line of dialogue from a female confederate. Subjects are allowed 10 sec to make a single response and are instructed to respond as they would in real life. The two forms were equated for mean ranked discomfort level. Seven measures—which include subject's ratings of anxiety; judges' global ratings of anxiety, adequacy, and likability; scoring of specific anxiety signs; latency of response; and number of words per response—are derived for performance. Although test–retest reliabilities have not been reported, parallel-form reliabilities vary from .05 for specific anxiety signs to a range of .67 to .87 for the other measures. Intrascorer reliability coefficients vary from .46 to .94; whereas interscorer reliabilities range from .47 to .69. The TST has been used to identify characteristics of high and low daters (Arkowitz *et al.*, 1975b). In addition, the TST or its variations have been used to assess treatment effects (Christensen *et al.*, 1975; Hall & Edelstein, 1977; Martinez & Edelstein, 1977a; Melnick, 1973; Rehm & Marston, 1968).

The Heterosocial Adequacy Test (HAT) for males (Perri & Richards, in press; Perri *et al.*, 1978) is a single-statement audiotaped test. The 22 situations and scoring criteria were derived empirically according to the Goldfried and D'Zurilla (1969) model. The situations are described by a narrator without confederate dialogue. Although test–retest reliability has not been reported, inter- and intrarater reliabilities of .85 and .93 have been reported for the verbal content variable yielded by the test. An internal consistency coefficient of .90 was obtained. The test discriminated between known groups. Limited evidence of convergent and discriminant validity has been reported (Perri & Richards, in press).

Tests with Live Confederates. The Heterosocial Skills Behavior Checklist (Barlow *et al.*, 1977) and Role-Played Dating Interactions (RPDI) test (Rhyne *et al.*, 1974) represent contrasting examples of per-

Table 2. Performance Tests

Test/source	Relevant sex	Response format	Focus and comments
Taped Situation Test (TST), Rehm & Marston (1968)[a]	Males	Two forms each consisting of 10 audiotaped situations. Single response is given to audiotaped stimulus statement.	1. Situations include initiating and/or maintaining a conversation with a female, asking out a date, and taking a date home. 2. A male voice presents the narration and a female voice is used for the line of dialogue.
Modified versions of the Taped Situation Test			
A. Situation Test, Melnick (1973)	Males		Situations are presented on videotape.
B. Taped Situation Tests			
1. Arkowitz *et al.* (1975b)	Males		10 social situations were modified from Rehm and Marston's (1968) TST.
2. Christensen *et al.* (1975)	Males and females		Situations were modified from the preceding study for use with males and females.
C. Hall & Edelstein (1977)	Males and females		Modified situations are for use in live interactions. Male and female subjects interact with each other.
Heterosocial Adequacy Test (HAT), Perri, Richards, & Goodrich (1978), Perri & Richards (in press)[a]	Males	22 items are presented on audiotape. Single response is required after bell is sounded.	1. Items include asking out a date; initiating, maintaining, and ending conversations; telling a girl how you feel about her; and discussing the

(Continued)

Table 2 (*Continued*)

Test/source	Relevant sex	Response format	Focus and comments
			need for some form of contraception. 2. HAT was empirically derived according to Goldfried & D'Zurilla's (1969) behavior analytic model.
Heterosocial Skills Behavior Checklist for Males, Barlow *et al.* (1977)[a]	Males	Five-min live interaction with a female confederate.	1. Checklist is used to assess social behaviors important in initiating a heterosocial interaction. 2. Four categories of behavior are rated: form of conversation (initiating, following up, and maintaining a conversation, and indicating interest in a female); affect (facial expression, eye contact, and laughter); voice (loudness, pitch, and inflection); and masculine vs. feminine motor behaviors while seated.

Role-Played Dating Interactions (RPDI), Rhyne *et al.* (1974)[a]	Males	3 four-min live encounters with a highly trained confederate following a contingency programmed script.	1. The three situations include a telephone call to ask for a date, a waiting-room interaction, and a double-date situation. 2. The RPDI assesses both skills and anxiety level.
Behavioral Test of Tenderness Expression (BTTE), Warren & Gilner (1978)	Males and females	15 audiotaped situations each requiring a single response.	1. The BTTE assesses the degree and quality of positive feelings toward an intimate, self-disclosures, and supportive expressions. 2. Subjects respond to each situation as if their spouse, boyfriend/girlfriend, or lover were in the situation with them. 3. The situations include receiving compliments, sharing personal feelings and experiences, and expressing concern for one another.

[a]Discussed in the text.

formance tests for males with live confederates. The checklist involves a 5-min interaction with a female confederate. Males are instructed to behave as if they wanted a date. The female's behavior is standardized and restricted regardless of the male's comments. The empirically derived checklist evaluates behaviors in four categories—voice, form of conversation, affect, and motor behavior while seated. Percentages of agreement reliabilities have typically been high (.85), and the checklist has differentiated known groups.

The RPDI, on the other hand, involves three discrete 4-min interactions (telephone, waiting room, double date). The confederates follow programmed scripts that allow for response variation contingent on the subject's comments. The situations and specific behaviors (skills) for each were derived by consensual agreement of three clinicians. The test yields scores for skill and anxiety (modified Paul's 1966, Timed Behavioral Checklist) that have shown good interjudge agreement (97.2% and 87.2%, respectively). The mean intersituation correlation is .43, suggesting that, although the situations are related, each taps unique heterosocial skills. The RPDI skill score differentiated males who had been accepted and rejected for treatment of heterosocial skill deficits.

A variety of performance tests other than role plays have also been used. These have included phone call procedures to ask for a date (Arkowitz *et al.*, 1975b), to practice getting to know women (Glass *et al.*, 1976), or to demonstrate willingness to call for a date (Twentyman & McFall, 1975); conversations with other subjects (Glasgow & Arkowitz, 1975); and interactions within the natural environment with "planted" confederates (Twentyman *et al.*, 1978). Many of these procedures are promising but unvalidated. Another major assessment need is the empirical development and validation of both laboratory and naturalistic performance tests which are designed for females.

Role-playing has been the principal means of sampling behavior, and the results of early validational efforts (Borgatta, 1955; Efran & Korn, 1969; Kreitler & Kreitler, 1968; Stanton & Litwack, 1955) indicated that it permits relatively direct assessment of behavior with good control and standardization of stimulus conditions. Recently, however, concern has been expressed about both the internal and external validity (Spencer, 1978) of role-playing. With respect to heterosocial skills, there is a need to validate performance in analogue situations against performance in the natural environment (Lick & Unger, 1977; Nay, 1977; Twentyman *et al.*, 1978) or to use nonreactive measures in a setting unrelated to treatment (Shepherd, 1977). It cannot be assumed that laboratory performance changes generalize to the natural environment. At present, such validational data are not available for the performance tests cited above.

The fact that both assertive and heterosocial behavior can be significantly altered by procedural variations in performance tests (Bellack,

Hersen, & Turner, 1978; Curran, 1978; M. D. Galassi & Galassi, 1976; Martinez & Edelstein, 1977b; Nietzel & Bernstein, 1976; Wessberg *et al.*, 1978; Westefeld, Galassi, & Galassi, 1978) weakens the confidence that can be placed in role-play assessments as well as the outcome of treatment studies based on them. The chapters by Bellack and Curran discuss these issues more completely.

A few comments are in order about the physiological measures that have been used in performance tests. Borkovec *et al.* (1974), Martinez and Edelstein (1977a), Miller and Arkowitz (1977), and Twentyman and McFall (1975) studied heart-rate differences among known groups of males, and Williams and Ciminero (1978) investigated similar differences for females. Only Borkovec *et al.* (1974) and Twentyman and McFall (1975) found any differences. Of the treatment studies (Geary & Goldman, 1978; Heimberg, 1977; O'Brien & Borkovec, 1977; Twentyman & McFall, 1975) that used heart rate as a dependent variable, only the first failed to find at least some group differences following treatment. Thus, although treatment can effect change in heart (or pulse) rate, it is not clear that such measures differentiate socially competent and incompetent individuals.

As noted by Eisler (1976), measurement of heterosocial competence is far more complicated than measurement in earlier areas of behavior therapy research such as phobic behavior. Social judgments are invariably involved with the former but not with the latter. However, as with phobic behavior, no one measure or assessment approach is sufficient, and multiple measures are required. As has been shown repeatedly (Arkowitz *et al.*, 1975b; Borkovec *et al.*, 1974; Farrell *et al.*, 1978; Geary & Goldman, 1978; Martinez & Edelstein, 1977a; Miller & Arkowitz, 1977; Steffen & Redden, 1977; Twentyman *et al.*, 1978; Wallander *et al.*, 1978), the interrelationships between measures of dating frequency, skill, and anxiety tend to be only low to moderate. These relationships are attenuated when either a small number of behavioral observations is made or measures from different assessment modalities (e.g., self-report vs. direct-performance samples) are involved. As such, we recommend that a battery of assessment procedures selected from those just discussed be used in both the clinical treatment and research of heterosocial difficulties. The particular instruments used would reflect the problems of the clients/subjects and be tailored to the goals of treatment. Clearly, further development of assessment procedures is needed.

TREATMENT

Treatments have generally been based on theoretical formulations about etiology and controlling variables or assumptions about techniques presumed to effect change with these variables. In this section we will discuss three major treatment approaches. These treatments have been labeled anxiety reduction, skills acquisition, and cognitive modification, based on face validity and theoretical extrapolation rather than on empirical evidence of how they effect change. The issue of treatment effectiveness, however, is distinct from causation and the mechanism responsible for change (Bandura, 1969; Davison, 1968; McFall, 1976). The fact that a so-called anxiety-reduction approach produces improvement neither establishes that change was the result of anxiety deconditioning nor substantiates the hypothesis that anxiety precipitated the problem initially. Treatment techniques are not necessarily the property of a single theoretical model (McFall, 1976; Trower *et al.*, 1978). Although treatment programs may be labeled as skill acquisition, their components often serve multiple functions. For example, hierarchically arranged behavior rehearsal not only facilitates skills acquisition but may also reduce anxiety. These distinctions between effectiveness, mechanism, and causation should be borne in mind in the final sections of this chapter.

Description of Treatment Approaches

Two major anxiety-reduction approaches, systematic desensitization and practice dating, have been used. Since desensitization is generally well known, we will describe only practice dating.

Practice dating was developed by Christensen and Arkowitz (1974) as a treatment that can be administered on a large scale simultaneously to males and females. Volunteers respond to advertisements for a program to increase dating comfort and activity. Following screening and an orientation in which the participants are told that the purpose is to provide practice dating and not an ideal partner, six weekly practice dates are completed. Pairings are determined with attention only to age, height, race, and distance from campus. All date arrangements are made by the participants. The only staff contacts that the participants have are weekly phone calls in which they receive the name and phone number of their next date. In one variation (Christensen *et al.*, 1975), feedback is exchanged by the partners after the date via the experimenter. Feedback consists of citing aspects of physical appearance, dress, and behavior that are positive and identifying one aspect of behavior for which change is suggested. The main objective is to reduce anxiety in dating situations. The advantages of practice dating are its economical use of professional time, large-scale applicability, and potential to enhance generalization by taking place in a naturalistic setting.

In a skills-training program, a variety of components are used to assist the individual to acquire skills assumed to be lacking. Such programs can include behavior rehearsal, modeling, instructions, coaching, audio- or videotape feedback, homework, lectures, discussions, and dating manuals. Behavior rehearsal is generally considered the primary change strategy and provides the client an opportunity to practice and receive feedback in simulated heterosocial situations.

Lindquist, Kramer, McGrath, MacDonald, & Rhyne (1975) described a multifaceted skills-training program that was used in one of the few studies (MacDonald *et al.*, 1975) of very-low-frequency daters (four or fewer dates in 12 months). The approach involves considerable therapist time compared with practice dating, and was used only with males.

The program consists of six 2-h group sessions with six low-dating males and a female leader. Although not empirically developed, the comprehensive program is organized according to a logical progression in dating relationships. Behavior rehearsal is used to practice initiating conversations, responding to female initiations, continuing and extending a conversation, and asking for a date. The program also includes modeling, cognitive restructuring, and self-, peer, female, and leader feedback. Group discussions focus on meeting women, reading cues concerning a woman's interest, making a date fun, evaluating a date, asking for a second date, conversing intimately, and becoming physically involved. Homework assignments are related to session content and range from identifying places to meet women to actual dating. Reports of successful performance are reinforced by the therapist and other group members, whereas unsuccessful performance meets with suggestions for improvement and encouragement to do better.

To date, cognitive treatment approaches for heterosocial difficulties have received limited attention. As such, they may not be as well developed as for other target behaviors (cf. Beck & Rush, 1975; Meichenbaum, 1973, 1977).

Glass *et al.* (1976) employed a four- and five-session semiautomated, audiotaped program which is administered individually to males by undergraduate trainers. The first session involves an orientation stressing the cognitive nature of heterosocial deficits and how to identify and change negative self-statements. Each subsequent session involves learning to recognize negative self-statements and practice in using them to cue more positive self-statements. The training sequence during each session is as follows: first, the subject listens to a taped description of a heterosocial situation and verbalizes what his self-statements might be in that situation. Then a tape is played in which a model verbalizes negative self-statements in the situation, uses them as a cue to substitute positive self-statements, and compliments himself for becoming aware of the negative and for substituting more positive statements. Following the modeling tape, the subject listens to the situation again, rehearses the statements aloud, and receives audiotaped feedback and coaching from the undergraduate trainer until he is satisfied with his performance and ready to proceed to the next situation. An important feature of the Glass *et al.* (1976) program is that all the problem situations and model cognitive responses were generated empirically.

Outcome Studies

Anxiety Reduction—Systematic Desensitization. Several studies have used systematic desensitization to reduce heterosocial anxiety. Comparing several desensitization treatments, Hokanson (1972) found that socially anxious males who received five group-treatment sessions demonstrated more positive behavior with a confederate, reported more reduction in dating anxiety, and were more likely to date (85% vs. 55%) than waiting-list control subjects. Subjects who relaxed during scene presentation were significantly superior to nonrelaxed subjects on behavioral measures. The results were interpreted as support for a counterconditioning explanation of therapeutic effect. However, unlike Hokanson (1972), Taylor (1972) found no significant differences between systematic desensitization, attention-placebo, and no-treatment control-group subjects in number of dates.

Using only self-report measures, Mitchell and Orr (1974) reported that both traditional and short-term desensitization groups for males achieved significant pre-follow-up (4-week) reductions in heterosexual anxiety in comparison to relaxation and no-treatment control groups, which did not. A significant reduction in avoiding women was achieved by the short-term desensitization group, whereas the effect for the traditional group only approached significance ($p < .06$).

For socially anxious females, no significant differences were found on self-report measures between desensitization with a standardized hierarchy, noncontiguous relaxation, hierarchy exposure, and a no-treatment control group (O'Brien & Borkovec, 1977). There was only one behavioral difference, with subjects in the combined treatment groups demonstrating significantly fewer overt signs of anxiety than untreated subjects. On a physiological measure (heart rate), subjects who received relaxation (SD and NCR) showed relatively lower anticipatory anxiety in performance situations compared to the HE and NT subjects, whereas SD subjects demonstrated relatively quicker and greater reduction in reactive anxiety than the other groups.

Although the data are not consistent, the above studies provide some support for using systematic desensitization with heterosocially anxious subjects. It should be noted that the following methodological limitations may have attenuated the effects of desensitization: brevity of treatment (five or fewer sessions), standardized rather than individualized hierarchies, and failing to determine whether anxiety was due to conditioning or skill deficits. Given anxiety due to skill deficits, desensitization alone would be expected to produce limited results.

Other investigations of desensitization have compared its effects to those of skill-based training programs. Stark (1971) reported that, in a performance situation, symbolic modeling and systematic desensitization appeared to be the most effective treatments for socially anxious males. At an 8-month follow-up, however, only relaxation and symbolic modeling subjects continued to improve. Results for an assertion-training group were mixed.

Using mostly males, 50% of whom had had 20 or fewer dates in their lives, Curran (1975) compared systematic desensitization (six 75-min sessions) to social skill training, an attention placebo (relaxation), and a waiting-list control group. Social skills training focused on giving and receiving compliments, listening skills, feeling talk, assertion, nonverbal methods of communication, training in planning dates, and methods of enhancing physical attractiveness. Training included video models, behavior rehearsal, video- and group feedback, and homework. Results were limited to both treatment groups exhibiting significant changes over time on global anxiety and skill ratings of performance. The control groups failed to do so. It appears that the treatment and control groups were never directly compared.

In a subsequent study incorporating a 6-month follow-up and lower frequency daters, Curran and Gilbert (1975) compared the effects of expanded versions of systematic desensitization (eight 90-min sessions over 9 weeks) and social skills training to a no-treatment control group. Both treatments differed significantly from the controls on self-report anxiety at posttest and follow-up, and on global anxiety ratings at follow-up. However, the treatments did not differ from each other. At follow-up, the skills group was rated significantly higher on skill than the other groups. Within-group changes in dating frequency were noted only for the treatment groups. The results were interpreted as supporting the hypothesis that only skills training could improve dating skill.

The superiority of skills training in the Curran and Gilbert (1975) as opposed to the Curran (1975) study may be due to greater heterosocial deficits in the former population (74% had had 10 or fewer dates in their lives), longer treatment, or the fact that assessment in the latter study (no follow-up) was conducted before differences were manifested. During the posttreatment period, subjects who underwent skills training may have refined their skills through further dating to the point at which a significant difference was realized at follow-up. Not surprisingly, superiority of skills training has been reported in other areas of social skills research (e.g., Marzillier *et al.*, 1976; Wright, 1976). Consistent with our position, Curran and Gilbert (1975) stressed the need for assessment instruments that differentiate anxiety from skill problems and for treatment to be tailored accordingly.

With respect to this last point, studies of general social skills training with psychiatric patients (Marzillier & Winter, 1978; Trower *et al.*, 1978) provide suggestive evidence about the value of such an approach. Patients who exhibited varying amounts of skill deficits, anxiety, and cognitive distortions responded differentially to different treatments. Although more refined classifications of heterosocial difficulties have been attempted (Curran *et al.*, 1977), such efforts have yet to be systematically integrated into outcome research.

The existing research suggests that systematic desensitization and social skills training may achieve similar amounts of anxiety reduction, but that skills training may produce greater changes in globally rated dating skills in subjects exhibiting undifferentiated heterosocial difficulties. These results seem consistent with the focus of systematic desensitization which, of course, is anxiety reduction. In comparison, skills training not only improves skills but also reduces anxiety by exposing subjects to anxiety-arousing dating situations in a nonpunitive atmosphere. Thus, it would be expected that desensitization would not add significantly to the effects of a skills-training program.

Two studies address this issue. Because of a general absence of treatment effects, one study (Bander *et al.*, 1975) represents a weak test of the additive effects of desensitization. Geary and Goldman (1978) compared the effects of five weekly, 50-min group sessions of systematic desensitization, behavior rehearsal, a combined SD-BR treatment, traditional insight-oriented therapy, and a no-treatment control group. They concluded that: behavior rehearsal was somewhat superior in reducing anxiety and enhancing social skills; nonspecific factors were as potent in reducing anxiety on most measures as behavioral treatments; and SD offered no advantages, either alone or in combination with BR. However, several problems with the study preclude such a straightforward interpretation of the results. Selection criteria are sketchy and suggest that subjects were only mildly anxious. With less than 5 treatment hours available, it may not have been possible to develop an effective combined treatment. Also, it is difficult to determine which comparisons were tested and which were not. Thus, results from these two studies provide weak but suggestive evidence that desensitization may not augment the effects of a skills training package.

Anxiety Reduction—Practice Dating. Arranged interactions or practice dating represents an alternate approach for reducing anxiety in heterosocial interactions. The approach assumes that low-frequency daters have at least minimally acceptable skills for successful heterosocial performance. The primary objective is to decrease anxiety rather than improve skills. Because this approach involves *in vivo* exposure, it should result in greater generalization than imagery and laboratory-based techniques.

Martinson and Zerface (1970) investigated the effects of a semistructured program of arranged interactions (discussions) with college coeds, individual counseling, and a delayed-treatment control group for males who had not dated in the last month. On a Specific Fear Index (telephoning for a date), the arranged-interactions treatment was significantly more effective than the other groups. This difference is probably due to the fact that males in the arranged-interactions treatment had to phone the coeds to schedule their assigned meetings. An overall χ^2 test failed to reveal significant differences in dating frequency among the groups. However, when each group was compared separately to the control group, a significantly greater number of subjects in the arranged-interactions group than the control group were dating at the 8-week, but not at the 3-week, follow-up.

Arkowitz and his colleagues (Arkowitz *et al.*, 1975a; Arkowitz *et al.*, 1978; Christensen & Arkowitz, 1974; Christensen *et al.*, 1975; Perl *et al.*, 1977; Royce & Arkowitz, 1978), with help from independent researchers (Kramer, 1975; Thomander, 1975), have produced a body of literature on practice dating. The initial study (Christensen & Arkowitz, 1974) lacked a no-treatment control group but suggested the potential of practice dating. Following six practice dates, significant pre–post changes for males and females were obtained and indicated decreases in self-reported anxiety and increases in self-reported skill, frequency of dates, and range of opposite-sex interactions.

In a subsequent study of lower frequency male and female daters (one date per month), combined practice dating and practice dating plus partner feedback treatments were significantly more effective than a delayed-treatment control group on self-report, self-monitoring, and behavioral composite scores that indicated increased skill, comfort, and frequency of heterosocial interactions (Christensen *et al.*, 1975). There were no differences on a peer-rating composite. The no-feedback treatment did significantly better than the feedback group on the behavioral composite. Treatments did not affect males and females differently. Process measures indicated significant decreases in self- and partner ratings of anxiety but not social skills from the first three to last three dates. Based on these findings, the authors suggested that anxiety reduction is the major factor in treatment. The fact that the majority of dates (77.6%) and casual interactions (96%) during the posttreatment assessment period were with persons other than the practice dating partner strongly indicated generalization of treatment. Because of attrition at a 3-month follow-up, data for treatment groups were combined. Results indicated significant change on the SAD and self-ratings of anxiety, social skill, and dating frequency. Although the results of the study are encouraging, the facts that differences were found only on composite scores with small actual differences on individual measures and that differences on judge and partner ratings were not obtained raise some questions about the potency of the approach. Finally, the effects of feedback seem mixed. Although the behavioral composite score clearly favored the no-feedback group, a trend emerged on the important self-monitoring composite score suggesting more frequent and a greater range of dates and casual interaction for the feedback group. These mixed results may be due to the fact that feedback appeared to encourage some subjects while making others more anxious.

Kramer (1975) found that practice dating, practice dating plus cognitive restructuring, and behavior rehearsal were about equally effective and superior to a waiting-list control group for males and females in increasing heterosocial interaction and decreasing heterosocial anxiety. There was only minimal support for the hypothesis that practice dating plus cognitive restructuring would effect greater changes than practice dating alone.

Perl *et al.* (1977) attempted a more stringent test of practice dating by employing subjects with more severe deficits than in past studies. From 3,800 students, they selected subjects who had a mean dating frequency of .45 in the last month and 2.07 in the last six months. The subjects were assigned to the following five groups: (1) waiting-list control, (2) group discussion (six 90-min group sessions about dating concerns), (3) practice dating with low-frequency dating partners, (4) practice dating with high-frequency dating partners (mean = 10 dates per month) plus knowledge of partner's dating frequency, and (5) practice dating with high-frequency dating partners but without knowledge of partner's high dating frequency. Despite significant pre–post improvement on a number of measures for all groups, no significant differences emerged between any of the groups on any of the measures. Previous studies had not demonstrated such improvements for controls. On the posttreatment to follow-up analyses (3 months), the low–low and low–high dating groups continued to increase their dating frequency (overall practice dating = 3.81 dates/month). Similar data were not available for the controls. The authors recommended the continued use of the standard practice-dating procedure (low–low) because it is more economical than the two other variations. Perl *et al.* emphasized that a replication was in order before conclusions could be drawn about practice dating with very-low-frequency daters.

In a study of differing amounts (6–8 h) of practice dating, dyadic interaction, conversational skills training, and group discussion, Thomander (1975) reported results somewhat at variance with the general finding of equal effects of practice dating for males and females. Although males showed similar improvement on dating frequency and self-concept regardless of treatment, change for females was related to a linear function of number of practice dates.

Royce and Arkowitz (1978) extended the practice interaction strategy to treating shyness in same-sex dyads. This study has two important implications. First, the effects of practice interactions with same-sex peers did not generalize to increases in either casual interactions or dates with the opposite sex. Second, social skills training did not add significantly to the effects of practice interactions. This last finding strengthened the belief (Arkowitz *et al.*, 1975a; Christensen *et al.*, 1975) that practice interactions are effective due to *in vivo* desensitization which reduces social anxiety. However, practice dating does not involve systematic graded exposure which is required for *in vivo* desensitization. As proposed by Davison and Wilson (1972), extinction resulting from nonreinforced exposure in the absence of aversive stimuli might constitute a more suitable explanation.

Several tentative conclusions may be offered with respect to anxiety-reduction approaches. Both systematic desensitization and practice dating have received some support for reducing heterosocial anxiety. Unfortunately, there has been no direct comparison of these approaches. Given existing data and results with other target behaviors, it seems likely that the direct-exposure procedure (Bandura, 1977) of practice dating would be the more effective treatment. However, the critical task appears to be assessing subjects' needs and tailoring treatment programs accordingly, rather than attempting to find the one treatment that is best for all heterosocial problems.

Skills-Acquisition Approaches. In the previous section, several studies were discussed that compared the effectiveness of skills-development approaches and systematic desensitization. This section presents more extensive data about other skills-development treatments.

Morgan (1971) found no significant differences on three self-report measures of date-initiating behavior between model exposure, behavior rehearsal, model exposure plus behavior rehearsal, and a control condition of focused counseling with a group of seldom-dating college males. On a fourth variable, degree of anxiety associated with date-initiating situations, only the behavior-rehearsal group differed significantly from pre- to posttest in reported anxiety. Yost (1974) reported that two social skills training programs (one of which included contracting for weekly homework assignments) for males did not differ from each other and resulted in significant improvement as compared to a waiting-list control group on four of five tests measuring self- and other-reported change in social activities and competence. On only one measure did the contracting group significantly outperform the other two groups.

In a study marred by the lack of minimal-dating (two or fewer dates/week) males, low interrater reliabilities, and the fact that the author served as experimenter for several groups and conducted the pre- and posttest assessments, Melnick (1973) compared the effects of a no-treatment control group to five treatment groups—therapy control (discussion), vicarious conditioning (video modeling), participant modeling, participant modeling and self-observation (video feedback), participant modeling plus self-observation and analogue reinforcement. Although no significant differences in dating frequency emerged, results on lab performance measures indicated that the males in the two self-observation treatments benefited the most. The basic modeling treatments had relatively little effect. Melnick (1973) speculated that the self-observation treatments effected change through one of two mechanisms; self-observation may have allowed for objective feedback which led to self-evaluative and corrective changes, or the knowledge that one is being videotaped activated self-monitoring which led to behavior change.

Twentyman and McFall (1975) used a skills-acquisition treatment with shy males who dated less than once a month and scored at least one standard deviation below the mean on the SHI. They compared the effectiveness of an analogue three-session treatment consisting of covert and overt rehearsal, coaching, modeling, and homework to an assessment control group. Treated subjects showed some indication of reduced physiological response in a performance situation, significant increases on the SHI, and less avoidance responding to a role-playing test than control subjects. In a variety of performance situations, self- and judges' ratings of anxiety did not consistently differentiate the two groups. Surprisingly, despite the focus on skills training, only one of two global ratings of skillfulness significantly differentiated the groups at posttest. Based on behavioral diaries, treated subjects reported more and longer interactions with women than control subjects. Treatment effects were no longer evident at a 6-month follow-up. However, considerable subject attrition may have obscured the differences.

In a study that lacked a control group and had disproportionate attrition, Curran *et al.* (1976) found social skills training superior to sensitivity training with heterosocially anxious students (mean = 1.08 dates in the preceding 2 months). Following eight 90-min group-training sessions over an 8-week period, the groups did not differ on self-report questionnaires of general social anxiety. In a live interaction test, the skills group rated themselves as significantly more skilled and judges rated them as significantly more skilled and less anxious than the sensitivity group. In fact, judges saw no improvement in the sensitivity group's skills or anxiety from pre- to posttesting. Significant within-group changes in dating frequency were found for the behavioral group but not for sensitivity training. This study, coupled with the research by Curran and Gilbert (1975), suggests the superiority of a skills-acquisition treatment as compared to at least two treatments, systematic desensitization and sensitivity training, developed from other conceptual perspectives.

Using the social skills program described earlier, MacDonald *et al.* (1975) evaluated two skills-training programs involving behavior rehearsal with and without homework, an attention placebo, and a waiting-list control group with low-frequency-dating males (four or fewer dates in the last 12 months). Results were limited primarily to a few within-group changes (RPDI skill and anxiety scores and Profile of Mood Scale scores), with the behavioral groups showing the majority of such changes. Homework did not significantly augment the effectiveness of behavioral rehearsal. The study failed to find differences in dating frequency among the groups.

In a study that lacked a behavioral measure and in which the senior author conducted the treatments, McGovern *et al.* (1975) compared the effects of group discussion (dating manual and contact with female confederates) to that of discussion plus behavior rehearsal in the office, discussion plus behavior rehearsal in the campus environment, and a waiting-list control group. Subjects were selected according to lenient criteria and consisted of males who had no more than three dates in the past month and/or no more than seven dates in the past 6 months. The treated groups performed significantly better than the control group on only two of five self-report measures. There were few differences among the treatments. Rehearsal in the natural environment did not significantly enhance the effects of social skills training.

Brown and Gilner (1978) used a skills-training approach (covert rehearsal, modeling, coaching, instructions, and self-feedback) to teach groups of psychiatric outpatients either to stand up for their rights or to express the type of positive feelings needed in close interpersonal relationships. The authors were interested in whether training in one aspect of self-expression generalized to improvement in the other. Treated subjects performed significantly better than controls in both trained and untrained areas of expression. One important contribution of this study is that subjects were trained in skills needed for the later stages of heterosocial interactions—expressing positive feelings toward an intimate, making self-disclosing statements, and providing supportive comments. These aspects of heterosocial interactions have rarely been investigated.

Finally, the effects of social skills training have also been investigated in intensive design experiments. Two studies (Edelstein & Eisler, 1976; Hall & Edelstein, 1977) using either modeling, instructions, and feedback of instructions and feedback provided mixed support for skills training with heterosocial deficits.

For a variety of reasons, it is premature to draw firm conclusions about the effectiveness of heterosocial skills training programs. As has been noted for social skills programs with other target behaviors (M. D. Galassi & Galassi, 1978), treatments vary substantially in the number and type of components that comprise them. As a result, it is difficult to discuss the effectiveness of heterosocial skills training *per se*. Other factors which mitigate against definitive conclusions in many of the studies include failure to employ severely incapacitated subjects, failure to increase dating frequency, and absence of adequate transfer measures and long-term follow-ups. Particularly striking is the paucity of skills-training research with females. Nevertheless, results obtained by a number of skills programs, although modest, are promising as they frequently surpass those obtained by alternative treatments, attention placebo, and no-treatment control groups.

Cognitive Modification Approaches. In addition to Kramer (1975), we are aware of only two other studies (Glass *et al.*, 1976; Rehm & Marston, 1968) that were conceptualized as modifying faulty cognitions or that used cognitive techniques. Rehm and Marston (1968) viewed social anxiety as involving negative self-evaluation resulting from a malfunction of self-reinforcement. A treatment in which males reinforced themselves contingent on performing specific behaviors *in vivo* with females was compared to nondirective therapy and a minimal urging toward self-help group. A series of multiple t-tests, which often followed nonsignificant F tests, indicated that the SR group as compared to the other groups showed significantly greater improvement on self-report questionnaires, dating frequency, and on self-rated anxiety and number of words in a behavioral performance test. At a 7- to 9-month follow-up, significant differences between the SR and combined control groups were found on a variety of self-report measures of anxiety and two items concerning self-understanding, but not on dating frequency. Although the results could not be attributed to the passage of time or to nonspecific therapy factors, we also cannot be certain that they are due to cognitive therapy as the SR group experienced the effects of *in vivo* desensitization in their graded performance assignments.

In an attempt to increase generalization and maintenance Glass *et al.* (1976) studied the effects of a combined cognitive self-statement/response-acquisition treatment, in comparison to a cognitive self-statement approach, a response-acquisition (social skill) treatment, and a waiting-list control group with girl-shy males. In addition, two lengthened treatment groups (enhanced response acquisition and enhanced cognitive self-statement modification) were used to control for the longer training time of the combined treatment group. Unlike many previous studies, problem situations, competent responses, and coping self-statements all were empirically derived. Results at posttest indicated that each of the three treatment groups (response acquisition, cognitive, and combined) did significantly better than the controls on trained situations in a performance test with the response acquisition and the combined treatment performing significantly better than the cognitive group. On untrained situations, the authors reported that the groups which included a cognitive component performed significantly better than those that did not. The cognitive self-statement group made significantly more phone calls and a better impression on women during the calls than subjects in the other groups. There were no significant differences between the regular and the enhanced treatment groups at posttest. At a 6-month follow-up, the response-acquisition, enhanced response-acquisition, cognitive self-statement, and combined groups scored significantly higher than the waiting-list group but did not differ from each other on trained situations. No differences were found on untrained situations. The cognitive self-statement group differed significantly from the waiting and enhanced response-acquisition groups on number of calls. The authors concluded that a cognitive self-statement treatment leads to greater transfer than a response-acquisition approach, and that treatment programs need to be tailored to individual needs. Surprisingly, there was little evidence of the superiority of the combined treatment. This result may be due partly to the limited amount of time available for each of the treatment components in the combined approach. We would recommend caution in generalizing the results of the study given that they were not unequivocal and that the enhanced cognitive group performed relatively poorly on the behavioral test at follow-up. In addition, the study is weakened by the failure to specify the degree of subject deficits initially, the use of undergraduates as therapists, and the lack of significant differences in dating frequency among the groups.

As we discussed in a previous section, there have been a number of studies that suggest the role of cognitive factors in the maintenance of heterosocial problems; however, the effects of cognitive-modification treatments remain essentially unexplored. Because of the emphasis on generalized coping skills, cognitive treatments ultimately may yield greater generalization of results than more traditional behaviorally oriented treatments. At this time, however, there is only limited data to support such a conclusion with heterosocial problems.

Generalization to Clinical Populations

It is customary to dismiss the results of treatment studies with college students as relatively unimportant because they may not generalize to a more severely disturbed but vaguely defined "clinical" population. It is as if college students and clinical patients represent mutually exclusive groups, with patients being depressed, nonverbal, incompetent, and maladjusted, and students being highly motivated, verbal, competent, and problem-free. Such a characterization ignores the fact that a sizable percentage of students seek help from college counseling and community mental health agencies each year.

Without studies that catalogue treatment-related differences between clearly defined groups of students and clinical patients or that replicate the effects of treatment in both populations, we will not know whether the results obtained with one population generalize to the other. Our bias is that, if an investigator is interested in the treatment of psychiatric inpatients, then research should not be performed with college students and excused as an analogue. Similarly, if the investigator is interested in college students, then psychiatric patients should not be used as analogue subjects.

We can probably generalize with minimum risk from the studies cited above to low-dating college students who respond to advertisements for treatment. However, the extent to which we can generalize to students who seek help from counseling facilities for problems in heterosocial interactions or for depression and other undesirable behaviors related to heterosocial difficulties is unknown.

A rather sizable body of literature (Apostal, 1968, Berdie & Stein, 1966; De Blassie, 1968; Doleys, 1964; J. P. Galassi & Galassi, 1973; Gaudet & Kulick, 1954; Lewis & Robertson, 1961; Mendelsohn & Kirk, 1962; Merrill & Heathers, 1954; Minge & Bowman, 1967; Parker, 1961; Rossman & Kirk, 1970) indicates that college students who seek personal-adjustment counseling differ significantly on factors that relate to treatment effectiveness from both those who seek vocational-educational counseling and those who do not seek counseling. It is those students who seek personal counseling to whom the question of generalization is most directly relevant.

As noted by Arkowitz, Levine, Grosscup, O'Neal, Youngren, and Royce (1976), it is easier to treat a single circumscribed problem and studies have typically not evaluated whether dating problems are relatively discrete or embedded in a network of other problems. In fact, it has been a common practice (Arkowitz *et al.*, 1978; Rehm & Marston, 1968; Twentyman & McFall, 1975) in treatment studies to screen out subjects who exhibit either severe deficits in the heterosocial area or other clinical problems. With respect to subject selection, Little, Curran, and Gilbert (1977) have shown that heterosocially anxious males recruited either from a newspaper ad for a therapy analogue study or an introductory psychology class for an experiment differ significantly on dating frequency and self- and judges ratings of skill and anxiety. In contrast, Royce and Arkowitz (1977) have shown that it is not the method of recruiting that leads to differences but whether subjects are recruited for an experiment or for treatment. When subjects were stringently screened on both criterion behaviors and motivation for treatment, there were no differences between the groups on any of 18 dependent variables. Neither study investigated differences between their groups and students who actually sought counseling because of heterosocial difficulties. Such comparisons are necessary in order to determine whether we can generalize our results to a clinical population of students.

We are not aware of any controlled studies that used a clinical population of college students. Such a population would include students who have severe dating problems and who independently seek help from a counseling facility. In an uncontrolled but clinically relevant study of outpatients from an unspecified population, Arkowitz *et al.* (1976) reported that 12–18 individual sessions of hierarchically arranged behavior rehearsal, feedback, coaching, modeling, and homework assignments led to results which were "far short of satisfactory." Even when improvement was achieved, it was still short of the treatment goals.

In contrast, Engeman *et al.* (1977) reported somewhat more encouraging results with seven 2 1/4 h sessions and a 1-month maintenance session of cognitive-behavioral social skills training with groups of college and community men and women referred for treatment by psychotherapists. Engeman *et al.* reported that clinical subjects were more resistant than analogue subjects to trying new behaviors both inside and outside the group, and suggested the following treatment modifications with clinical subjects: greater emphasis on cognitive change procedures and expectancy of change; more group support and attention to group process; specific methods to deal with resistance to role-playing; and systematic efforts to effect transfer of training. At present, there is a clear need for controlled studies of the effectiveness of treatment procedures with clinical populations of college students experiencing heterosocial problems and for studies of treatment-related differences between analogue and clinical groups of college students. In addition, treatment studies with other outpatient adults and studies of a preventive nature designed to teach heterosocial skills to beginning daters are needed.

CONCLUSIONS

As our review has shown, heterosocial skills have been an extremely popular topic for behavior therapists during the last 10 years. Since 1975, in particular, research activity has increased remarkably. In many ways, however, our knowledge of the development, assessment, and treatment of heterosocial problems does not appear to have increased at the same rate. In this final section, we will identify problems impeding further progress, present conclusions that can be drawn from existing research, and suggest methodological changes and new directions.

1. Equating heterosocial problems with skill deficits has biased research efforts. Heterosocial problems are not necessarily synonymous with heterosocial skill deficits. Although globally defined skill deficits have often been recognized, heterosocial problems may exist even though no skill deficits are evident. Equating heterosocial problems with skill deficits or with anxiety, for example, results in equating the problem with a presumed cause or explanatory model and may lead to overlooking other important variables.

Curiously, no one has yet referred to this area of research as heterosocial cognitive deficits. In fact, there is more evidence that heterosocial problems are characterized (not necessarily caused) by deficits in cognitive rather than behavioral skills. In addition, no one has referred to the area as heterosocial attractiveness deficits. Progress might be accelerated if conceptualization is guided by such terms as heterosocial problems and heterosocial competence/incompetence, with competent performance being influenced by a number of variables.

2. As noted by McFall (1976), social skills research continues to be unduly influenced by trait-like conceptions of behavior. Contrasting subjects high and low in skill, anxiety, cognitive evaluation, or attractiveness in order to identify treatment targets obscures the differential

importance of these variables in different dating stages and situations. What may be more helpful is to take a developmental-interactional view of heterosocial competence by first identifying crucial stages of dating and salient variables for each. Once these stages/situations have been identified, individuals can be observed in either laboratory analogues or naturalistic settings, and measurements of skills, anxiety, cognitions, physical attractiveness, and other potentially important variables can be taken. The relevance of these variables to situationally defined competent performance can then be determined from either peer judgments or empirical validation. In this manner, heterosocial problems can be more precisely identified.

3. Being careful not to overemphasize skill deficits, we should not abandon the search for these deficits without first conducting studies that use the structural approach to investigating dyadic interactions. If judges can routinely detect global skill differences between crudely defined groups of competent and incompetent performers, it seems likely that they are responding to observable differences that current methodology has been unable to detect.

4. The role of physical attractiveness needs to be more fully appreciated and incorporated into our conceptualization and treatments of heterosocial incompetence.

5. Studies of beginning daters, of people who resume dating after a divorce or separation, and of people who differ in age, education, race, and social class may provide us with a better understanding of dating competence. For example, longitudinal studies of beginning daters can provide information on the development of dating patterns and perhaps on the etiology of these problems as well.

As was mentioned earlier, heterosocial competence involves more than just successful dating. Research could explore the development of social but nondating (friendship) relationships between members of the opposite sex at a variety of points across the life span. Such studies would broaden our knowledge of heterosocial competence.

6. As with other areas of behavior therapy research, assessment remains relatively underdeveloped as compared to treatment. Improving assessment represents an important key to progress. Treatment targets need to be more clearly specified for interventions to be maximally successful.

7. Some progress in assessment has been evident in the last few years. Attempts at empirically deriving and validating situationally specific measures (e.g., Barlow *et al.*, 1977; Glass & Merluzzi, 1978) are promising and represent the type of increased specificity likely to enhance understanding of heterosocial competence.

8. Behavioral performance tests, although promising, require vali-

dation against performance in the naturalistic setting. In general, the focus of assessment should move from the laboratory to the subject's environment. Such an approach would involve increasing the reliability and validity of diary reports and other transfer of training measures, and might include the use of confederates as dating partners who retrospectively analyze the subject's dating behavior on empirically derived behavioral checklists.

9. Heterosocially incompetent females have received much less attention than their male counterparts. Research on the parameters of heterosocially competent performance, development of relevant assessment instruments, treatments designed especially for females, and outcome studies with females are needed.

10. Existing research suggests that behavior in heterosocial interactions may be sex-role stereotyped. If future research continues to indicate such a pattern, then the counselor/therapist may face the dilemma of trying to help clients be more effective in opposite-sex interactions without simultaneously reinforcing sex-role stereotypes.

11. Treatments for heterosocial problems have generally yielded modest results. Effects have often been limited to within-group changes and frequently have not been maintained. Controlled studies with clinical populations of heterosocially incompetent subjects have not been reported.

12. Outcome research suggests that systematic desensitization reduces heterosocial anxiety, whereas various skills-acquisition programs not only reduce anxiety but also produce changes in globally rated heterosocial skills. In general, a variety of treatments that facilitate exposure (e.g., practice dating) to avoided heterosocial situations have resulted in at least some improvement. In theory, cognitive-acquisition treatments hold promise for increasing the generalization, transfer, and maintenance of treatment effects. Whether this is actually the case has not been satisfactorily investigated.

13. The search for the single best treatment package seems misdirected. What is needed is careful assessment of client problems and goals followed by investigations of which treatments are most effective for which problems and for which clients.

14. The issues of causation, treatment effectiveness, and treatment mechanism must be studied separately. Data on treatment effectiveness, for example, cannot be used as support for a particular explanatory model of the causation and maintenance of heterosocial problems. Further, although multifaceted treatment programs probably effect change in multiple ways, we actually know little about these mechanisms. One necessary, but not sufficient, condition for identifying the mechanisms responsible for treatment effectiveness is demonstrating change on rele-

vant dependent variables. A skills-acquisition treatment must effect changes in related skills and a cognitive treatment must produce changes in cognitions as a prerequisite for concluding that skills acquisition or cognitive modification were functioning as hypothesized.

In conclusion, achieving heterosocial competence is a fundamental task that has life-long implications. Clearly, the work on heterosocial problems to date has provided some basic knowledge necessary for understanding and treating incompetence. However, much more remains to be accomplished.

ACKNOWLEDGMENTS

We thank Kathy Fee Fulkerson and Mary Lay for their suggestions and help in preparing the manuscript.

REFERENCES

Abel, G. G. Assessment of sexual deviation in the male. In M. Hersen & A. S. Bellack (Eds.), *Behavioral assessment: A practical handbook*. New York: Pergamon, 1976.

Abel, G. G., Blanchard, E. G., & Becker, J. V. An integrated treatment program for rapists. In R. T. Raba (Ed.), *Clinical aspects of the rapist*. New York: Grune & Stratton, 1977.

Apostal, R. A. Comparison of counselees and noncounselees with type of problem controlled. *Journal of Counseling Psychology*, 1968, *15*, 407–410.

Argyle, M. *Social interaction*. Chicago: Aldine, 1969.

Argyle, M., & Kendon, A. The experimental analysis of social performance. In L. Berkowitz (Ed.), *Advances in experimental social psychology* (Vol. 3). New York: Academic, 1967.

Argyle, M., & Williams, M. Observer or observed? A reversible perspective in person perception. *Sociometry*, 1969, *32*, 396–412.

Arkowitz, H. Measurement and modification of minimal dating behavior. In M. Hersen (Ed.), *Progress in behavior modification* (Vol. 5). New York: Academic, 1977.

Arkowitz, H., Christensen, A., & Royce, S. *Treatment strategies for social inhibition based on real-life practice*. Paper presented at the annual meeting of the Association for the Advancement of Behavior Therapy, San Francisco, December 1975. (a)

Arkowitz, H., Lichtenstein, E., McGovern, K., & Hines, P. The behavioral assessment of social competence in males. *Behavior Therapy*, 1975, *6*, 3–13. (b)

Arkowitz, H., Levine, A., Grosscup, S., O'Neal, A., Youngren, M., Royce, W. S., & Largay, D. *Clinical applications of social skill training: Issues and limitations in generalization from analogue studies*. Paper presented at the annual meeting of the Association for the Advancement of Behavior Therapy, New York, December 1976.

Arkowitz, H., Hinton, R., Perl, J., & Himadi, W. Treatment strategies for dating anxiety in college men based on real-life practice. *The Counseling Psychologist*, 1978, *7*, 41–46.

Bander, K. W., Steinke, G. V., Allen, G. J., & Mosher, D. L. Evaluation of three dating-specific treatment approaches for heterosexual dating anxiety. *Journal of Consulting and Clinical Psychology*, 1975, *43*, 259–265.

Bandura, A. *Principles of behavior modification*. New York: Holt, Rinehart and Winston, 1969.

Bandura, A. *Social learning theory*. Englewood Cliffs, N.J.: Prentice-Hall, 1977.

Barlow, D. H., Abel, G. G., Blanchard, E. B., Bristow, A. R., & Young, L. D. A heterosocial skills behavior checklist for males. *Behavior Therapy*, 1977, *2*, 229–239.

Barocas, R., & Karoly, P. Effects of physical appearance on socialness. *Psychological Reports*, 1972, *31*, 495–500.

Beck, A. T., & Rush, J. *Cognitive behavioral theory of depression: Treatment manual*. University of Pennsylvania, 1975.

Bellack, A. S., & Hersen, M. Self report inventories in behavioral assessment. In J. D. Cone & R. P. Hawkins (Eds.), *Behavioral assessment: New directions in clinical psychology*. New York: Brunner/Mazel, 1977.

Bellack, A. S., Hersen, M., & Turner, S. M. Role-play tests for assessing social skills: Are they valid? *Behavior Therapy*, 1978, *9*, 448–461.

Berdie, R. F., & Stein, J. A comparison of new university students who do and do not seek counseling. *Journal of Counseling Psychology*, 1966, *13*, 310–317.

Berscheid, E., & Walster, E. Physical attractiveness. In L. Berkowitz (Ed.), *Advances in experimental social psychology* (Vol. 7). New York: Academic, 1973.

Berscheid, E., Dion, K., Walster, E., & Walster, G. W. Physical attractiveness and dating choice: A test of the matching hypothesis. *Journal of Experimental Social Psychology*, 1971, *7*, 173–189.

Borgatta, E. F. Analysis of social interaction: Actual, role playing and projective. *Journal of Abnormal and Social Psychology*, 1955, *51*, 394–405.

Borkovec, T. D., Stone, N. M., O'Brien, G. T., & Kaloupek, D. G. Evaluation of a clinically relevant target behavior for analog outcome research. *Behavior Therapy*, 1974, *5*, 503–513.

Brislin, R. W., & Lewis, S. A. Dating and physical attractiveness. *Psychological Reports*, 1968, *22*, 976.

Brown, H. M., & Gilner, F. H. *Modifying unassertive behavior with psychiatric outpatients*. Unpublished manuscript, 1978.

Bryant, B. M., & Trower, P. E. Social difficulty in a student sample. *British Journal of Educational Psychology*, 1974, *44*, 13–21.

Bryant, B. M., Trower, P., Yardley, K., Urbieta, H., & Letemendia, F. A survey of social inadequacy among psychiatric outpatients. *Psychological Medicine*, 1976, *6*, 101–112.

Byrne, D. Attitudes and attraction. In L. Berkowitz (Ed.), *Advances in experimental social psychology* (Vol. 4). New York: Academic, 1969.

Byrne, D., London, O., & Reeves, K. The effects of physical attractiveness, sex, and attitude similarity on interpersonal attraction. *Journal of Personality*, 1968, *36*, 259–271.

Byrne, D., Ervin, C. R., & Lamberth, J. The continuity between the experimental study of attraction and "real life" computer dating. *Journal of Personality and Social Psychology*, 1970, *16*, 157–165.

Christensen, A., & Arkowitz, H. Preliminary report on practice dating and feedback as treatment for college dating problems. *Journal of Counseling Psychology*, 1974, *21*, 92–95.

Christensen, A., Arkowitz, H., & Anderson, J. Practice dating as treatment for college dating inhibitions. *Behaviour Research and Therapy*, 1975, *13*, 321–331.

Clark, J. V., & Arkowitz, H. Social anxiety and self-evaluation of interpersonal performance. *Psychological Reports*, 1975, *36*, 211–221.

Coons, F. W. The resolution of adolescence in college. *Personnel and Guidance Journal*, 1970, *48*, 533–541.

Cross, J. F., & Cross, J. Age, sex, race, and the perception of facial beauty. *Developmental Psychology*, 1971, *5*, 433–439.

Curran, J. P. Correlates of physical attractiveness and interpersonal attraction in the dating situation. *Social Behavior and Personality*, 1973, *1*, 153–157.

Curran, J. P. Social skills training and systematic desensitization in reducing dating anxiety. *Behaviour Research and Therapy*, 1975, *13*, 65–68.

Curran, J. P. Skills training as an approach to the treatment of heterosexual-social anxiety: A review. *Psychological Bulletin*, 1977, *84*, 140–157.

Curran, J. P. Comments on Bellack, Hersen and Turner's paper on the validity of role-play test. *Behavior Therapy*, 1978, *9*, 462–468.

Curran, J. P., & Gilbert, F. S. A test of the relative effectiveness of a systematic desensitization program and an interpersonal skills training program with date anxious subjects. *Behavior Therapy*, 1975, *6*, 510–521.

Curran, J. P., & Lippold, S. The effects of physical attraction and attitude similarity on attraction in dating dyads. *Journal of Personality*, 1975, *43*, 528–538.

Curran, J. P., Gilbert, F. S., & Little, L. M. A comparison between behavioral replication training and sensitivity training approaches to heterosexual dating anxiety. *Journal of Counseling Psychology*, 1976, *23*, 190–196.

Curran, J. P., Wallander, J. L., & Fishetti, M. *The role of behavioral and cognitive factors in heterosexual-social anxiety*. Paper presented at the annual meeting of the Midwestern Psychological Association, Chicago, May 1977.

Curran, J. P., Little, L. M., & Gilbert, F. S. Reactivity of males of differing heterosexual social anxiety to female approach and non-approach cue conditions. *Behavior Therapy*, 1978, *9*, 961.

Davison, G. C. Systematic desensitization as a counterconditioning process. *Journal of Abnormal Psychology*, 1968, *73*, 91–99.

Davison, G. C., & Wilson, G. T. Critique of "Desensitization: Social and cognitive factors underlying the effectiveness of Wolpe's procedure." *Psychological Bulletin*, 1972, *78*, 28–31.

De Blassie, R. R. Diagnostic usefulness of the 16PF in counseling. *Journal of College Student Personnel*, 1968, *9*, 378–381.

Dion, K., Berscheid, E., & Walster, E. What is beautiful is good. *Journal of Personality and Social Psychology*, 1972, *24*, 285–290.

Doleys, E. J. Differences between clients and non-clients on the Mooney Problem Check List. *Journal of College Student Personnel*, 1964, *6*, 21–24.

Duncan, S., Jr. Nonverbal communication. *Psychological Bulletin*, 1969, *72*, 118–137.

Duncan, S., Jr., & Fiske, D. W. *Face to face interaction: Research, methods, and theory*. Hillsdale, N.J.: Erlbaum, 1977.

Edelstein, B. A., & Eisler, R. M. Effects of modeling and modeling with instructions and feedback on the behavioral components of social skills. *Behavior Therapy*, 1976, *7*, 382–389.

Efran, J. S., & Korn, P. R. Measurement of social caution. *Journal of Consulting and Clinical Psychology*, 1969, *33*, 78–83.

Eisler, R. M. Behavioral assessment of social skills. In M. Hersen & A. S. Bellack (Eds.), *Behavioral assessment: A practical handbook*. New York: Pergamon, 1976.

Eisler, R. M., Frederiksen, L. W., & Peterson, G. L. The relationship of cognitive variables to the expression of assertiveness. *Behavior Therapy*, 1978, *9*, 419–427.

Endler, W. S., & Magnusson, D. *Interactional psychology and personality*. New York: Hemisphere, 1976.

Endler, W. S., Hunt, J. McV., & Rosenstein, A. J. An S-R Inventory of Anxiousness. *Psychological Monographs*, 1962, *76* (17, Whole No. 536).

Engeman, J., Campbell, M., & Steffen, J. *Development and evaluation of a cognitive-behavioral social skills training program with a clinical population.* Paper presented at the annual meeting of the Association for the Advancement of Behavior Therapy, Atlanta, December 1977.

Farrell, A. D., III, Mariotto, M. J., Conger, A. J., Curran, J. P., & Wallander, J. L. *Self and judges' ratings of heterosexual-social anxiety and skill: A generalizability study.* Unpublished manuscript, 1978.

Fischetti, M., Curran, J. P., & Wessberg, H. W. Sense of timing: A skill deficit in heterosexual-socially anxious males. *Behavior Modification,* 1977, *1,* 179–194.

Galassi, J. P., & Galassi, M. D. Alienation in college students: A comparison of counseling seekers and nonseekers. *Journal of Counseling Psychology,* 1973, *20,* 44–49.

Galassi, J. P., & Galassi, M. D. A comparison of the factor structure of an assertion scale across sex and population. *Behavior Therapy,* 1979, *10,* 117–128.

Galassi, M. D., & Galassi, J. P. The effects of role playing variations on the assessment of assertive behavior. *Behavior Therapy,* 1976, *7,* 343–347.

Galassi, M. D., & Galassi, J. P. Assertion: A critical review. *Psychotherapy: Theory, Research and Practice,* 1978, *15,* 16–29.

Galassi, M. D., & Galassi, J. P. Similarities and differences between two assertion measures: Factor analyses of the College Self-Expression Scale and the Rathus Assertiveness Schedule. *Behavioral Assessment,* in press.

Gaudet, F. J., & Kulick, W. Who comes to a vocational guidance center? *Personnel and Guidance Journal,* 1954, *33,* 211–214.

Geary, J. M., & Goldman, M. S. Behavioral treatment of heterosexual social anxiety: A factorial investigation. *Behavior Therapy,* 1978, *9,* 971–972.

Glasgow, R. E., & Arkowitz, H. The behavioral assessment of male and female social competence in dyadic heterosexual interactions. *Behavior Therapy,* 1975, *6,* 488–498.

Glass, C. R., & Merluzzi, T. V. *Approaches to the cognitive assessment of social anxiety.* Paper presented at the annual meeting of the Association for the Advancement of Behavior Therapy. Chicago, November 1978.

Glass, C. R., Gottman, J. M., & Shmurak, S. H. Response-acquisition and cognitive self-statement modification approaches to dating-skills training. *Journal of Counseling Psychology,* 1976, *23,* 520–526.

Goffman, E. *The presentation of self in everyday life.* Garden City, N.Y.: Doubleday, 1959.

Goldfried, M. R., & D'Zurilla, T. J. A behavioral-analytic model for assessing competence. In C. D. Spielberger (Ed.), *Current topics in clinical psychology* (Vol. 1). New York: Academic, 1969.

Greenwald, D. P. The behavioral assessment of differences in social skill and social anxiety in female college students. *Behavior Therapy,* 1977, *8,* 925–937.

Greenwald, D. P. Self-report assessment in high- and low-dating college women. *Behavior Therapy,* 1978, *9,* 297–299.

Hall, D., & Edelstein, B. *The remediation of heterosexual social skills deficits in a single-subject experiment.* Paper presented at the annual meeting of the Association for the Advancement of Behavior Therapy, Atlanta, December 1977.

Havighurst, R. J. *Developmental tasks and education* (3rd ed.). New York: David McKay, 1972.

Heimberg, R. G. Comment on "Evaluation of a clinically relevant target behavior for analog outcome research." *Behavior Therapy,* 1977, *8,* 492–493.

Hersen, M., & Bellack, A. J. Assessment of social skills. In A. R. Ciminero, K. R. Calhoun, & H. E. Adams (Eds.), *Handbook of behavioral assessment.* New York: Wiley, 1976.

Hokanson, D. T. Systematic desensitization and positive cognitive rehearsal treatment of social anxiety. (Doctoral dissertation, University of Texas at Austin, 1971.) *Dissertation Abstracts International,* 1972, *32,* 6649B–6650B. (University Microfilms No. 72-15, 775)

Huston, T. L. Ambiguity of acceptance, social desirability, and dating choice. *Journal of Experimental Social Psychology*, 1973, *9*, 32–42.

Kaats, C. R., & Davis, K. E. The dynamics of sexual behavior of college students. *Journal of Marriage and the Family*, 1970, *32*, 390–399.

Kazdin, A. E. Assessing the clinical or applied importance of behavior change through social validation. *Behavior Modification*, 1977, *1*, 427–452.

Kiesler, D. J. Some myths of psychotherapy research and the search for a paradigm. *Psychological Bulletin*, 1966, *65*, 110–136.

Klaus, D., Hersen, M., & Bellack, A. S. Survey of dating habits of male and female college students: A necessary precursor to measurement and modification. *Journal of Clinical Psychology*, 1977, *33*, 369–375.

Kramer, S. R. Effectiveness of behavior rehearsal and practice dating to increase heterosexual social interaction. (Doctoral dissertation, University of Texas, 1975.) *Dissertation Abstracts International*, 1975, *36*, 913B–914B. (University Microfilms No. 75–16, 693)

Kreitler, H., & Kreitler, S. Validation of psychodramatic behaviour against behaviour in life. *British Journal of Medical Psychology*, 1968, *41*, 185–192.

Lanyon, R. I. Measurement of social competence in college males. *Journal of Consulting Psychology*, 1967, *31*, 495–498.

Levenson, R. W., & Gottman, J. M. Toward the assessment of social competence. *Journal of Consulting and Clinical Psychology*, 1978, *46*, 453–462.

Lewis, H., & Robertson, M. Socio-economic status and a university psychological clinic. *Journal of Counseling Psychology*, 1961, *8*, 239–242.

Libet, J. M., & Lewinsohn, P. M. Concept of social skill with special reference to the behavior of depressed persons. *Journal of Consulting and Clinical Psychology*, 1973, *40*, 304–312.

Lick, J. R., & Unger, T. E. The external validity of behavioral fear assessment: The problem of generalizing from the laboratory to the natural environment. *Behavior Modification*, 1977, *1*, 283–306.

Lindquist, C. V., Kramer, J. A., McGrath, R. A., MacDonald, M., & Rhyne, L. D. Social skills training: Dating skills. JSAS *Catalog of Selected Documents*, 1975, *5*, 279. (Ms. No. 1009)

Linehan, M. M. Issues in behavioral interviewing. In J. D. Cone & R. P. Hawkins (Eds.), *Behavioral assessment: New directions in clinical psychology*. New York: Brunner/Mazel, 1977.

Little, L. M., Curran, J. P., & Gilbert, F. S. The importance of subject recruitment procedures in therapy analogue studies on heterosexual-social anxiety. *Behavior Therapy*, 1977, *8*, 24–29.

Lowe, M. R., & Cautela, J. R. A self-report measure of social skill. *Behavior Therapy*, 1978, *9*, 535–544.

MacDonald, M. L., Lindquist, C. U., Kramer, J. A., McGrath, R. A., & Rhyne, L. D. Social skills training: Behavior rehearsal in groups and dating skills. *Journal of Counseling Psychology*, 1975, *22*, 224–230.

McFall, R. M. Behavioral training: A skill-acquisition approach to clinical problems. In J. T. Spence, R. C. Carson, & J. W. Thibaut (Eds.), *Behavioral approaches to therapy*. Morristown, N.J.: General Learning, 1976.

McFall, R. M. Analogue methods in behavioral assessment: Issues and prospects. In J. D. Cone & R. P. Hawkins (Eds.), *Behavioral assessment: New directions in clinical psychology*. New York: Brunner/Mazel, 1977.

McGovern, K. B., Arkowitz, H., & Gilmore, S. K. Evaluation of social skills training programs for college dating inhibitions. *Journal of Counseling Psychology*, 1975, *22*, 505–512.

Martinez, J. A., & Edelstein, B. A. *Heterosocial competence: A validity study*. Paper presented at the annual meeting of the Association for the Advancement of Behavior Therapy, Atlanta, December 1977. (a)

Martinez, J. A., & Edelstein, B. A. *The effects of demand characteristics on the assessment of heterosocial competence in college males*. Paper presented at the annual meeting of the Association for the Advancement of Behavior Therapy, Atlanta, December 1977. (b)

Martinson, W. D., & Zerface, J. P. Comparison of individual counseling and a social program with nondaters. *Journal of Counseling Psychology*, 1970, *17*, 36–40.

Marzillier, J. S., & Winter, K. Success and failure in social skills training. *Behaviour Research and Therapy*, 1978, *16*, 67–84.

Marzillier, J. S., Lambert, C., & Kellett, J. A controlled evaluation of systematic desensitization and social skills training for socially inadequate psychiatric patients. *Behaviour Research and Therapy*, 1976, *14*, 225–238.

Mathes, W. D., & Kahn, A. Physical attractiveness, happiness, neuroticism, and self-esteem. *Journal of Psychology*, 1975, *90*, 27–30.

Meichenbaum, D. *Therapist manual for cognitive behavior modification*. University of Waterloo, Waterloo, Canada, 1973.

Meichenbaum, D. *Cognitive-behavior modification*. New York: Plenum Press, 1977.

Melnick, J. A comparison of replication techniques in the modification of minimal dating behavior. *Journal of Abnormal Psychology*, 1973, *81*, 51–59.

Mendelsohn, G. A., & Kirk, B. A. Personality differences between students who do and do not use a counseling facility. *Journal of Counseling Psychology*, 1962, *9*, 341–346.

Merrill, R. M., & Heathers, L. B. The use of an adjective checklist as a measure of adjustment. *Journal of Counseling Psychology*, 1954, *1*, 137–143.

Miller, W. R., & Arkowitz, H. Anxiety and perceived causation in the social success and failure experiences: Disconfirmation of an attribution hypothesis in two experiments. *Journal of Abnormal Psychology*, 1977, *86*, 665–668.

Minge, M. R., & Bowman, T. F. Personality differences among nonclients and vocational-educational and personal counseling clients. *Journal of Counseling Psychology*, 1967, *14*, 137–139.

Mitchell, K. R., & Orr, T. E. Note on treatment of heterosexual anxiety using short-term massed desensitization. *Psychological Reports*, 1974, *35*, 1093–1094.

Morgan, J. The effect of model exposure and behavior rehearsal on the initiation of dating experiences by seldom dating college men. (Doctoral dissertation, University of Colorado, 1970.) *Dissertation Abstracts International*, 1971, *31*, 3275A. (University Microfilms No. 70–26, 943)

Nay, W. R. Analogue measures. In A. R. Ciminero, K. S. Calhoun, & H. E. Adams (Eds.), *Handbook of behavioral assessment*. New York: Wiley, 1977.

Nietzel, M. T., & Bernstein, D. A. The effects of instructionally-mediated demand upon the behavioral assessment of assertiveness. *Journal of Consulting and Clinical Psychology*, 1976, *44*, 500.

O'Banion, K., & Arkowitz, H. Social anxiety and selective memory for affective information about the self. *Social Behavior and Personality*, 1977, *5*, 321–328.

O'Brien, G. T., & Borkovec, T. D. The role of relaxation in systematic desensitization: Revisiting an unresolved issue. *Journal of Behavior Therapy and Experimental Psychiatry*, 1977, *8*, 359–364.

O'Brien, T. P. Further comments on "Evaluation of a clinically relevant target behavior for analog outcome research." *Newsletter of the Association for the Advancement of Behavior Therapy*, 1977, *4*, 19, 22.

Parker, C. A. The predictive use of the MMPI in a college counseling center. *Journal of Counseling Psychology*, 1961, *8*, 154–158.

Paul, G. L. *Insight vs. desensitization in psychotherapy*. Stanford, Calif.: Stanford University Press, 1966.

Perl, J., Hinton, R., Arkowitz, H., & Himadi, W. *Partner characteristics and the effectiveness of practice dating procedures in the treatment of minimal dating*. Paper presented at the annual meeting of the Association for the Advancement of Behavior Therapy, Atlanta, December 1977.

Perri, M. G. Behavior modification of heterosocial difficulties: A review of conceptual, treatment and assessment considerations. JSAS *Catalog of Selected Documents in Psychology*, 1977, *7*, 75. (Ms. No. 1530)

Perri, M. G., & Richards, C. S. An investigation of naturally occurring episodes of self-controlled behaviors. *Journal of Counseling Psychology*, 1977, *3*, 178–183.

Perri, M. G., & Richards, C. S. The empirical development of a behavioral role-playing test for the assessment of heterosocial skills in male college students. *Behavior Modification*, in press.

Perri, M. G., Richards, C. S., & Goodrich, J. D. The heterosocial adequacy test (HAT): A behavioral role-playing test for the assessment of heterosocial skills in male college students. JSAS *Catalog of Selected Documents in Psychology*, 1978, *8*, 16. (Ms. No. 1650)

Perrin, F. A. Physical attractiveness and repulsiveness. *Journal of Experimental Psychology*, 1921, *4*, 203–217.

Rehm, L. P., & Marston, A. R. Reduction of social anxiety through modification of self-reinforcement: An instigation therapy technique. *Journal of Consulting and Clinical Psychology*, 1968, *32*, 565–574.

Rhyne, L. D., MacDonald, M. L., McGrath, R. A., Lindquist, C. U., & Kramer, J. A. The roleplayed dating interactions (RPDI): An instrument for the measurement of male social dating skills. JSAS *Catalog of Selected Documents in Psychology*, 1974, *4*, 42. (Ms. No. 615)

Richardson, F. C., & Tasto, D. L. Development and factor analysis of a social anxiety inventory. *Behavior Therapy*, 1976, *7*, 453–462.

Rossman, J. E., & Kirk, B. A. Comparison of counseling seekers and nonseekers. *Journal of Counseling Psychology*, 1970, *17*, 184–188.

Royce, W. S., & Arkowitz, H. Clarification of some issues concerning subject recruitment procedures in therapy analog studies. *Behavior Therapy*, 1977, *8*, 64–69.

Royce, W. S., & Arkowitz, H. Multi-modal evaluation of practice interactions as treatment for social isolation. *Journal of Consulting and Clinical Psychology*, 1978, *46*, 239–245.

Schwartz, R. M., & Gottman, J. M. A task analysis approach to clinical problems: A study of assertive behavior. *Journal of Consulting and Clinical Psychology*, 1976, *44*, 910–920.

Shepherd, G. Social skills training: The generalization problem. *Behavior Therapy*, 1977, *8*, 1008–1009.

Shmurak, S. H. *A comparison of types of problems encountered by college students and psychiatric inpatients in social situations*. Unpublished manuscript, 1973.

Sigall, H., & Landy, D. Radiating beauty: Effects of having a physically attractive partner on person perception. *Journal of Personality and Social Psychology*, 1973, *28*, 218–224.

Smith, R. E. Social anxiety as a moderator variable in the attitude-similarity-attraction relationship. *Journal of Experimental Research in Personality*, 1972, *6*, 22–28.

Smith, R. E., & Jeffrey, R. W. Social-evaluative anxiety and the reinforcement properties of agreeing and disagreeing attitude statements. *Journal of Experimental Research in Personality*, 1970, *4*, 276–280.

Smith, R. E., & Sarason, I. G. Social anxiety and the evaluation of negative interpersonal feedback. *Journal of Consulting and Clinical Psychology*, 1975, *43*, 429.

Spencer, C. D. Two types of role playing: Threats to internal and external validity. *American Psychologist,* 1978, *33,* 265–268.

Stanton, H. R., & Litwak, E. Toward the development of a short form test of interpersonal competence. *American Sociological Review,* 1955, *20,* 668–674.

Stark, J. E. The comparative efficacy of three behavior modification techniques in the treatment of interpersonal anxiety. (Doctoral dissertation, University of Georgia, 1970.) *Dissertation Abstracts International,* 1971, *31,* 6914B. (University Microfilms No. 71-13, 132)

Steffen, J. J., & Reckman, R. F. *Selective perception and interpretation of interpersonal cues in dyadic interactions.* Unpublished manuscript, 1978.

Steffen, J. J., & Redden, J. Assessment of social competence in an evaluation-interaction analogue. *Human Communication Research,* 1977, *4,* 30–37.

Steffen, J. J., Greenwald, D. P., & Langmeyer, D. *A factor analytic study of social competence in women.* Unpublished manuscript, 1978.

Stevenson, I., & Wolpe, J. Recovery from sexual deviation through overcoming nonsexual neurotic responses. *American Journal of Psychiatry,* 1960, *116,* 737–742.

Stroebe, W., Insko, C. A., Thompson, V. D., & Layton, B. D. Effects of physical attractiveness, attitude, similarity, and sex on various aspects of interpersonal attraction. *Journal of Personality and Social Psychology,* 1971, *18,* 79–91.

Taylor, R. *Systematic desensitization of dating anxiety.* Unpublished doctoral dissertation, Arizona State University, 1972.

Tesser, A., & Brodie, M. A note on the evaluation of a "computer date." *Psychonomic Science,* 1971, *23,* 200.

Thomander, L. D. The treatment of dating problems: Practice dating, dyadic interaction, and group discussion. (Doctoral dissertation, Michigan State University, 1975.) *Dissertation Abstracts International,* 1975, *36,* 461B. (University Microfilms No. 75-14, 850)

Trower, P., Yardley, K., Bryant, B. M., & Shaw, P. The treatment of social failure: A comparison of anxiety-reduction and skills-acquisition procedures of two social problems. *Behavior Modification,* 1978, *2,* 41–60.

Twentyman, C. T., Boland, T., & McFall, R. M. *Five studies exploring the problem of heterosocial avoidance in college males.* Unpublished manuscript, 1978.

Twentyman, C. T., & McFall, R. M. Behavioral training of social skills in shy males. *Journal of Consulting and Clinical Psychology,* 1975, *43,* 384–395.

Wallander, J. L., Conger, A. J., Mariotto, M. J., Curran, J. P., & Farrell, A. D. *An evaluation of selection instruments in the heterosexual-social anxiety paradigm.* Unpublished manuscript, 1978.

Walster, E. The effects of self esteem on liking for dates of various social desirabilities. *Journal of Experimental Social Psychology,* 1970, *6,* 248–253.

Walster, E., Aronson, V., Abraham, D., & Rottman, L. Importance of physical attractiveness in dating behavior. *Journal of Personality and Social Psychology,* 1966, *4,* 508–516.

Warren, N. J., & Gilner, F. H. Measurement of positive assertive behaviors: The behavioral test of tenderness expression. *Behavior Therapy,* 1978, *9,* 169–177.

Watson, D., & Friend, R. Measurement of social-evaluative anxiety. *Journal of Consulting and Clinical Psychology,* 1969, *33,* 448–457.

Weiss, R. L. Operant conditioning techniques in psychological assessment. In P. McReynolds (Ed.), *Advances in psychological assessment.* Palo Alto, Calif.: Science and Behavior, 1968.

Wessberg, H. W., Mariotto, M. J., Conger, A. J., Farrell, A. D., & Conger, J. C. *The ecological validity of role plays in the assessment of heterosexual-social skill and anxiety.* Unpublished manuscript, 1978.

Westefeld, J. S., Galassi, J. P., & Galassi, M. D. *Effects of instructional variations during role playing on the behavioral assessment of assertion.* Unpublished manuscript, 1978.

Williams, C. L., & Ciminero, A. R. *Further investigations of a new heterosocial skills inventory for females.* Paper presented at the annual meeting of the Association for the Advancement of Behavior Therapy, Atlanta, December 1977.

Williams, C. L., & Ciminero, A. R. Development and validation of a heterosocial skills inventory: The Survey of Heterosexual Interactions for Females. *Journal of Consulting and Clinical Psychology,* 1978, *46,* 1547–1548.

Wright, C. A comparison of systematic desensitization and social skill acquisition in the modification of a social fear. *Behavior Therapy,* 1976, *7,* 205–210.

Yost, E. J. B. The development and evaluation of a social skills training program for college males. (Doctoral dissertation, University of Oregon, 1973.) *Dissertation Abstracts International,* 1974, *34,* 5649A. (University Microfilms No. 74-6, 920)

Zeichner, A., Pihl, R. O., & Wright, J. C. A comparison between volunteer drug-abusers and non-drug-abusers on measures of social skills. *Journal of Clinical Psychology,* 1977, *33,* 585–590.

Zeichner, A., Wright, J. C., & Herman, S. Effects of situation on dating and assertive behavior. *Psychological Reports,* 1977, *40,* 375–381.

Zigler, E., & Phillips, L. Case history data and psychiatric diagnosis. Unpublished manuscript, 1959. Available from Worcester State Hospital, Worcester, Mass.

Zigler, E., & Phillips, L. Social competence and outcome in psychiatric disorders. *Journal of Abnormal and Social Psychology,* 1961, *63,* 264–271.

CHAPTER 6

Modification of Skill Deficits in Psychiatric Patients

Michel Hersen

INTRODUCTION

The relationship between social competence and psychiatric disorder is well documented in both the psychological and psychiatric literatures (cf. Bellack & Hersen, 1978; Hersen & Bellack, 1976b, 1977; Paul, 1969; Sylph, Ross, & Kedward, 1978). For close to two decades, Zigler and his colleagues (Levine & Zigler, 1973; Phillips & Zigler, 1961, 1964; Zigler & Levine, 1973; Zigler & Phillips, 1960, 1961, 1962) have completed a series of retrospective studies, essentially showing that level of premorbid social competence in a hospitalized psychiatric patient is the best predictor of his posthospital adjustment. This relationship seems to hold irrespective of the patient's diagnostic label (i.e., schizophrenia, alcoholism, depression) and regardless of the treatment regimen carried out during the course of hospitalization. Indeed, only recently has the relationship between social skill deficits and particular diagnostic entities been carefully evaluated. Among the disorders so considered are: schizophrenia (Bellack & Hersen, 1978; Bellack, Hersen, & Turner, 1976; Hersen & Bellack, 1976a), depression (Libet & Lewinsohn, 1973; Wells, Hersen, Bellack, & Himmelhoch, 1977), alcoholism (Miller & Eisler, 1977; O'Leary, & Donovan, 1976; Van Hasselt, Hersen, & Milliones, 1978), drug addiction (Callner & Ross, 1976; Van Hasselt *et al.*, 1978), hysterical neurosis (Blanchard & Hersen, 1976), and sexual deviation (Barlow, Abel, Blanchard, Bristow, & Young, 1977). Undoubtedly, as further research evidence is adduced, other disorders will be added to this growing list.

Michel Hersen • Department of Psychiatry, Western Psychiatric Institute and Clinic, University of Pittsburgh School of Medicine, Pittsburgh, Pennsylvania 15261.

Despite the fact that the precise relationship of skill deficits and many specific diagnostic groupings has not yet been identified, the social disability of chronic psychiatric patients is a well-established fact (cf. Bellack & Hersen, 1978; Hersen & Bellack, 1976a,b; Paul, 1969). Perhaps, however, the extent of the disabilities of such patients has been underestimated. In a recent study, using the Adaptive Behavior Scale as the primary assessment tool, Sylph *et al.* (1978) found that social skill deficits in 147 chronic psychiatric patients were greater and more extensive than in retardates who are institutionalized. More specifically, in this study it was found that:

> One third of the patients had distasteful table manners. Almost all could dress and undress, but without supervision many would attract unfavorable attention in the community. About half were unkempt, were inappropriately attired or groomed, or would go about dirty unless reminded to wash. A slightly smaller proportion could not take a bath unaided, and 30% were unable to wash their hands and face adequately. A quarter wet or soiled themselves at least occasionally, and about 20% needed help at the toilet. Over three-quarters could not manage money satisfactorily, 57% had difficulties with public transportation, and almost half were unable to use the telephone or find their way about outside the hospital.
>
> Impairment of work capacity was marked. Poor performance of domestic chores was universal, and only a third of the patients managed other types of work. Common types of work failure are suggested by the prevalence of slow and sluggish movements and lack of initiative, perseverance, and organizing capacity among 58%–75% and unreliability in carrying out assignments in 96%.
>
> Between 10% and 30% of the patients had defects of sight, hearing, mobility, or manual control, and the body balance of 79% was rated as impaired. Language and communication handicaps were common: 10% of the patients were mute or unintelligible and a third had blocked or unclear speech. Vocabulary and sentence construction were severely limited in 60%–75%. One-third could not understand complex instructions, and twice as many had difficulty reading. (pp. 1391–1392)

Many other skill deficits in this sample were identified by the investigators. Suffice it to say, however, the number of problems found was astoundingly high. Also, given that this sample was drawn from two independent mental institutions, it would appear that the survey adequately depicts the kind of patient typically found in large understaffed state and federal psychiatric institutions.

With the advent of behavior therapy in general and assertive training in particular (cf. Bellack & Hersen, 1977, chap. 5; Salter, 1949; Wolpe, 1973a; Wolpe & Lazarus, 1966), considerable attention has been focused on teaching skill-deficient patients the adaptive behaviors needed for successful extrahospital living. However, nonbehaviorists also have recognized the importance of adequate interpersonal functioning in their patients. For example, Gladwin (1967) and Frank (1974) implicitly and explicitly note that following effective psychotherapy a patient's social functioning should be markedly improved. But in spite of such recognition, it is only the behavior therapists who have precisely identified socially deficient behaviors (e.g., Eisler, Hersen, Miller, & Blanchard, 1975; Hersen, Bellack, & Turner, 1978) in their patients as well as developing systematic procedures for remediating such deficiencies (e.g., Bellack, Hersen, & Turner, 1976; Frederiksen & Eisler, 1977; Hersen & Bellack, 1976a; Hersen, Turner, Edelstein, & Pinkston, 1975; Liberman, Lillie, Falloon, Vaughn, Harpin, Leff, Hutchinson, Ryan, & Stoute, 1977; Wells *et al.*, 1977).

Our primary concern in this chapter will be with the modification of skill deficits in psychiatric patients. Many different kinds of psychiatric patients have received social skills training as the major treatment strategy or as one element in a more comprehensive behavioral approach. The case histories and single-case and group-comparison studies to be reviewed here have dealt with the following diagnostic categories: schizophrenia, depression, alcoholism, drug addiction, passive-aggressive personality, mental retardation, epilepsy, sexual deviation, self-mutilation, marital dysfunction, and hysterical neurosis.

The first section will be devoted to a discussion of social skills assessment with psychiatric patients. Although there are some common denominators to the assessment of social skill deficits in all populations (psychiatric or otherwise), to avoid needless duplication (see Chapter 3) we will not discuss in great detail the general issues facing the field. Rather, we will highlight the special assessment problems that are confronted on a day-to-day basis by the practitioner or researcher working in the psychiatric setting. Next, we will review a number of case histories where skills-training procedures were carried out. Then, we will examine single-case research in which social skill approaches have been evaluated. This is to be followed by a survey of the short-term treatment analogues that have appeared in the literature in the last decade or so. We will then proceed to a review of controlled outcome studies in which social skills treatment (administered individually or in group format) has been contrasted with other therapeutic modalities. Finally, we will discuss the current status of the art as applied to psychiatric populations, as well as indicating possible future directions.

ASSESSMENT

Assessment of social skill in psychiatric patients has followed a course that typifies the behavior therapy field in general. The majority of assessments involve tasks that range from analogues to observations of actual responses in the natural environment (see Table 1). Most of the analogues consist of role-playing of situations requiring assertive behavior (e.g., Eisler *et al.*, 1975; Goldsmith & McFall, 1975; Hersen, Bellack, & Turner, 1978). These analogue assessments are presumed to mirror how the patient would actually behave in the community if confronted with similar real-life situations. However, in at least one study (see Bellack, Hersen, & Turner, 1978) this assumption has been challenged. Thus, at this time the external validity of role play to ascertain level of assertion or social skill has yet to be established. Therefore, until further research on this topic emerges, this assessment strategy should neither be accepted on the basis of its face validity alone nor automatically discarded as useless (see Chapter 3).

It should be noted that many of the behavioral tasks used for assessing social skill were developed out of expediency rather than on the basis of sound psychometric principles. However, as noted in Chapter 3, this unfortunately is representative of a large proportion of the work done in the social skills area. On the other hand, the particular strategy developed by King, Liberman, Roberts, and Bryan (1977) for directly observing the patient in his/her natural environment (i.e., observing of patient completing preprogrammed homework assignments such as bargaining with a salesperson about price in a department store) definitely represents a major stride forward as to technology. Of course, the problem with this kind of assessment is that the observer serves as an ever-present stimulus for the targeted behaviors to occur (i.e., reactivity to being observed) (Hersen, 1977). The question still remains as to whether or not such behaviors would be seen in the absence of the observer.

One way out of this dilemma is to observe the patient's behavior in an unobtrusive fashion. However, serious ethical objections have been raised and undoubtedly would result from this approach. Perhaps, then, the patient might be informed right from the inception of treatment that his/her behavior may or may not be monitored in the natural environment. Although this certainly does establish an expectation on the part of the patient, the unknown feature (as to where and when) might circumvent excessive reactivity to being observed. Moreover, such observation would not take place unless the patient consented to the strategy in the first place. Thus, the surreptitious aspects of being so observed would be greatly diminished.

A second major assessment strategy has involved construction of self-report scales and inventories (see Table 2). However, here too, with few exceptions, the psychometric properties of the scales are not well researched. A recent exception is the study by Hersen, Bellack, Turner,

Table 1. Behavioral Tasks for Psychiatric Populations

Reference	Task	Number of items	Ratings
Argyle *et al.* (1974)	Videotaped conversations	Information conversation with male and female	17 verbal and nonverbal measures
Barlow *et al.* (1977)	Videotaped conversations	Structured conversation with female	11 verbal and nonverbal measures
Eisler *et al.* (1973)	Behavioral Assertiveness Test	Responses to 14 role-played situations	10 verbal and nonverbal measures
Eisler *et al.* (1975) Hersen *et al.* (1978a)	Behavioral Assertiveness Test—Revised	Responses to 32 role-played situations	12 verbal and nonverbal measures
Frederiksen *et al.* (1976)	On-Ward Situations	3 contrived situations to elicit assertive or aggressive responses	1–5 point ratings
Goldsmith & McFall (1975)	Interpersonal Behavior Role-Playing Test	Responses to 25 role-played situations	0–2 point ratings per situation on competence
Goldstein *et al.* (1973)	Tape-recorded Interpersonal Situations	Responses to 50 audiotaped situations	5-point rating scale for independence-dependence
Gutride *et al.* (1973)	Interaction with experimental "accomplice"	Responses to 6 questions posed by experimental confederate	20 ratings on a social interaction checklist
King *et al.* (1977)	Homework assignments	50 assignments in the community	Percentage successfully completed
Weinman *et al.* (1972)	Behavior in critical situations	Responses to four contrived situations	2- to 5-point rating scales on the four critical situations

Table 2. Self-Report Measures for Psychiatric Populations

Reference	Scale	Number of items	Format	Normative population
Callner & Ross (1976)	Assertion Questionnaire	40	1- to 4-point ratings	Male addicts and male veterans
Eisler *et al.* (1978)	Social Alternatives Test	8	Assertive, aggressive, passive	Psychiatric patients
Goldsmith & McFall (1975)	Interpersonal Situation Inventory	55	1- to 5-point ratings	Male psychiatric patients
Lazarus (1971)	Assertive Questionnaire	20	Yes–no	None originally reported but used clinically with psychiatric patients
Wolpe & Lazarus (1966)	Assertiveness Scale	30	Yes–no	None originally reported; male and female psychiatric patients (Hersen, Bellack, & Turner, 1978)

Harper, Watts, and Williams (1979), in which the internal consistency, reliability, and validity of the Wolpe–Lazarus Assertiveness Scale were examined.

A third strategy of assessment has involved evaluation of the patient's physiological reactivity before, during, and after role-playing interpersonal situations calling for expression of positive and negative assertion (e.g., Hersen, Bellack, & Turner, 1978). However, there definitely are many problems when measuring physiological reactivity, not the least of which is that different patients respond to stress with totally distinct physiological patterns (i.e., some are heart rate, respiration rate, pulse rate, or GSR responders, etc.). Therefore, unless the investigator is able to carry out a multichannel assessment, resulting data will be misleading and actually may yield statistical insignificance. It is in part for these reasons that some workers in the field (e.g., Eisler, 1976) have discouraged the use of physiological measurement in social skills research.

Aside from the aforementioned, there are some particular assessment problems that are prevalent in the psychiatric setting (i.e., the inpatient service) when making social skill determinations. The first of these concerns *drugs*. That is, most psychiatric inpatients are prescribed pharmacological agents by their physicians, especially to control psychotic or depressive symptomatology. Indeed, many psychiatric outpatients also are prescribed tranquilizers (major and minor) as well as antidepressants and soporifics. The problem for social skill assessors is not that drugs *per se* are being administered. Rather, the primary issue is that during the inpatient or outpatient course of psychiatric management the dosage and combination of drugs often are changed. If repeated social skill assessments are taken, either in single-case or group-comparison studies (see Hersen & Barlow, 1976), a confound is obviously introduced. Thus, any modification in the drug regime taking place during the phase of social skills training will not allow for an unequivocal interpretation as to the specific effects of that behavioral procedure.

Of course, several solutions to the above do present themselves. One is to delay social skill assessment and treatment procedures until the patient's medication has been fully regulated. This strategy has been rigorously adhered to in a number of single-case experimental anaylses (cf. Bellack, Hersen, & Turner, 1976; Hersen & Bellack, 1976a; Hersen *et al.*, 1975). Nevertheless, even if social skill assessment and treatment are purposely delayed, exigencies of psychiatric treatment may, at times, require a medication change during the course of the behavioral intervention. That being the case, the investigator then has an opportunity to evaluate the effects of medication change by systematically withdrawing and reintroducing the drug change (cf. Liberman & Davis, 1975; Stern,

1978; Turner, Hersen, & Bellack, 1978) while concurrently monitoring social skill responses.

Also with regard to drugs and the assessment of social skill, it should be underscored that, in the absence of pharmacological management, many acutely disturbed psychiatric patients simply would be too psychotic to attend to the requirements of a social skill assessment task (whether it be of the behavioral or self-report variety). Moreover, in the case of schizophrenia, there is an ample literature documenting that unless relatively symptom free (i.e., the absence of delusions, hallucinations, and thought disturbances), such patients will not be able to learn new material and retain it (cf. Hersen & Bellack, 1976b). In that sense, then, administration of drugs and social skills treatment may have a synergistic effect, with drugs controlling psychotic symptoms and the behavioral approach teaching the patient effective coping strategies.

The second of the assessment issues peculiar to psychiatric patients relates to *ward observations* of their behavior. In several studies, the effects of individually administered training to severely disturbed and aggressive psychiatric patients have been evaluated by systematically observing their interpersonal behavior in ward situations (Elder, Edelstein, & Narick, in press; Frederiksen & Eisler, 1977; Frederiksen, Jenkins, Foy, & Eisler, 1976; Matson & Stephens, 1978). Aside from some of the ethical dilemmas posed by staging incidents on the ward that require the expression of assertion (negative or positive) in otherwise aggressive patients, a considerable amount of administrative control and available personnel are needed to carry out such assessments successfully. As argued elsewhere by Hersen and Bellack (1978), in the absence of necessary control over the ward environment, many behavioral assessment and treatment possibilities remain unfulfilled. Therefore, although we label this problem as one of assessment, in practice it is representative of a larger administrative-political issue. That is, who controls what and where at which particular time on a given psychiatric ward?

CASE HISTORIES

An inspection of Table 3 clearly indicates that case histories detailing the use of social skill approaches with psychiatric patients abound in the literature. In some of the reports, theoretical and clinical considerations are fully discussed (e.g., Edwards, 1972; Wolpe, 1973a, b); in others, skill approaches are used in innovative fashion with some hard data reported (e.g., Laws & Serber, 1971; Wells *et al.*, 1977); and in still others, there appears to be greater concern for assessing maintenance of therapeutic gains posttreatment (cf. Blanchard & Hersen, 1976; Eisler &

Table 3. Case Histories Involving Skills-Training Procedures

Reference	Cases	Treatments	Outcome	Follow-ups
Barnard *et al.* (1966)	1 female depressive with urinary retention	Faradic aversion, assertive training	Symptom removal	18 months
Blanchard & Hersen (1976)	4 hysterical neurotics	Differential attention, extinction, behavioral family therapy, assertive training	Successful in all cases, minimal recurrence of symptoms	4, 4, 3, and 12 months
Bloomfield (1973)	Group of chronic schizophrenics	Group assertive training	Not indicated	Not indicated
Edwards (1972)	Homosexual pedophile	Thought stopping, assertive training	Improvement of sexual pathology	4 months
Eisler & Hersen (1973)	3 families in crisis	Behavioral contracting, communications training	Improved interpersonal functioning	3, 0, and 3 months
Eisler *et al.* (1974a)	3 unassertive psychiatric patients	Assertive training	Improvement on targeted behaviors and marital interaction	0, 0, and 6 weeks
Fensterheim (1972a)	3 socially isolated males	Group assertive training	Moderate improvement	None reported
Fensterheim (1972b)	1 dysfunctional couple	Communication and assertive training	Improved marital and sexual relationship	1 year
Goldstein *et al.* (1970)	3 fearful outpatients	Induced anger	Symptomatic improvement	6, 0, and 6 months

(*Continued*)

Table 3 (*Continued*)

Reference	Cases	Treatments	Outcome	Follow-ups
Hersen & Luber (1977)	100 chronic day hospital patients	Group social skills	Improved target behavior	None reported
Hersen & Miller (1976)	2 unipolar depressives	Assertive training	Decrease in depression	5 and 5 months
Katz (1971)	1 chronically passive male	Assertive training in a hierarchical presentation, modeling	Decrease in passivity	Treatment not complete at time of report
Laws & Serber (1971)	1 homosexual pedophile	Assertive training	Improvement in targeted behavior	None reported
Liberman *et al.* (1976)	1 chronic psychiatric inpatient	Inpatient and day hospital skills training	Improvements in targeted behaviors and social functioning	None reported
Luber *et al.* (1979)	2 epileptics	Day hospital and individual skills training	Improved social functioning	None reported
Macpherson (1972)	1 female anxiety-hysteric	Differential reinforcement of assertive behavior	Symptomatic improvement	2 years
Patterson (1972)	1 dependent child with crying spells	Time-out, assertive training	Symptomatic improvement	9 months
Piaget & Lazarus (1969)	1 chronically anxious woman, particularly in crowds	Rehearsal-desensitization	Improved assertiveness and interpersonal functioning	6 months
Rimm (1967)	1 male schizophrenic with chronic crying spells	Assertive training	Symptomatic improvement	10 months

Roback *et al.* (1972)	1 self-mutilating female patient	Chemotherapy, emotional labeling, modeling, role playing, assertive training, psychodrama, group psychotherapy, contingency management	Symptomatic improvement	4 months
Wallace *et al.* (1973)	1 aggressive, organically impaired male	Contingency management, assertive training	Symptomatic improvement	9 months
Wells *et al.* (1977)	4 unipolar depressive females	Social skills training	Improved assertion, diminished depression	0, 0, 3, and 3 months
Wolpe (1973a)	1 unassertive male and 1 unassertive female	Assertive training	Not indicated (supervision transcript)	Not indicated (supervision transcript)
Wolpe (1973b)	1 unassertive male	Assertive training	Not indicated (supervision transcript)	Not indicated (supervision transcript)
Wolpe & Lazarus (1966)	1 depressed passive male, 1 depressed female with hyperventilation syndrome, asthmatic	Systematic desensitization, assertive training	Symptomatic improvement and increased assertiveness	5 years, 1 year, and 18 months
Wood *et al.* (1975)	17 female chronic psychiatric patients	Assertive training	72% of goal attainment	8 months

Hersen, 1973; Fensterheim, 1972b; Rimm, 1967; Wallace, Teigen, Liberman, & Baker, 1973). Examination of Table 3 also indicates the wide divergence of patient populations seen, methods of treatment used, and differing outcomes resulting from the particular skills-training procedures or combinations of treatment procedures carried out. Given such wide divergence, it obviously is very difficult to reach any definitive conclusions on the basis of one clinical case or even as a function of a series of cases.

In spite of the limitations of the clinical single-case approach,[1] the importance of such reports for the researcher should never be discounted. That is, the single-case method often provides a rich source of hypotheses that later can be subjected to more rigorous empirical verification by enterprising researchers (see Lazarus & Davison, 1971). For example, on the basis of their clinical work with hysterical neurotics, Blanchard and Hersen (1976) describe a tripartite treatment model for this diagnostic group. Included are: (1) extinction procedures during the inpatient phase of decrease rate of hysterical symptom presentation, (2) restructuring of the natural environment to reinforce the patient's behavior differentially, and (3) instruction of the patient, using a social skills approach, to enable him to obtain greater gratification from the environment. Although as yet untested, a group-comparison research design could easily be designed to evaluate, say, the contribution of social skill training to the overall symptomatic remission and maintenance of treatment in hysterical neurotics.

The suggestion that assertive training in schizophrenics might improve their interpersonal skills, thus protecting them from relapse (cf. Bloomfield, 1973; Hersen & Luber, 1977; Liberman *et al.*, 1976; Rimm, 1967), is now being examined more systematically in two, large, funded grants from the National Institute of Mental Health (Hogarty, 1978; Liberman, Vaughn, Aitchison, & Falloon, 1977). Similarly, very moderate success in treating two unipolar depressives with assertive training procedures (Hersen & Miller, 1976) led to the development of a more comprehensive research approach (cf. Wells *et al.*, 1977) for evaluating both initial and maintenance social skill training for such patients. Indeed, this research project is also in progress and funded by the National Institute of Mental Health (Hersen, Himmelhoch, & Bellack, 1978).

Let us now consider two of the case studies that were presented in the literature, and then consider in greater detail the skills approach followed by Wells *et al.* (1977) in the treatment of depressed women.

[1] This is in contradistinction to the single-case experimental design approach (cf. Hersen & Barlow, 1976), where the controlling effects of therapeutic variables may be documented.

Eisler, Miller, Hersen, and Alford (1974) describe the short-term treatment of a 52-year-old married male who was a manager of a large automotive service station. He had a 6-year drinking history, with alcoholic consumption being especially heavy following periods of marital discord. Marital arguments centered around three issues: (1) discipline of the 20-year-old retarded daughter, (2) time spent with the wife after work hours, and (3) questions about his social drinking (i.e., a few beers) in the evening. The patient and his wife were initially videotaped for 20 min while discussing their marital problems. Based on retrospective ratings of this videotape, several of the husband's deficit behaviors were targeted for modification during assertive training: (1) eye contact, (2) speech duration, (3) latency of response, and (4) requests for his wife to change her behavior. In addition to the above, 6 weeks prior to receiving assertive training the patient's blood/alcohol levels (obtained in the natural environment) were taken weekly.

Using 10 simulated marital encounters reflecting the typical interchanges occurring between the patient and his wife, assertive training was conducted in four 45-min sessions. A female research technician enacted the part of the wife during role-played scenes. An example of one of the 10 scenes follows:

> You've just come home from work about an hour later than usual. You had to stay later than usual to finish a difficult job. As soon as you come in the door, your wife rushed up to you very upset. She says, "Don't tell me you've been working all this time. I know you've been drinking again." (Eisler *et al.*, 1974, pp. 644–647)

Assertive training for the husband consisted of instructions, behavior rehearsal, and performance feedback. Subsequent to training, the husband and his wife were reevaluated by having them once again discuss the marital situation while being videotaped. Analysis of the posttreatment videotape not only indicated positive changes in all of the specific behaviors targeted for change in the husband, but there were reciprocal changes in some of the specific behaviors seen in the wife as well. This took place despite the fact that the wife did not receive any form of treatment during this time period. Moreover, weekly blood/alcohol levels taken for 6 weeks posttreatment showed marked decreases when contrasted to pretreatment levels. Thus, treatment in this case resulted in improved assertiveness for the patient, a concomitant improvement in the marital interaction, and a substantial decrease in drinking.

Wallace, *et al.* (1973) describe the behavioral treatment of a 22-year-old male inpatient who had a long history of violent behavior. Treatment lasted 37 days and involved the concurrent administration of contingency contracting and assertive training. Contingency contracting was introduced to reduce episodes of aggressive behavior on the ward. Specifically, a signed contract (between the patient and the inpatient unit staff) was drawn up, stating that home visits would be contingent on the total absence of assaultive behavior for the prior 7 days. In addition to suppressing assaultive behavior by contingency management, assertive training was instituted in order to teach the patient more appropriate ways of interacting with nursing staff and other patients. Twenty-five scenes comprising four specific hierarchies were constructed. Thus, treatment was conducted in graduated fashion, with each scene in each hierarchy representing "ascending potential frustration" for this patient. Following 37 days of such treatment the patient was discharged from the hospital. A 9-month follow-up revealed that only one aggressive incident had been recorded. In summary, a twofold treatment approach (one resulting in the extinction of aggressive behavior and one resulting in the increase in appropriate behavior) proved beneficial for this patient whose aggressive disorder dated some 3 years.

Treatment Protocol

We will examine the social skills treatment protocol for depressed women followed by Wells *et al.* (1977) that currently is being investigated on a larger scale by Hersen, Himmelhoch, and Bellack (1978).

Initial Treatment. Social skills training consists of 12 1-hour sessions held over a 12-week time period. The treatment procedure consists of four distinct elements: skills training *per se,* social perception training, practice, and self-evaluation and self-reinforcement.

Skills training first entails assessment of the particular skill deficits of the depressed patient. Based on review of the videotapes of an initial behavioral assessment (Behavioral Assertiveness Test–Revised; Hersen *et al.*, 1978a) and a therapist-conducted interview, specific verbal and nonverbal deficiencies are identified in each of four content areas: family interactions (or heterosocial and/or parent interactions for unmarried patients), work interactions, interactions with friends, and interactions with strangers. A minihierarchy of four to eight situations is developed for each of the relevant categories. The situations represent interpersonal interactions in which the patient did or is likely to encounter difficulty in expressing feelings and receiving reinforcement. For example, work-related situations might involve responding to undue criticism, asking for a day off, or requesting a raise. A family situation might entail exerting control over children or expressing annoyance to a spouse who neglects to share a responsibility. Scenes are selected to allow for the expression of positive feelings as well as negative feelings. The hierarchical arrangement reflects the degree of difficulty the patient experiences in each situation. In addition, the four content areas are hierarchically arranged as well.

Training is conducted for each specific behavior, in each situation, in each content area in order of increasing difficulty. Skills training consists of instructions, feedback, role-playing, modeling, and positive reinforcement. The therapist first instructs the patient in the appropriate manner of performing the first target response in the first situation (i.e., the least difficult situation in the least difficult content area). For example, "We are going to begin with the situation in which you tell a salesperson at a crowded counter that you are next. It is important that you speak in a loud voice if you are to make your point. Try to speak out loudly." The therapist and patient then role-play the situation. The patient is instructed to respond as if he/she is really in the situation and the therapist provides an initiating prompt (e.g., "Who is next?").

Following the patient's response, the therapist provides feedback which emphasizes the positive aspects of the response (positive reinforcement) and suggests needed modifications. If instructions and feedback are not sufficient, the therapist then models appropriate responses (often this is required with the exception of the simplest responses that occur low on the hierarchy). This sequence is repeated until the patient reaches a clinically determined criterion of adequacy, at which time training shifts to a different target behavior. When the response to the situation is adequate in all respects, training focuses on the next situation in the hierarchy. Training in subsequent areas focuses more on verbal aspects and the more difficult nonverbal components such as intonation and speech rate. Treatment, of course, proceeds at a rate determined by the patient's unique improvement.

Social perception training assumes that for effective interpersonal functioning to occur the ability accurately to perceive feelings and intentions of others as well as the ability to make appropriate responses are required. Training in social perception is incorporated in the modeling component of social skills training. As previously suggested, there is an expectation that the patient will gain some competency by the conclusion of the first content area. From that point on, the patient is requested to evaluate and interpret the therapist's modeled responses. This involves periodic modeling of poor responses as well as effective responses. The therapist also begins to react (via role-playing) to the patient's role-played responses; the patient is required to interpret such reactions and to determine what is being communicated and what should follow.

The gains made in sessions have a low probability of being maintained if the patient fails to *practice* the newly acquired responses in the natural environment. Homework assignments to insure practice, therefore, are an integral part of the treatment. At each session the patient is specifically directed to perform a designated number of responses for the succeeding week. Particular responses are identified so as to insure the likelihood of success, as well as increasing the probability that they will be practiced.

Inasmuch as depressed individuals have been shown to make inordinately negative evaluations of their behavior (cf. Fuchs & Rehm, 1977; Rehm, 1977), it is conceivable that they might well fail to perceive environmental reinforcement of their homework performance. Or they may fail to judge such performance as adequate. Thus, patients are trained to evaluate their responses more objectively (i.e., correct *self-evaluation*) and to use self-reinforcement (cf. Bellack, 1976; Bellack, Glanz, & Simon, 1976; Fuchs & Rehm, 1977).

Beginning with the second content area, patients are requested to evaluate each of their role-played responses by assigning a letter grade (i.e., A, B, C, D, F). Grades of A or B are to be followed by emission of a positive self-statement. The therapist corrects any inappropriately low evaluations and requests that positive self-statements be made even for corrected evaluations. Patients similarly are requested to perform self-evaluative and self-reinforcing operations for homework responses. Thus, they are asked to keep a self-monitoring diary of their homework performance, which is reviewed by the therapist at each session.

Maintenance Treatment. During a 6-month maintenance phase, social skills training is primarily directed toward reinforcing and consolidating improvements obtained during initial treatment. Thus, maintenance can alternatively be conceived as "booster" sessions. Generally, six to eight maintenance sessions are given, with treatment strategies paralleling those administered in the initial phase.

SINGLE-CASE EXPERIMENTAL DESIGNS

In the single-case experimental designs, the controlling effects of social skill procedures for specified target behaviors have been documented for a number of diagnostic categories. Skills approaches evaluated in these experimental analyses have primarily been applied to schizophrenics, alcoholics, and patients evidencing aggressive and explosive behaviors. In some of the studies, only the controlling effects of treatment on targeted behaviors are shown. In others, there appears to be evidence that generalization from trained to untrained items takes place. And, more important, in still other studies there is documentation that skills training carries over to the natural environment.

In almost all of the studies to be reviewed, the multiple-baseline design across behaviors, patients, or situations have been employed (see Hersen & Barlow, 1976, Chapter 7, for a complete description of this design). This design strategy is especially useful for the study of skills approaches in psychiatric populations for a number of reasons. *First,* inasmuch as most psychiatric patients are deficient in several target behaviors, the multiple-baseline design across behaviors allows for the systematic treatment of each deficiency in sequential and cumulative fashion. *Second,* for schizophrenic patients in particular, the treatment of one behavior at a time (until mastered to criterion level) counteracts their pronounced difficulty in attending to compound stimuli and their general inability to process new information with success. *Third,* the multiple-baseline design across behaviors permits a clear demonstration that the treatment is only effective when it is directly applied to the targeted behavior or that generalization to different behaviors and novel situational contexts is possible. *Fourth,* in the case of the multiple-baseline design across situations, the strategy indicates whether treatment is situation specific or whether the possibility exists for generalized training to a number of situational contexts.

Schizophrenia

Hersen *et al.* (1975) report one of the first single-case analyses in which social skills training procedures were applied to a chronic schizophrenic patient. The patient was a 27-year-old black male (diagnosed schizophrenic, catatonic type) who had a long history of withdrawn and seclusive behavior. For 3 years prior to the current inpatient hospitalization he had been seen in "supportive" psychotherapy and was maintained on phenothiazines. However, his behavior had deteriorated to the point where his mother initiated hospitalization on the psychiatric service in a Veterans Administration hospital. For the first 7–8 weeks of this hospitalization the patient's medication was regulated, and an individualized token economy program was instituted to improve his physical hygiene. Only then was social skills training begun, following a behavioral analysis of skill deficits based on retrospective videotape ratings of the patient's responses to Behavioral Assertiveness Test (BAT) (Eisler, Miller, & Hersen, 1973) scenes.

Using four scenes from the BAT as a training vehicle, social skills treatment (instructions, behavior rehearsal, feedback, and modeling) was applied in a multiple-baseline design in sequential fashion to eye contact, response latency, and requests for new behavior. Three other behaviors were monitored concurrently (overall assertiveness, voice trails, speech disruptions), but remained untreated. In addition, the patient's verbal initiations were recorded in group psychotherapy sessions.

Social skills treatment required a total of 5 weeks, with three sessions administered per week. Probe sessions conducted throughout treatment revealed substantial improvements in eye contact and response latency, with correlated improvements noted for requests when latency was specifically treated. Concurrent improvements in the three untreated behaviors (i.e., overall assertiveness, voice trails, speech disruptions) appeared, as well as increased verbal initiation in group psychotherapy sessions. However, despite the patient's generalized improvement, the experimental data were not at all clear-cut. Following social skills training the patient was sufficiently improved clinically to warrant vocational counseling and subsequent job placement. At the 22-week postdischarge follow-up the patient had a steady job in the community, an accomplishment that he had been unable to attain in the prior 3 years.

In a very similar social skills training approach with two chronic schizophrenics, Hersen and Bellack (1976a) demonstrated more convincingly the controlling effects of the treatment over targeted behaviors. In one case treatment consisted of instructions and feedback; in the second case instructions, feedback, and modeling were used. Again, targeted behaviors were treated sequentially and cumulatively in multiple-baseline fashion, using Behavioral Assertiveness Test–Revised (BAT-R) scenes (Eisler *et al.*, 1975) as both the assessment and training vehicle. Also, training was directed toward improving both positive and negative assertiveness in these patients. In each case training began after phenothiazines were regulated, with drug dosage maintained at a constant level in baseline, training, and follow-up phases. As can be seen from inspection of Figures 1 and 2, the results were positive for both patients. Not only did specific targets improve when treatment was specifically directed toward them, but gradual gains in overall assertiveness were recorded by independent observers. In addition, treatment gains seemed to hold at the 2-, 4-, 6-, and 8-week posttreatment assessments.

In a continuation of their research program, Bellack, Hersen, and Turner (1976) examined generalization effects of social skills training in three chronic schizophrenics (one male and two females). Skills training here involved instructions, feedback, and modeling. There were 25, 26, and 31 sessions of treatment for the three patients. Eight scenes from the BAT-R were used to assess behavioral deficiencies. From this analysis, five to seven target behaviors were identified for each patient. Training

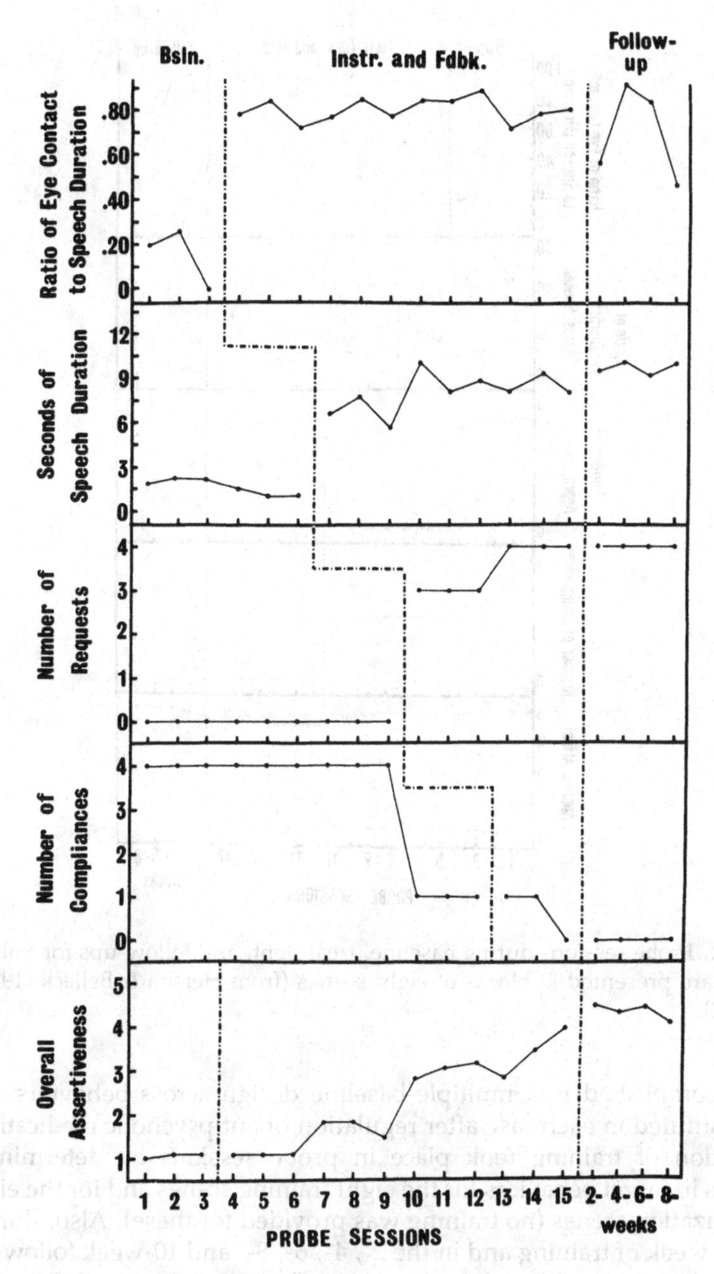

Figure 1. Probe sessions during baseline, treatment, and follow-up for Subject 1. Data are presented in blocks of eight scenes (from Hersen & Bellack, 1976a, Figure 1).

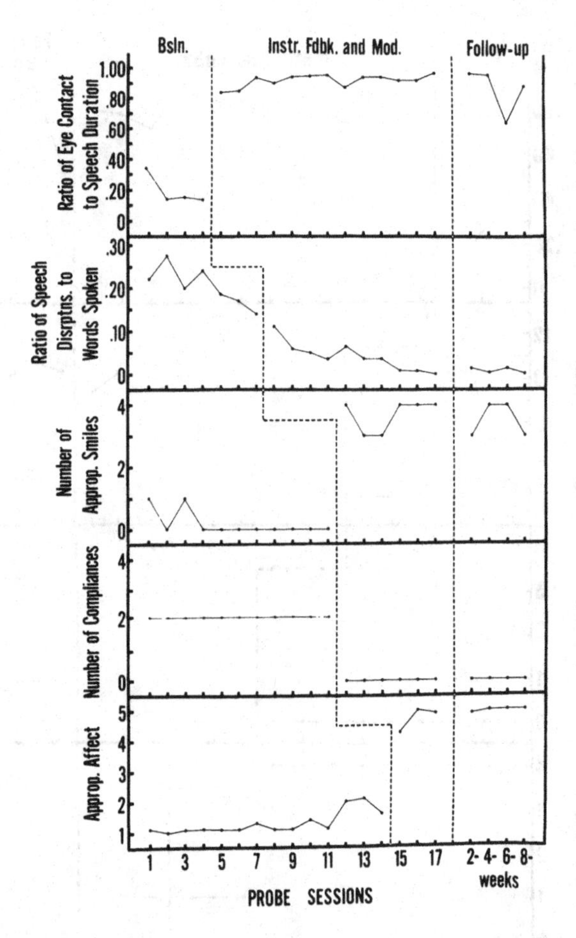

Figure 2. Probe sessions during baseline, treatment, and follow-ups for Subject 2. Data are presented in blocks of eight scenes (from Hersen & Bellack, 1976a, Figure 2).

was accomplished in a multiple-baseline design across behaviors and was instituted in each case after regulation of antipsychotic medication. Evaluation of training took place in probe sessions by determining changes in target behaviors for the eight training scenes and for the eight generalization scenes (no training was provided for these). Also, during the last week of training and in the 2-, 4-, 6-, 8-, and 10-week follow-up probes, a third set of eight scenes (novel scenes) was introduced to assess further the effects of generalization.

In general, the results showed that social skills training was highly successful for the two female schizophrenics but only partially successful for the male schizophrenic. In the two female patients there was substantial improvement in targeted behaviors as evidenced in the training, generalization, and novel scenes (see Figure 3, 25-year-old female outpatient schizophrenic, twice divorced, and enrolled in a partial hospitalization program). Examination of Figure 3 shows that baselines were independent and that when treatment was applied to each target, improvements occurred (an exception is ratio of words spoken to speech duration in which no treatment effects appeared). Overall assertiveness, rated independently, tended to increase gradually as treatment progressed. Follow-up data indicate that treatment gains were maintained.

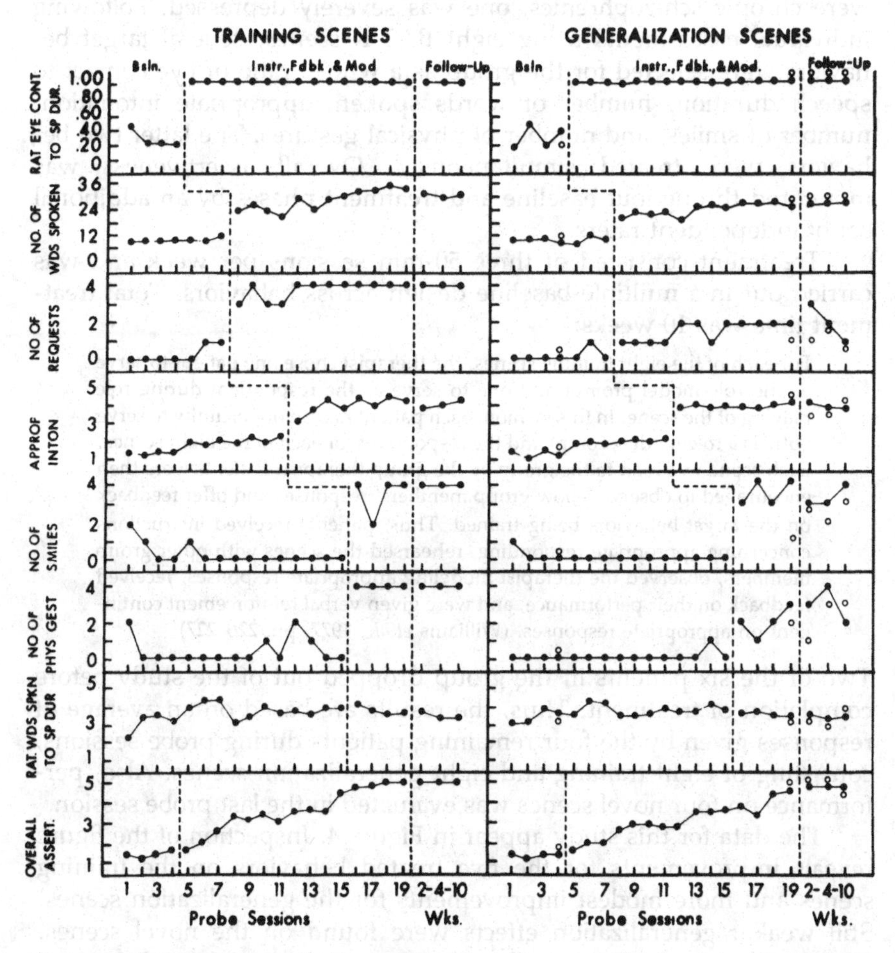

Figure 3. Probe sessions during baseline, treatment, and follow-ups for Subject 3 (from Bellack *et al.*, 1976b, Figure 3).

Data for generalization scenes parallel those for the training scenes. However, treatment gains did not hold up as well for number of requests in follow-up. Also, evidence for generalization was not as great for the novel scenes.

Now that the social skills approach developed by Hersen, Bellack, and Turner and their colleagues appeared to effect changes in chronic schizophrenics when applied on an individual basis, we decided to evaluate the value of such training when conducted in a group format (Williams, Turner, Watts, Bellack, & Hersen, 1977). Initially, six of the most severely debilitated patients from our partial hospitalization service were selected for group social skills training. Five of the six patients were chronic schizophrenics; one was severely depressed. Following individual assessments using eight BAT-R scenes, several target behaviors were selected for the group as a whole: ratio of eye contact to speech duration, number of words spoken, appropriate intonation, number of smiles, and number of physical gestures. The latter two behaviors were treated simultaneously. Overall assertiveness was monitored throughout baseline and treatment phases by an additional set of independent raters.

Treatment consisted of three 50-min sessions per week and was carried out in a multiple-baseline design across behaviors. Total treatment time was 10 weeks:

> For each of the eight training scenes, the therapist chose one patient to serve as the role model prompt and one to serve as the respondent during role playing of the scene. In this fashion, each patient had an opportunity to serve both as a role model prompt and the respondent for each of the eight scenes. Each scene was read to the group by the group therapist. Patients were then encouraged to observe fellow group members' responses and offer feedback on the target behaviors being trained. Thus, patients received instructions concerning appropriate responding, rehearsed the scenes with other group members, observed the therapist modeling appropriate responses, received feedback on their performance, and were given verbal reinforcement contingent on appropriate responses. (Williams *et al.*, 1977, pp. 226–227)

Two of the six patients in the group dropped out of the study before completion of treatment. Thus, the results are based on an average of responses given by the four remaining patients during probe sessions, consisting of eight training and eight generalization scenes. Also, performance on four novel scenes was evaluated in the last probe session.

The data for this study appear in Figure 4. Inspection of the figure reveals improvements for the five treated behaviors on the training scenes and more modest improvements for the generalization scenes. Still weaker generalization effects were found on the novel scenes. Slight improvements in overall assertiveness appeared for training and generalization scenes.

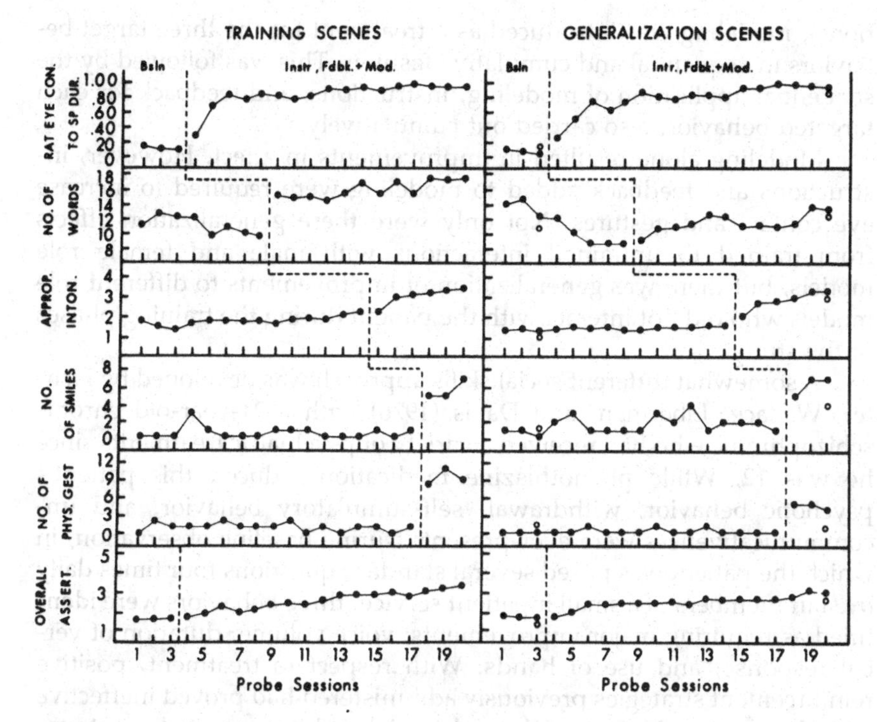

Figure 4. Group data for training, generalization, and novel scenes. Performance on novel scenes is indicated by the open circle (from Williams *et al.*, 1977, Figure 4).

When contrasted to application of social skill procedures at the individual level (cf. Bellack *et al.*, 1976b; Hersen & Bellack, 1976a; Hersen *et al.*, 1975), the results of Williams *et al.* (1977) are somewhat less striking. However, from a cost-effectiveness analysis standpoint, the group-training format for these four patients represents savings of 70 treatment hours when contrasted to the time requirement for individual treatment of four such patients on the same five targeted behaviors.

Edelstein and Eisler (1976) were concerned with the comparative effects of modeling alone versus modeling, instructions, and feedback on three component behaviors in a 32-year-old paranoid schizophrenic who was a social isolate on a psychiatric ward. Despite the fact that administration of phenothiazines led to a remission of psychotic symptoms, the patient's social withdrawal persisted. Using a complex multiple-baseline design across behaviors, three targets were selected for modification (eye contact, gestures, affect). Following baseline assessment (consisting of retrospective videotape ratings of the patient's interactions with male and female role models in interpersonal situations), modeling was introduced as a treatment for the three target behaviors in sequential and cumulative fashion. This was followed by the sequential application of modeling, instructions, and feedback for each targeted behavior, also carried out cumulatively.

Modeling alone resulted in improvements in affect. However, instructions and feedback added to modeling were required to increase eye contact and gestures. Not only were there generalization effects from trained to untrained interactions with male and female role models, but there was generalization of improvements to different role models who did not interact with the patient during the training phases of the study.

A somewhat different social skills approach was developed by Fichter, Wallace, Liberman, and Davis (1976) with a 21-year-old chronic schizophrenic who had received a variety of psychiatric treatments since he was 12. While phenothiazine medication reduced this patient's psychotic behavior, withdrawal, self-stimulatory behavior, and uncommunicativeness were ever-present. During baseline observation, in which the patient was posed several standard questions four times daily by staff members of a small inpatient service, three behaviors were identified as requiring major improvements: voice volume, duration of verbal response, and use of hands. With respect to treatment, positive reinforcement strategies previously administered had proved ineffective with this patient. However, through a trial-and-error procedure, it was discovered that "nagging" (i.e., verbal prompts for each behavior until it met a preset criterion) was a potent treatment. That is, when the patient improved a given behavior (e.g., voice volume), only then would the staff person stop "nagging" (i.e., using repeated prompts). Thus, the effects of "nagging" were evaluated in a combined multiple-baseline design across behaviors and withdrawal design.

The sequential application of "nagging" for each behavior, followed by withdrawal and reinstitution of the contingency, demonstrated the controlling effects of the treatment for this patient. Subsequent to treatment the patient was discharged from the inpatient service and followed in a day treatment center and a residential care home. Improvements in two of the targets (voice volume, duration) were maintained in these new settings. However, there appeared to be a renewal of the inappropriate use of hands in these new environments. Fichter *et al.* (1976) conclude that "given some degree of compliance a 'nagging' procedure may be useful for establishing appropriate social and independent behavior in patients who are unresponsive to the more 'typical' positive reinforcement contingencies" (p. 385).

Probably the most ambitious applied behavioral analysis as yet conducted with schizophrenic patients is that of Liberman *et al.* (1977a). In this study three chronic schizophrenics at high risk for relapse were selected for social skills training. First, in multiple-baseline format (i.e., multiple-baseline design across behaviors), several deficit behaviors were specifically targeted for modification, sequentially and cumulatively. Once these specific behaviors reached acceptable levels, patients were asked to carry out interpersonal assignments in several different settings (hospital, family, community). This part of the study was evaluated with a multiple-baseline design across settings. That is, the training of patients to carry out assignments in the three independent settings was done in sequence and cumulatively.

The results of this multiple-baseline analysis within yet another multiple-baseline design indicates that after specific behaviors improved, then each of the patients was able to carry out assignments in each of the designated settings. That is, the controlling effects of the treatment were documented both for specific behaviors and for performance in specific settings. The authors argue that their results suggest "that high risk schizophrenic patients can be taught to function with a greater degree of social competence" (Liberman *et al.*, 1977).

As part of the training to perform interpersonal assignments in the natural environment, self-reinforcement strategies were taught to the patients. However, following successful performance of these interpersonal tasks, the patients failed to continue using self-reinforcement strategies. Also, there were very few attempts by patients to complete new interpersonal assignments at their own initiative. Although there was good evidence of generalization during the inpatient training phase, when patients were discharged and training sessions diminished in frequency, levels of social competence tended to return to baseline levels.

Alcoholism

There are relatively few single-case experimental designs in which social skill procedures with alcoholics have been evaluated (cf. Eisler, Hersen, & Miller, 1974; Foy, Miller, Eisler, & O'Toole, 1976; Martorano, 1974). This is somewhat surprising in light of the fact that alcoholics (as opposed to most social drinkers) tend to increase their consumption following interpersonal stress (e.g., Miller, Hersen, Eisler, & Hilsman, 1974). Moreover, assertive training for many alcoholics would appear indicated considering that the more unassertive the alcoholic is rated on a behavioral role-playing test, the more alcohol he is likely to consume on a laboratory drinking task ($r = -0.63$; Miller & Eisler, 1977).

In one single-case analysis, Eisler *et al.* (1974b) identified four deficit behaviors (eye contact, compliance, affect, behavioral requests) in a 34-year-old twice-divorced male with a history of alcoholism. These deficits were identified through role-playing of six interpersonal encounters that were typical of the patient's difficulties. Using a multiple-baseline design across behaviors, assertive training resulted in marked improvements of all targets both in trained and untrained scenes. However, a postdischarge follow-up was disappointing, inasmuch as the patient eventually was arrested for driving while intoxicated and then admitted to a state hospital for further treatment.

Foy *et al.* (1976) used social skill training procedures to teach alcoholics effectively to refuse drinks from their drinking peers. The patients were two chronic alcoholics, with 15- and 25-year drinking histories, respectively. In each case, a multiple-baseline design across behaviors (requests, offer alternative, changed subject, duration of looking, affect) was used to evaluate the effects of training. Treatment, applied sequentially and cumulatively to the five targets, consisted of modeling and focused instructions. An example of one of the training scenes follows:

> You're at your brother's house. It's a special occasion and your whole family and several friends are there. Your brother says, "How about a beer?" In each session throughout the three phases scenes were described by the fourth author; then 2 minutes were allowed for verbal interaction between the patient and the first and second authors who role-played the patient's social acquaintances ("pushers") who attempted to coerce him into taking a drink. During each 2-min interaction the patient was directly offered a drink at least three times. The pusher's goal was to persuade the patient to accept verbally an offered drink. The pushers used various rebuttals to counter the patient's initial refusals to accept offered drinks (e.g., "What kind of a man are you who won't drink with his friends?" "One drink won't hurt you"). (Foy *et al.*, 1976, pp. 1341–1342)

The results of the study show that modeling alone led to minor improvements in some of the targeted behaviors, but when focused instructions were added to modeling, substantial changes appeared in all of the targets. These changes were maintained at a 3-month follow-up. With respect to changes in the natural environment, at the 3-month follow-up one of the two patients was abstinent. However, the second patient resumed drinking and had been returned to inpatient status for additional treatment. In spite of the partial success, Foy *et al.* (1976) conclude that "these preliminary findings suggest that many alcoholics might benefit from refusal training. Detailed self-report data obtained from the patients indicated that both used their refusal skills successfully while on leave from hospital and when they were first discharged. For example, the first patient reported that he was able to refuse when several of his former drinking companions came to his home with alcohol and invited him to join them in drinking and playing pool" (p. 1344).

In a within-subject analysis with four chronic inpatient alcoholics, Martorano (1974) examined the effects of assertive training during baseline, nondrinking, and drinking phases (when the patients had free access to alcohol). The results indicated that during nondrinking phases, assertive training appeared to improve alcoholics' "social desirability," activity, and ability to suppress anger. However, during drinking periods the effects of assertive training seemed to be greater personal discomfort as well as social rejection from peers.

Aggressivity

In the last few years a number of single-case analyses have appeared in the literature in which skills approaches have been applied to patients evidencing aggressive and explosive behaviors (Eisler *et al.*, 1974b; Elder *et al.*, 1979; Foy, Eisler, & Pinkston, 1975; Frederiksen & Eisler, 1977; Frederiksen *et al.*, 1976; King *et al.*, 1977; Matson & Stephens, 1978; Turner, Hersen, & Bellack, 1978). Not only do these studies represent an important extension of the social skills strategy to a population of patients previously ignored with respect to this approach, but they clearly document the important distinction between assertion and aggression. Recently, this distinction has received some attention in the behavioral literature (cf. DeGiovanni & Epstein, 1978; Hollandsworth, 1977).

Before describing several prototypical examples of the aforementioned single-case studies, let us first consider our own distinction between assertive and aggressive behavior. In both instances the behavior (be it assertive or aggressive) may have its desired effect on the environment. That is, the interpersonal partner who is the object of assertiveness or aggressiveness generally changes his behavior. However, it also is quite obvious that in the case of assertiveness (i.e., negative assertiveness), the behavior is deemed socially appropriate. By contrast, aggressiveness or explosiveness *is* socially inappropriate. Thus, in summary, both assertiveness and aggressiveness can be effective ways of getting others to alter their behavior. Nonetheless, only assertiveness is both effective and appropriate.

One of the first published studies describing the effects of modeled assertion in a case of explosive rages was presented by Foy *et al.* (1975). The patient was a 56-year-old twice-married carpenter who dealt with "unreasonable demands from others" by responding with verbal abuse and assaultiveness at times. The patient's marriage (separated at the time of the study) had been characterized by strife and repeated beatings of his wife.

Baseline assessment of this patient consisted of his role-playing seven work-related situations that depicted conflictual relationships. Analysis of the patient's role-played responses revealed the following: numerous hostile comments, considerable compliance to unreasonable requests, a moderate number of irrelevant comments, and no requests for his interpersonal partners to change their behavior. Thus, these four targets were treated in a multiple-baseline design across behaviors.

The results of this study appear in Figure 5. Inspection of the figure indicates that modeling effected changes in all four targets, which were consolidated by the addition of instructions. Five follow-up assessments over a 6-month period revealed maintenance of gains. Moreover, by the patient's self-report, it appears that his work-related aggressiveness and interpersonal discord had improved considerably. In addition, he indicated that he experienced an improved relationship with his son.

In an extension of the work of Foy *et al.* (1975), Frederiksen *et al.* (1976) used social skills training to reduce abusive verbal outbursts in two hospitalized psychiatric patients. Both patients had long histories of aggressive behavior, some of which was observed on the ward before skills training was introduced. In this study, social skills training con-

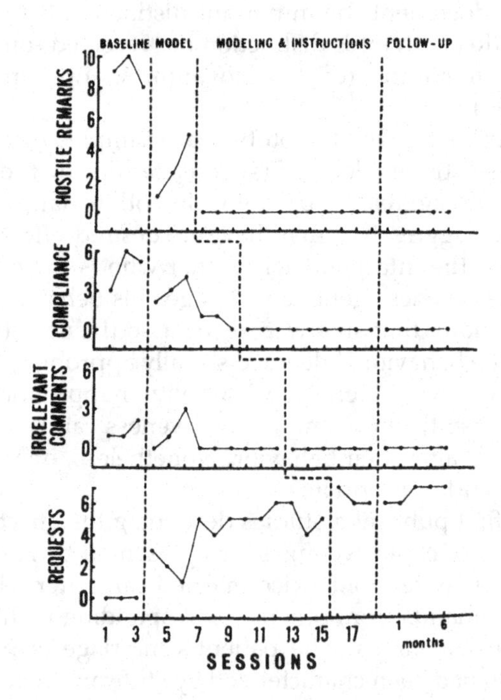

Figure 5. Target assertive behaviors during the four phases (from Foy *et al.*, 1975, Figure 1).

sisted of behavior rehearsal, modeling, focused instructions, and feedback. This treatment was evaluated in a multiple-baseline design across subjects. Specific targets selected on the basis of role-played scenes included: looking, irrelevant comments, hostile comments, inappropriate requests, and appropriate requests.

As a function of skills training all targeted behaviors improved substantially both in training and generalization scenes. Also, training generalized from the original role model to a new role model with whom the patients had not had any prior interaction. Perhaps more important, however, training generalized to staged *in vivo* siutations on the ward that these two patients previously had responded to in hostile fashion. After training, their responses to these staged interactions were rated by independent staff members as "socially appropriate."

King *et al.* (1977) applied assertive training (i.e., personal effectiveness training) to a 25-year-old married male who evidenced episodes of rage and anger in spite of his rather "docile and passive personality." Personal effectiveness training was conducted twice weekly for a 6-week period. Hierarchies of items related to three separate settings (in which the patient experienced difficulty interacting) were identified (see Table 4). Training was instituted for each of the settings in a multiple-baseline design. Assessment of treatment involved the therapist's accompanying the patient into the environment to observe how many of the 10 items from each of the three hierarchies were successfully performed.

Results of this study clearly document the utility of personal effectiveness therapy for this particular patient. Indeed, as training was specifically directed to each of the settings, the number of items successfully completed in each rose dramatically. Of course, this study can easily be criticized on methodological grounds (e.g., Hersen, 1977) as to the reactive effects of being observed. That is, the observer (the therapist in this instance) obviously serves as a discriminative stimulus for the trained behavior to occur. However, in spite of this limitation the study *did evaluate actual behavior in the natural environment*. And that in itself "is certainly pointing us in the right direction" (Hersen, 1977, p. 93).

Matson and Stephens (1978) evaluated the effects of social skills training with four severely debilitated female patients who had spent the last 4, 3, 10, and 11 years, respectively, in a psychiatric hospital. These four patients were selected for treatment inasmuch as they "had been characterized as loud, uncooperative, hostile, and combative" (Matson & Stephens, 1978, p. 64). Social skills training was assessed in a combined multiple-baseline design across behaviors and subjects. Training consisted of instructions, modeling, role-playing, and feedback, and was implemented using role-played scenes typical of each patient's interactive difficulties. The effects of treatment were highly positive, in

Table 4. Hierarchies of Assertive Behaviors for Three Community Situations[a]

I. Department store
   1. Ask where to buy something that store doesn't have
   2. Ask for information about some product without buying
   3. Try on clothes without buying
   4. Return item with receipt for credit
   5. Return item with receipt for cash
   6. Return item without receipt for credit
   7. Return item without receipt for cash
   8. Ask to use the phone
   9. Ask where the restroom is
   10. Bargain with salesperson on price

II. Service stations
   1. Without buying gas, ask directions
   2. Without buying gas, ask for key to restroom
   3. Without buying gas, ask for free map
   4. Ask for estimate on repairs
   5. When buying gas, ask attendent to check oil, water, tire pressure, and wash windows
   6. Go to unknown station and ask for advice about doing own repairs
   7. Ask to borrow tools
   8. Buy $1.00 worth of gas and write a check for it
   9. Buy $1.00 worth of gas and charge it
   10. Without buying gas, ask attendent to check oil, water, tire pressure, and wash windows

III. Restaurants
   1. Compliment waitress on service
   2. Request substitutions on menu
   3. Request special preparation of food
   4. Request bill at time of being served
   5. Ask for a separate check
   6. Change mind about order right after waitress has taken it
   7. Request order be hurried
   8. Question a bill
   9. Send food back to be recooked or warmed
   10. Criticize waitress on service

[a]From King *et al.* (1977), Table 2.

that generalization to the ward setting was observed. Also, treatment gains were maintained at the 3-month follow-up.

Finally, Elder *et al.* (in press) describe their treatment of four aggressive adolescent psychiatric inpatients. Three targets were selected (responses to negative communication, requests for behavior change, interruptions) for modification in a multiple-baseline design across behaviors. Not only did the effects of social skill treatment generalize from trained to untrained scenes, but generalization to the psychiatric ward was also obtained. Indeed, there was a significant reduction of the number of token economy fines and need for seclusion for these four patients. Moreover, at the 3-month follow-up, three of the four patients had been discharged to the community and subsequently remained there for 9 months.

SHORT-TERM TREATMENT ANALOGUES

Although analogue studies in the behavioral literature are primarily short-term treatment evaluations of theoretical or clinical import with nonclinical volunteer populations (usually college students), in the social skills literature parallel investigations have been carried out on a short-term basis with psychiatric patients as well. Of course, these studies cannot be classified as legitimate outcome studies for at least three major reasons: (1) most involve a treatment regime of four sessions or less; (2) many are concerned more with theoretically oriented issues, such as an analysis of the active components in assertive training (cf. Eisler *et al.*, 1974a; Hersen, Eisler, & Miller, 1974; Hersen, Eisler, Miller, Johnson, & Pinkston, 1973), rather than comparative efficacy of different short-term strategies; (3) some do involve comparisons with other therapeutic modalities, but a full test of any of the therapies evaluated is not the case.

Despite the fact that these analogues do not yield conclusive data as to comparative efficacy and durability of changes, their importance should not be discounted. This is especially true as many of the investigations have considerable theoretical importance, and they do point to future directions that research and practice might assume. In short, these studies often serve in the capacity of pilot investigations.

In the earliest study subsumed under this rubric, Lazarus (1966) compared four 30-min sessions of behavior rehearsal, direct advice, and nondirective therapy, with 25 patients experiencing interpersonal difficulties being assigned to each of the three treatment conditions. Specific criteria for improvement were established, with greatest change seen in the behavior-rehearsal group, followed by the direct-advice and nondirective groups. However, as Lazarus was both therapist and assessor of change for patients in all three conditions, the possibility of experimenter and observer bias obviously is apparent.

Eisler *et al.* (1974a) matched unassertive male psychiatric patients on several variables, including self-reported assertiveness, and assigned 10 to each of three conditions: (1) modeling, (2) practice-control, and (3) test–retest. Patients were videotaped pre- and posttraining responding to five BAT situations. Patients in the modeling group were exposed to four sessions of a videotaped model responding assertively in the five BAT scenes. Following each viewing of the modeling tape, patients then practiced the scenes. Practice-control patients received four sessions merely consisting of practice on the scenes. No modeling was given. Test–retest patients neither received modeling nor did they have an opportunity to practice the scenes.

The results of this study showed that the modeling condition led to significant improvement on five of the eight components of assertiveness evaluated on the BAT. Such improvements were not noted for either the practice-control or test–retest groups. There were no significant differences between these two latter conditions. Eisler *et al.* (1974a) argue, on the basis of their results, "that in cases where response deficits exist (lack of assertiveness), repeated exposure to the difficult situation alone does not change the behavior" (p. 5).

In a partial replication of Eisler *et al.* (1974a), Hersen *et al.* (1973b) contrasted five groups, with 10 unassertive male psychiatric patients assigned to each: (1) test–retest, (2) practice-control, (3) instructions, (4) modeling, and (5) modeling plus instructions. As in the Eisler *et al.* investigation, there were no differences between test–retest and practice-control conditions. Modeling and instructions and the two conditions combined proved to be the most powerful strategies for bringing about behavioral change. For simple targets (e.g., loudness of speech), instructions were as effective as modeling plus instructions. However, for more complicated targets, such as requesting that the interpersonal partner change her behavior, modeling plus instructions was superior. None of the conditions seemed effective in altering patients' responses to a self-report measure of assertiveness (i.e., the Wolpe–Lazarus Assertiveness Scale). Nonetheless, this kind of change in self-report is rarely seen in short-term behavioral treatment directed toward motoric targets in the first place (see Hersen, 1973).

In a third study in this series of investigations, Hersen, Eisler, and Miller (1974) examined the effects of training with regard to generalization. There were five conditions, with 10 unassertive male psychiatric patients assigned to each: (1) test–retest, (2) practice-control, (3) practice-control with generalization instructions, (4) modeling and instructions, and (5) modeling and instructions with generalization instructions. Using 10 BAT scenes in pre- and posttraining assessments, patients in conditions 2–5 received treatment for only five of the scenes. The remaining five served as a measure of generalization. Test–retest patients, of course, did not receive any training.

The results of this study showed that the modeling and instructions groups led to significant improvements on seven of the eight targeted behaviors on training scenes, whereas significant changes on only five of the eight targets were noted for generalization scenes. Generalization instructions appeared to have minimal effects. Also, as in the two prior studies, there were no differences or treatment effects for either of the practice-control groups (with or without generalization instructional sets) or for the test–retest group. In addition, there was very little evidence of generalization to an *in vivo* ("short-change") task. Thus, generalization effects were only obtained from trained to untrained items on the BAT.

Goldstein, Martens, Hubben, van Belle, Schaaf, Wiersma, and Goedhart (1973) conducted a series of three interrelated studies to examine the effects of modeling with respect to improving independent behavior in psychiatric patients. Ninety neurotic outpatients were divided into three groups in the first study: (1) independence modeling, (2) dependence modeling, and (3) no modeling–control. For the independence modeling condition, patients were exposed to one session in which a model responded to 40 situations in independent fashion. For the dependence modeling condition, patients were exposed to the same 40 situations, except that the model here responded in dependent fashion. The results of this study, based on differences between 10 base-rate and 10 posttest situations, indicated that male and female patients in the independence modeling group showed an increase in independent responses. For the dependence modeling condition, only the female patients showed the effect.

In the second study, 60 neurotic outpatients listened to audiotapes of the independence model. Prior to this, the group was divided into three preconditions involving different instructional sets: (1) warm structuring, (2) cold structuring, and (3) no structuring. Cold structuring resulted in diminished modeling, but warm structuring, contrary to expectation, did not lead to enhanced modeling effects.

The third study was quite similar to the one performed by Hersen *et al.* (1973b). Fifty-three chronic schizophrenic inpatients were assigned to one of four conditions: (1) modeling and instructions, (2) modeling, (3) instructions, and (4) no modeling, no instructions. The results of the pre–post analysis indicated that the three treatment groups (1–3) improved in independence responding. However, these results are at variance with those reported by Hersen *et al.* (1973b), where modeling plus instructions generally yielded the most potent effects. Of course, different dependent measures were used by Goldstein *et al.* and Hersen *et al.*, as well as the fact that there were four times as many training sessions in the Hersen *et al.* investigation. Thus, the results are truly not comparable.

Goldsmith and McFall (1975) contrasted an empirically derived interpersonal skill training condition to a pseudotherapy group and an assessment-only control group. Twelve male psychiatric inpatients were assigned to each of the groups. Three sessions of training were offered to the two treatment groups. Training sessions were 1 h in duration and were conducted over a 5-day period. The results of this study indicated significantly greater pre–post differences on behavioral and self-report measures for the interpersonal skill training patients when contrasted to the other two groups. Behavioral changes were obtained on laboratory measures and a "simulated real-life behavior test." Also, at the 8-month follow-up (based on 72% of the original sample), there was a slight albeit nonsignificant superiority as to recidivism rates for the interpersonal skill training patients.

In a somewhat more recent study, Chartier, Ainley, and Voss (1976) evaluated the effects of vicarious reward and punishment on the social imitation of chronic psychotics. The patients were 18 males and 24 females who had been hospitalized for 3 of the 5 prior years. In each of six conditions, the patients viewed videotapes in which a model (i.e., confederate) was being evaluated for a job in the hospital by an interviewer (also a confederate). There were two types of models (*appropriate* and *inappropriate*) and three types of responses to the model (*reward, punishment, no consequences*), thus yielding a 2 × 3 factorial design. Subsequent to exposure to one of these six conditions, each patient was escorted to an adjacent room and interviewed (i.e., an imitation test). The patient's verbal and nonverbal behavior in response to the interviewer was rated. In general, the results of this study confirmed the suppressing "effect of observed punishment of appropriate behavior." By contrast, observing the appropriate model led to improved behavior. However, observing the appropriate model being rewarded did not lead to significantly greater improvement than observing a model receiving no consequences for his behavior.

Using a different subject population, Berger and Rose (1977) compared the effects of interpersonal skill training with a discussion control and an assessment-only control condition for improving social skills in elderly institutionalized patients. Patients in the interpersonal skill training group received three treatment sessions (consisting of modeling, coaching, feedback, and practice), with eight role-played scenes employed both as the assessment and training vehicle. The results of this short-term study indicated that the interpersonal skill training group was superior to the discussion control and assessment-only conditions on the training scenes. However, the effects of training failed to generalize to eight untrained scenes. Also, no differences were found among the groups for the self-report measures used in this study.

In the most recent of the short-term studies to be published, Eisler, Blanchard, Fitts, and Williams (1978), in a 3 × 2 analysis of variance design, compared the effectiveness of social skills training with modeling, social skills training without modeling, and a behavior rehearsal control group with 24 schizophrenics and 24 nonpsychotic hospitalized psychiatric patients. Treatment in each of the conditions involved six 30-min sessions, with 12 BAT-R scenes used for training purposes. Analysis of pre–post BAT-R data and the behavioral data obtained from the extended interaction test showed that both social skill conditions led to significantly improved performance. However, it should be noted that the addition of modeling was necessary for changing performance in the schizophrenics. This was not the case for the nonpsychotic patients. Not only were there significant pre–post differences on materials that were used in training sessions, but there was clear evidence for generalization on the extended interaction test on five of the eight dependent measures assessed. Eisler *et al.* (1978) point out that "the results . . . demonstrated that the different training packages had differential effects on two broad classifications of psychiatric patients. Thus, there appears to be some utility in making this distinction" (p. 169).

OUTCOME STUDIES

By our count, there are some 14 investigations at this time which generally may be categorized as outcome studies. Some of these studies are very nicely designed, yield meaningful pre–post differences, and present sufficiently lengthy follow-up data. Others are less meticulously designed, do not lead to any conclusive results, and do not present follow-up data. Still others have too small an *N* and, in fact, more closely resemble pilot investigations. Since the entire area is relatively new (the oldest studies dating 10 years or less: Lomont, Gilner, Spector, & Skinner, 1969; Serber & Nelson, 1971), it probably would be unfair to criticize severely any one study in particular. However, we will review several of the studies in some detail and will point out methodological deficiencies if present. Since space will not permit a critical appraisal of each of the investigations, some of the salient features of each study are depicted in tabular form (see Table 5).

In an early report (see Table 5), Serber and Nelson (1971) described the comparative evaluation of systematic desensitization and assertive training for hospitalized schizophrenic patients. Systematic desensitization was applied with methohexital as the relaxing agent. Assertive training was carried out in hierarchical fashion as recommended by Wolpe (1969). Relatively few details are provided as to how patients were treated and who their therapists were. Moreover, criteria for improvement at the 6-month follow-up did not appear to be very rigor-

Table 5. Outcome Studies

Reference	Treatment and length	Patients	Conditions	*N*	Follow-up	Best treatment
Argyle *et al.* (1974)	Individual 18 sessions	Personality disorders	Social skill	8	6 months	Social skill
			Psychotherapy	8	1 year	
Booraem & Flowers (1972)	Group, 12 sessions	Inpatients	Assertion training and milieu	7	None	Assertion training
			Milieu	7		
Field & Test (1975)	Group Number of sessions not indicated	Inpatients	Role playing	5	10 months	Role playing
			No role playing	5		
Finch & Wallace (1977)	Group, 12 sessions	Male schizophrenic inpatients	Skills training,	8	None	Skills training
			Control group	8		
Gutride *et al.* (1973)	Group, 12 sessions	Asocial inpatients	Structured learning therapy	45	None	Structured learning therapy in the absence of additional psychotherapy
			No structured learning therapy	42		
Gutride *et al.* (1974)	Group, 15 sessions	Chronic inpatients	Structured	24	None	Structured learning therapy with transfer training
			Structured learning therapy (continued)	24		
			Structured learning therapy (transfer training)	24		
			Social companionship therapy	24		
			No treatment	24		

Jaffe & Carlson (1976)	Individual 12 sessions	Chronic inpatients	Modeling	8	None	Modeling or instructions
			Instructions	8		
			Attention	8		
Lomont *et al.* (1969)	Group, 30 sessions	Chronic inpatients	Group assertion	7	None	Slight evidence for assertion group
			Group insight	5		
Marzillier *et al.* (1976)	Individual 15 sessions	Outpatients (1 exception)	Systematic desensitization	7	6 months	Mixed findings; no treatment supremacy
			Social skills	7		
			Waiting list control	7		
Monti *et al.* (1978)	Individual 10 sessions	Inpatients	Social skills	10	10 months	Social skills
			Bibliotherapy	10		
			Control	10		
Percell *et al.* (1974)	Group, 8 sessions	Outpatients	Assertive training	12	None	Assertive training
			Relationship (control)	12		
Serber & Nelson (1971)	Individual 18 sessions	Inpatient schizophrenics	Systematic desensitization	16	6 months	Assertive training (only minimal improvement)
			Assertive training	14		
Trower *et al.* (1978)	Individual 10 sessions	Outpatient	Systematic desensitization	20	6 months	Social skill training for unskilled patients
			Social skill training	20		Both therapies equal success with social phobics
Weinman *et al.* (1972)	Individual 36 sessions	Chronic inpatient schizophrenics	Socioenvironmental treatment	21	None	Socioenvironmental more effective with older patients
			Systematic desensitization	21		
			Relaxation training	21		

ously defined (i.e., much improved, slightly improved, unimproved). In any case, neither of the two treatment approaches seemed to be particularly effective for changing behavior. However, assertive training was described as leading to some improvement while it was being administered. But following treatment, "signs of assertive behavior regressed." In considering Serber and Nelson's disappointing results, it should be underscored that the multiple-baseline strategy for treating schizophrenics with social skill techniques (cf. Bellack *et al.*, 1976b; Hersen & Bellack, 1976a; Hersen *et al.*, 1975; King *et al.*, 1977) had not yet been developed at the time of their study. Thus, the results of Serber and Nelson (1971), on that basis and in light of other methodological considerations, can best be designated as tentative.

In a more recent and somewhat better controlled and designed study, Finch and Wallace (1977) evaluated the effects of interpersonal skills training with a group of schizophrenic patients. Treatment was conducted in group format and involved 12 sessions over a 4-week period. The effects of training were contrasted to a no-treatment control condition, with patients matched on relevant demographic variables. Assessment of pre–post differences was accomplished by evaluating patient performance (several verbal and nonverbal behaviors as identified by Eisler *et al.*, 1974a) on role-played and spontaneously enacted interpersonal encounters. Results of the study indicated superiority of the interpersonal skills training condition on a variety of measures: "loudness, fluency, affect, latency, eye contact, content . . ." (p. 885). Also, improvements were noted on the Wolpe–Lazarus Assertiveness Scale. Such improvements appeared both on trained and untrained items. Indeed, to ensure generalization, patient dyad teams were formed and instructed to complete homework assignments.

In spite of the apparent success of the treatment, a number of limitations regarding the generality of conclusions reached must be noted. *First,* although there was evidence of generalization from trained to untrained scenes, there was no assessment of *in vivo* interactions such as in King *et al.* (1977). *Second,* from Finch and Wallace's terse description, it does not appear that an active comparison treatment was being offered here. Therefore, one cannot fully discount the attentional aspects of the experimental condition (interpersonal skills training), although in analogue studies (cf. Eisler *et al.*, 1974a; Hersen *et al.*, 1973b, 1974), no evidence was adduced for change in target behaviors with a similar subject population in the absence of active treatment (i.e., instructions, modeling, feedback, etc.). *Third,* no follow-up data are reported. Nonetheless, when the results of this study are taken together with those of Gutride *et al.* (Gutride, Goldstein, & Hunter, 1973; Gutride, Goldstein, Hunter, Carrol, Clark, Furia, & Lower, 1974), Jaffe and

Carlson (1976), and Weinman, Gelbart, Wallace, and Post (1972), it certainly seems that social skills approaches with schizophrenics do at least yield significant pre–post treatment effects. Of course, maintenance of gains and generalization to extrahospital settings still is another question.

In an interesting comparison of a group assertive approach with that of a group relationship therapy control, Percell, Berwick, and Beigel (1974) randomly assigned 12 psychiatric outpatients to each of the two conditions. Each of the groups was given eight sessions of treatment and evaluated pre–post on several dependent measures. Measures included the Breger Self-Acceptance Scale, the Taylor Manifest Anxiety Scale, the Interpersonal Behavior Test, and a behavioral rating scale of assertiveness that was scored blindly by independent observers. Pre–post analyses revealed significantly greater changes for assertive training patients than relationship therapy controls on all measures. Assertive training patients were less anxious, more self-accepting, and were rated as more assertive, aggressive, empathic, spontaneous, and outgoing. However, despite the support this study tends to lend to the efficacy of group assertive training, it must be clearly noted that the evaluations were primarily of the self-report variety. No extratreatment behaviors were directly observed nor were any follow-up data obtained.

Argyle, Bryant, and Trower (1974) compared the effects of social skills training and psychotherapy in a crossover design, which was used in an attempt to control for the effects of drugs and spontaneous improvements. Subjects for the study were outpatients (ages 17–50) exhibiting a variety of interpersonal difficulties. Excluded from participation were psychotic and brain-damaged patients, drug addicts, alcoholics, and retardates. Following an initial 3-week assessment period, four patients each were assigned to 6 weeks of social skill treatment (1 session a week) or 6 weeks of psychotherapy (3 sessions a week). Then, 6 weeks later, following the initial assessment, another two sets of four patients each were assigned to the two treatments. Psychotherapy was psychoanalytically oriented, while social skills training consisted of role-playing, behavior, rehearsal, modeling, social perception training, and homework assignments.

The dependent measure for this study was a 17-item social skill rating scale developed by Argyle (1969), and it was used to evaluate patients' responses in "semistandardized" interpersonal situations. "The results provide evidence that both social skills training and brief psychotherapy were effective in improving the social behavior of patients, compared with no improvements during a 'no treatment' control period" (Argyle *et al.*, 1974, p. 44). Of course, these results are difficult to interpret, particularly in light of the small *N* employed. However, if one considers the findings in terms of a cost-effectiveness factor (cf. Kazdin & Wilson, 1978), the social skill condition must be deemed superior. That is, six sessions of social skill yielded effects equal to or better than those of 18 sessions of psychotherapy.

In one of the more theoretically interesting of the outcome studies recently completed, Trower, Yardley, Bryant, and Shaw (1978) contrasted the effects of systematic desensitization and social skills training with 20 socially unskilled patients and 20 socially phobic patients. Socially unskilled patients were those considered to exhibit specific deficits, whereas socially phobic patients were described as those whose anxiety apparently interfered with their adequate social expression. This distinction was made reliably on the basis of the Symptom Rating Scale (Gelder & Marks, 1966).

Ten patients in each of the diagnostic categories received either 10 sessions of systematic desensitization or 10 sessions of social skills training. Thus, there were four treatment groups. However, because of initial differences on most measures between socially unskilled and socially phobic patients, data for each category were analyzed separately. The results indicated that social phobics responded equally well to both systematic desensitization or social skills training. On the other hand, with socially unskilled patients, the social skill approach was clearly superior. These latter data definitely confirm the notion that, when response deficits are present, the patient must be taught a new behavioral repertoire (see Hersen & Bellack, 1976b).

Of course, Trower *et al.* (1978) are well aware that the distinction between a social deficit and a social phobia is not always crystal clear. Moreover, the distinction between a skills-acquisition approach and an anxiety-reduction approach also has its limitations. Certainly, in the case of social skills training, by virtue of exposing patients to presumed phobic situations, that in itself may lead to anxiety reduction. Nonetheless, despite the limitations of the study, the results undoubtedly have theoretical importance that should stimulate further research in this area.

CRITICAL APPRAISAL

As one of our colleagues once said, "Now that the easy studies have been done it is probably time to begin doing the difficult ones." This statement perhaps best captures the status of the art at the time of writing this chapter. However, before attempting to point to future directions that the field might take, let us briefly take stock of what has been accomplished to date.

Whereas in the late 1960s and early 1970s the positive effects of assertive training appeared limited to a circumscribed group of outpatient neurotics (cf. Serber & Nelson, 1971; Wolpe, 1969), the range of applicability of what is now termed social skills training has extended to many other diagnostic categories (e.g., elderly institutionalized patients, explosive patients, alcoholics, drug addicts, schizophrenics, etc.). Although the specific protocols for dealing with each diagnostic entity are not fully worked out, a relatively effective methodology has been devised to deal with severely disturbed psychiatric patients whose medication is controlling psychotic symptomatology (e.g., Bellack & Hersen, 1978; Hersen & Bellack, 1976a, b).[2] In this connection, the active ingredients that comprise social skill approaches have been identified (e.g., Hersen *et al.*, 1973b), thus permitting a highly effective treatment package that combines the specific strategies to maximize therapeutic effectiveness (e.g., Liberman *et al.*, 1977a).

However, despite the strides that have been made in the area in a relatively short time span (about 10 years), a number of problems need to be tackled by researchers and practitioners in the decade to come. *First,* a definition of terms is warranted. For what essentially amounts to the same training approach, a variety of descriptors have been attached: assertive training, assertion training, social skills training, personal effectiveness training, structured learning therapy, etc. Perhaps social skills training represents the most generic term, in that the particular skills taught a given patient probably exceed those dealt with, for example, in assertive training. Nonetheless, the plethora of terms used serves only to confuse the reader and adds little of value to the field. Thus, we are suggesting that investigators in this area refrain from attaching new labels to already existing therapies.

Second, and obviously more important, the continued development of *in vivo* measures to assess the patient in his/her natural environment is mandatory. The hallmark of behavior therapy has always been the evaluation of actual behavior when possible. In its absence in the social skills arena, false leads based on laboratory observations could conceivably result in erroneous conclusions as to efficacy of clinical strategems being carried out. *Third,* and related to assessment, is the critical issue of generalizability of results. The continued documentation that our cherished techniques facilitate in laboratory improvement of designated behaviors will be of theoretical significance in some cases. But if our

[2] The importance of having severely disturbed psychiatric patients under good pharmacological control cannot be sufficiently underscored. In the absence of such control, the schizophrenic patients in particular will be unable to learn and retain new material (*cf.* Hersen & Bellack, 1976b).

social skill techniques are to have an impact in the psychiatric arena in general, ample documentation of extrahospital utility (i.e., in the patient's home, work, and recreational environments) should be forthcoming. *Fourth,* the equally pertinent issue of durability of results will require attention with regard to the variety of the populations being treated. Along these lines, the use of booster training sessions to maintain and consolidate treatment gains will need to be carefully researched. *Fifth,* and finally, the cost effectiveness of the social skill approaches still need to be determined. At this juncture, most of the skill procedures being tested with psychiatric patients tend to require an enormous amount of therapeutic time and energy, with results that are not quite commensurate with the initial input. This is especially the case when considering the number of staff people needed to assess and treat any given patient in the skills modality. Thus, the comparative evaluation of treatments using group and individual formats will not only provide data as to overall efficacy, but the increasingly important cost-effectiveness factor will emerge as well. In view of the large number of patients who are socially unskilled and require remediation, the widespread application of social skill strategies will not materialize if the techniques remain costly and impractical to apply.

SUMMARY

This chapter has dealt with the research and practice involving modification of skill deficits in psychiatric patients. Following a general discussion of the issues, we examined some of the assessment problems peculiar to psychiatric patients. Then we examined how skills-training procedures have been applied alone and in combination with other techniques in selected cases. Next we looked in some detail at a treatment protocol developed for use with depressive women. This was followed by surveys of research in the field, with single-case experimental design, analogue, and outcome studies considered. Finally, the current status of skills approaches with psychiatric patients was critically appraised. Specific attention was directed to delineation of the field's needs for the future.

REFERENCES

Argyle, M. *Social interaction*. London: Methuen, 1969.

Argyle, M., Bryant, B., & Trower, P. Social skills training and psychotherapy. *Psychological Medicine*, 1974, *4*, 435–443.

Barlow, D. H., Abel, G. G., Blanchard, E. B., Bristow, A. R., & Young, L. D. A heterosocial skills behavior checklist for males. *Behavior Therapy*, 1977, *8*, 229–239.

Barnard, G. W., Flesher, C. K., & Steinbrook, R. M. The treatment of urinary retention by aversive stimulus cessation and assertive training. *Behaviour Research and Therapy*, 1966, *4*, 232–237.

Bellack, A. S. A comparison of self-monitoring and self-reinforcement in weight reduction. *Behavior Therapy*, 1976, *7*, 68–75.

Bellack, A. S., & Hersen, M. *Behavior modification: An introductory textbook*. Baltimore: Williams & Wilkins, 1977.

Bellack, A. S., & Hersen, J. Chronic psychiatric patients: Social skills training. In M. Hersen & A. S. Bellack (Eds.), *Behavior therapy in the psychiatric setting*. Baltimore: Williams & Wilkins, 1978.

Bellack, A. S., Glanz, L., & Simon, R. Covert imagery and individual differences in self-reinforcement style in the treatment of obesity. *Journal of Consulting and Clinical Psychology*, 1976, *44*, 490–491. (a)

Bellack, A. S., Hersen, M., & Turner, S. M. Generalization effects of social skills training in chronic schizophrenics: An experimental analysis. *Behaviour Research and Therapy*, 1976, *14*, 391–398. (b)

Bellack, A. S., Hersen, M., & Turner, S. M. Role-play tests for assessing social skills: Are they valid? *Behavior Therapy*, 1978, *9*, 448–461.

Berger, R. M., & Rose, S. D. Interpersonal skill training with institutionalized elderly patients. *Journal of Gerontology*, 1977, *32*, 346–353.

Blanchard, E. B., & Hersen, M. Behavioral treatment of hysterical neurosis: Symptom substitution and symptom return reconsidered. *Psychiatry*, 1976, *39*, 118–129.

Bloomfield, H. H. Assertive training in an outpatient group of chronic schizophrenics: A preliminary report. *Behavior Therapy*, 1973, *4*, 277–281.

Booream, C. D., & Flowers, J. V. Reduction of anxiety and personal space as a function of assertion training with severely disturbed neuropsychiatric inpatients. *Psychological Reports*, 1972, *30*, 923–929.

Callner, D. A., & Ross, S. M. The reliability and validity of three measures of assertion in a drug addict population. *Behavior Therapy*, 1976, *7*, 659–667.

Chartier, G. M., Ainley, C., & Voss, J. Effects of vicarious reward and punishment on social imitation in chronic psychotics. *Behaviour Research and Therapy*, 1976, *14*, 303–304.

DeGiovanni, I. S., & Epstein, N. Unbinding assertion and aggression in research and clinical practice. *Behavior Modification*, 1978, *2*, 173–192.

Edelstein, B., & Eisler, R. M. Effects of modeling and modeling with instructions and feedback on the behavioral components of social skills of a schizophrenic. *Behavior Therapy*, 1976, *7*, 382–389.

Edwards, N. B. Case conference: Assertive training in a case of homosexual pedophilia. *Journal of Behavior Therapy and Experimental Psychiatry*, 1972, *3*, 55–63.

Eisler, R. M. The behavioral assessment of social skills. In M. Hersen & A. S. Bellack (Eds.), *Behavioral assessment: A practical handbook*. New York: Pergamon, 1976.

Eisler, R. M., & Hersen, M. Behavioral techniques in family-oriented crisis intervention. *Archives of General Psychiatry*, 1973, *28*, 111–116.

Eisler, R. M., Hersen, M., & Miller, P. M. Effects of modeling on components of assertive behavior. *Journal of Behavior Therapy and Experimental Psychiatry*, 1973, *4*, 1–6. (a)

Eisler, R. M., Miller, P. M., & Hersen, M. Components of assertive behavior. *Journal of Clinical Psychology*, 1973, *29*, 295–299. (b)

Eisler, R. M., Hersen, M., & Miller, P. M. Shaping components of assertiveness with instructions and feedback. *American Journal of Psychiatry*, 1974, *131*, 1344–1347. (b)

Eisler, R. M., Miller, P. M., Hersen, M., & Alford, H. Effects of assertive training on marital interaction. *Archives of General Psychiatry*, 1974, *30*, 643–649.

Eisler, R. M., Hersen, M., Miller, P. M., & Blanchard, E. B. Situational determinants of assertive behaviors. *Journal of Consulting and Clinical Psychology*, 1975, *43*, 330–340.

Eisler, R. M., Blanchard, E. B., Fitts, H., & Williams, J. G. Social skill training with and without modeling on schizophrenic and non-psychotic hospitalized psychiatric patients. *Behavior Modification*, 1978, *2*, 147–172.

Elder, J. P., Edelstein, B. A., & Narick, M. M. Social skills training in the modification of aggressive behavior of adolescent psychiatric patients. *Behavior Modification*, in press.

Fensterheim, H. Assertive methods and marital problems. In R. D. Rubin, H. Fensterheim, J. D. Henderson, & L. P. Ullmann (Eds.), *Advances in behavior therapy*. New York: Academic Press, 1972. (a)

Fensterheim, H. The initial interview. In A. A. Lazarus (Ed.), *Clinical behavior therapy*. New York: Brunner/Mazel, 1972. (b)

Fichter, M. M., Wallace, C. J., Liberman, R. P., & Davis, J. R. Improving social interaction in a chronic psychotic using discriminated avoidance ("nagging"): Experimental analysis and generalization. *Journal of Applied Behavior Analysis*, 1976, *9*, 377–386.

Field, G. D., & Test, M. A. Group assertiveness training for severely disturbed patients. *Journal of Behavior Therapy and Experimental Psychiatry*, 1975, *6*, 129–134.

Finch, B. E., & Wallace, C. J. Successful interpersonal skills training with schizophrenic inpatients. *Journal of Consulting and Clinical Psychology*, 1977, *45*, 885–890.

Foy, D. W., Eisler, R. M., & Pinkston, S. G. Modeled assertion in a case of explosive rage. *Journal of Behavior Therapy and Experimental Psychiatry*, 1975, *6*, 135–137.

Foy, D. W., Miller, P. M., Eisler, R. M., & O'Toole, D. H. Social skills training to teach alcoholics to refuse drinks effectively. *Journal of Studies on Alcohol*, 1976, *37*, 1340–1345.

Frank, J. D. Therapeutic components of psychotherapy: A 25-year progress report of research. *Journal of Nervous and Mental Disease*, 1974, *159*, 325–342.

Frederiksen, L. W., & Eisler, R. M. The control of explosive behavior: A skill-development approach. In D. Upper (Ed.), *Perspectives in behavior therapy*. Kalamazoo, Mich.: Behaviordelia, 1977.

Frederiksen, L. W., Jenkins, J. O., Froy, D. W., & Eisler, R. M. Social skills training in the modification of abusive verbal outbursts in adults. *Journal of Applied Behavior Analysis*, 1976, *9*, 117–125.

Fuchs, C. Z., & Rehm, L. P. A self-control behavior therapy program for depression. *Journal of Consulting and Clinical Psychology*, 1977, *45*, 206–215.

Gelder, M. G., & Marks, I. M. Severe agoraphobia: A controlled prospective trial of behaviour therapy. *British Journal of Psychiatry*, 1966, *112*, 309–319.

Gladwin, T. Social competence and clinical practice. *Psychiatry*, 1967, *30*, 30–43.

Goldsmith, J. B., & McFall, R. M. Development and evaluation of an interpersonal skill-training program for psychiatric patients. *Journal of Abnormal Psychology*, 1975, *84*, 51–58.

Goldstein, A. P., Martens, J., Hubben, J., van Belle, H. A., Schaaf, W., Wiersma, H., & Goedhart, A. The use of modeling to increase independent behaviour. *Behaviour Research and Therapy*, 1973, *11*, 31–42.

Goldstein, A. J., Serber, M., & Piaget, G. Induced anger as a reciprocal inhibitor of fear. *Journal of Behavior Therapy and Experimental Psychiatry*, 1970, *1*, 67–70.

Gutride, M. E., Goldstein, A. P., & Hunter, G. F. The use of modeling and role playing to increase social interaction among asocial psychiatric patients. *Journal of Consulting and Clinical Psychology*, 1973, *40*, 408–415.

Gutride, M. E., Goldstein, A. P., Hunter, G. F., Carrol, S., Clark, L., Furia, R., & Lower, W. Structured learning therapy with transfer training for chronic inpatients. *Journal of Clinical Psychology*, 1974, *30*, 277–279.

Hersen, M. Self assessment of fear. *Behavior Therapy*, 1973, *4*, 241–257.

Hersen, M. Comment to L. W. King, R. P. Liberman, J. Roberts, & E. Bryan, "Personal effectiveness: A structured therapy for improving social and emotional skills." *European Journal of Behavioural Analysis and Modification*, 1977, *2*, 92–94.

Hersen, M., & Barlow, D. H. *Single case experimental designs: Strategies for studying behavior change*. New York: Pergamon, 1976.

Hersen, M., & Bellack, A. S. A multiple-baseline analysis of social-skills training in chronic schizophrenics. *Journal of Applied Behavior Analysis*, 1976, *9*, 239–245. (a)

Hersen, M., & Bellack, A. S. Social skills training for chronic psychiatric patients: Rationale, research findings, and future directions. *Comprehensive Psychiatry*, 1976, *17*, 559–580. (b)

Hersen, M., & Bellack, A. S. Assessment of social skills. In A. R. Ciminero, K. S. Calhoun, & H. E. Adams (Eds.), *Handbook for behavioral assessment*. New York: John Wiley & Sons, 1977.

Hersen, M., & Bellack, A. S. *Behavior therapy in the psychiatric setting*. Baltimore: Williams & Wilkins, 1978.

Hersen, M., & Luber, R. F. Use of group psychotherapy in a partial hospitalization service: The remediation of basic skill deficits. *International Journal of Group Psychotherapy*, 1977, *27*, 361–376.

Hersen, M., & Miller, P. M. Social skills training for unipolar (non-psychotic) depressed patients. *World Journal of Psychosynthesis*, 1976, *8*, 15–17.

Hersen, M., Eisler, R. M., & Miller, P. M. Development of assertive responses: Clinical, measurement and research considerations. *Behaviour Research and Therapy*, 1973, *11*, 505–521. (a)

Hersen, M., Eisler, R. M., Miller, P. M., Johnson, M. B., & Pinkston, S. G. Effects of practice, instructions, and modeling on components of assertive behavior. *Behaviour Research and Therapy*, 1973, *11*, 443–451. (b)

Hersen, M., Eisler, R. M., & Miller, P. M. An experimental analysis of generalization in assertive training. *Behaviour Research and Therapy*, 1974, *12*, 295–310.

Hersen, M., Turner, S. M., Edelstein, B. A., & Pinkston, S. G. Effects of phenothiazines and social skills training in a withdrawn schizophrenic. *Journal of Clinical Psychology*, 1975, *34*, 588–594.

Hersen, M., Bellack, A. S., & Turner, S. M. Assessment of assertiveness in female psychiatric patients: Motor and autonomic measures. *Journal of Behavior Therapy and Experimental Psychiatry*, 1978, *9*, 11–16. (a)

Hersen, M., Himmelhoch, J. M., & Bellack, A. S. *Pharmacological and social skills treatment for unipolar (non-psychotic depression)*. Funded grant from the National Institute of Mental Health, 1978. (b)

Hersen, M., Bellack, A. S., Turner, S. M., Harper, K. S., Watts, J. G., & Williams, M. T. Psychometric properties of the Wolpe–Lazarus Assertiveness Scale. *Behaviour Research and Therapy*, 1979, *17*, 63–69.

Hogarty, G. *Environmental-personal treatment indicators in schizophrenics*. Funded grant from the National Institute of Mental Health, 1978.

Hollandsworth, J. G. Differentiating assertion and aggression: Some behavioral guidelines. *Behavior Therapy*, 1977, *8*, 347–352.

Jaffe, P. G., & Carlson, P. M. Relative efficacy of modeling and instructions in eliciting social behavior from chronic psychiatric patients. *Journal of Consulting and Clinical Psychology*, 1976, *44*, 200–207.

Katz, R. Case conference: Rapid development of activity in a case of chronic passivity. *Journal of Behavior Therapy and Experimental Psychiatry*, 1971, *2*, 187–193.

Kazdin, A. E., & Wilson, G. T. Criteria for evaluating psychotherapy. *Archives of General Psychiatry*, 1978, *35*, 407–416.

King, L. W., Liberman, R. P., Roberts, J., & Bryan, E. Personal effectiveness: A structured therapy for improving social and emotional skills. *European Journal of Behavioural Analysis,* 1977, *2,* 82–91.

Laws, R., & Serber, M. *Measurement and evaluation of assertive training.* Paper presented at the meeting of the Association for Advancement of Behavior Therapy, Washington, D.C., September 1971.

Lazarus, A. A. Behaviour rehearsal vs. non-directive therapy vs. advice in effecting behaviour change. *Behaviour Research and Therapy,* 1966, *4,* 209–212.

Lazarus, A. A. *Behavior therapy and beyond.* New York: McGraw-Hill, 1971.

Lazarus, A. A., & Davison, G. C. Clinical innovation in research and practice. In A. E. Bergin & S. L. Garfield (Eds.), *Handbook of psychotherapy and behavior change.* New York: John Wiley & Sons, 1971.

Levine, J., & Zigler, E. The essential-reactive distinction in alcoholism: A developmental approach. *Journal of Abnormal Psychology,* 1973, *81,* 242–249.

Liberman, R. P., & Davis, J. Drugs and behavior analysis. In M. Hersen, R. M. Eisler, & P. M. Miller (Eds.), *Progress in behavior modification* (Vol. 1). New York: Academic Press, 1975.

Liberman, R. P., McCann, M. J., & Wallace, C. J. Generalization of behavior therapy with psychotics. *British Journal of Psychiatry,* 1976, *129,* 490–496.

Liberman, R. P., Lillie, F., Falloon, I., Vaughn, C., Harper, E., Leff, J., Hutchison, W., Ryan, P., & Stoute, M. *Social skills training for schizophrenic patients and their families.* Unpublished manuscript, 1977. (a)

Liberman, R. P., Vaughn, C., Aitchison, R. A., & Falloon, I. *Social skills training for relapsing schizophrenics.* Funded grant from the National Institute of Mental Health, 1977. (b)

Libet, J. M., & Lewinsohn, P. M. Concept of social skill with special reference to the behavior of depressed persons. *Journal of Consulting and Clinical Psychology,* 1973, *40,* 304–312.

Lomont, J. R., Gilner, F. H., Spector, N. J., & Skinner, K. K. Group assertion training and group insight therapies. *Psychological Reports,* 1969, *25,* 463–470.

Luber, R. F., Adebimpe, V. R., & Trocki, D. A. *The role of partial hospitalization and social skills training in the treatment of epileptic patients.* Unpublished manuscript.

Macpherson, E. L. R. Selective operant conditioning and deconditioning of assertive modes of behaviour. *Journal of Behavior Therapy and Experimental Psychiatry,* 1972, *3,* 99–102.

Martorano, R. D. Mood and social perception in four alcoholics. *Quarterly Journal of Studies on Alcohol,* 1974, *35,* 445–457.

Marzillier, J. S., Lambert, C., & Kelett, J. A controlled evaluation of systematic desensitization and social skills training for socially inadequate psychiatric patients. *Behaviour Research and Therapy,* 1976, *14,* 225–238.

Matson, J. L., & Stephens, R. M. Increasing appropriate behavior of explosive chronic psychiatric patients with a social-skills training package. *Behavior Modification,* 1978, *2,* 61–76.

Miller, P. M., & Eisler, R. M. Assertive behavior of alcoholics: A descriptive analysis. *Behavior Therapy,* 1977, *8,* 146–149.

Miller, P. M., Hersen, M., Eisler, R. M., & Hilsman, G. Effects of social stress on operant drinking of alcoholics and social drinkers. *Behaviour Research and Therapy,* 1974, *12,* 65–72.

Monti, P. M., Fink, E., Norman, W., Curran, J., Hayes, S., & Caldwell, A. *The effect of social skills training groups and social skills bibliotherapy with psychiatric patients.* Unpublished manuscript, 1978.

Nydegger, R. V. The elimination of hallucinatory and delusional behavior by verbal conditioning and assertive training: A case study. *Journal of Behavior Therapy and Experimental Psychiatry*, 1972, *3*, 225–228.

O'Leary, D. E., O'Leary, M. R., & Donovan, D. M. Social skill acquisition and psychosocial development of alcoholics: A review. *Addictive Behaviors*, 1976, *1*, 111–120.

Patterson, R. L. Time-out and assertive training for a dependent child. *Behavior Therapy*, 1972, *3*, 466–468.

Paul, G. L. Chronic mental patients: Current status–future direction. *Psychological Bulletin*, 1969, *71*, 81–94.

Percell, L. P., Berwick, P. T., & Beigel, A. The effects of assertive training on self-concept and anxiety. *Archives of General Psychiatry*, 1974, *31*, 502–504.

Phillips, L., & Zigler, E. Social competence: The action–thought parameter and vicariousness in normal and pathological behaviors. *Journal of Abnormal and Social Psychology*, 1961, *63*, 137–146.

Phillips, L., & Zigler, E. Role orientation, the action–thought dimension and outcome in psychiatric disorder. *Journal of Abnormal and Social Psychology*, 1964, *68*, 381–389.

Piaget, G. W., & Lazarus, A. A. The use of rehearsal desensitization. *Psychotherapy: Theory, Research and Practice*, 1969, *6*, 264–266.

Rehm, L. P. A self-control model of depression. *Behavior Therapy*, 1977, *8*, 787–804.

Rimm, D. C. Assertive training used in treatment of chronic crying spells. *Behaviour Research and Therapy*, 1967, *5*, 373–374.

Roback, H., Frayn, D., Gunby, L., & Tuters, K. A multifactorial approach to the treatment and ward management of a self-mutilating patient. *Journal of Behavior Therapy and Experimental Psychiatry*, 1972, *3*, 189–193.

Salter, A. *Conditioned reflex therapy*. New York: Capricorn, 1949.

Serber, M., & Nelson, P. The ineffectiveness of systematic desensitization and assertive training in hospitalized schizophrenics. *Journal of Behavior Therapy and Experimental Psychiatry*, 1971, *2*, 107–109.

Stern, R. S. Behavior therapy and psychotropic medication. In M. Hersen & A. S. Bellack (Eds.), *Behavior therapy in the psychiatric setting*. Baltimore: Williams & Wilkins, 1978.

Sylph, J. A., Ross, H. E., & Kedward, H. B. Social disability in chronic psychiatric patients. *American Journal of Psychiatry*, 1978, *134*, 1391–1394.

Trower, P., Yardley, K., Bryant, B. M., & Shaw, P. The treatment of social failure: A comparison of anxiety-reduction and skills-acquisition procedures on two social problems. *Behavior Modification*, 1978, *2*, 41–60.

Turner, S. M., Hersen, M., & Bellack, A. S. *Pharmacological approaches to the treatment of obsessive-compulsive neurosis*. Unpublished manuscript, 1978.

Turner, S. M., Hersen, M., & Bellack, A. S. Social skills training to teach prosocial behavior in an organically impaired and retarded ambulatory patient. *Journal of Behavior Therapy and Experimental Psychiatry*, 1978, *9*, 253–258.

Van Hasselt, V. B., Hersen, M., & Milliones, J. Social skills training in alcoholics and drug addicts: A review. *Addictive Behaviors*, 1978, *3*, 221–233.

Wallace, C. J., Teigen, J. R., Liberman, R. P., & Baker, V. Destructive behavior treated by contingency contracts and assertive training: A case study. *Journal of Behavior Therapy and Experimental Psychiatry*, 1973, *4*, 273–274.

Weinman, B., Gelbart, P., Wallace, M., & Post, M. Inducing assertive behavior in chronic schizophrenics: A comparison of socio-environmental, desensitization, and relaxation therapies. *Journal of Consulting and Clinical Psychology*, 1972, *39*, 246–252.

Wells, K. C., Hersen, M., Bellack, A. S., & Himmelhoch, J. M. *Social skills training for unipolar depressive females*. Paper presented at the meeting of the Association for Advancement of Behavior Therapy, Atlanta, December 1977.

Williams, M. T., Turner, S. M., Watts, J. G., Bellack, A. S., & Hersen, M. Group social skills training for chronic psychiatric patients. *European Journal of Behavioural Analysis and Modification*, 1977, *1*, 223–229.

Wolpe, J. *The practice of behavior therapy*. New York: Pergamon, 1969.

Wolpe, J. The instigation of assertive behavior: Transcripts from two cases. *Journal of Behavior Therapy and Experimental Psychiatry*, 1973, *1*, 145–151. (a)

Wolpe, J. Supervision transcript: V. Mainly about assertive training. *Journal of Behavior Therapy and Experimental Psychiatry*, 1973, *4*, 141–148. (b)

Wolpe, J., & Lazarus, A. A. *Behavior therapy techniques*. New York: Pergamon, 1966.

Wood, D. D., Lenhardt, S., Maggiani, M. R., & Campbell, M. D. The nurse as therapist: Assertion training to rehabilitate the chronic mental patient. *Journal of Psychiatric Nursing and Mental Health Services*, 1975, *13*, 41–46.

Zigler, E., & Levine, J. Premorbid adjustment and paranoid-nonparanoid status in schizophrenia. *Journal of Abnormal Psychology*, 1973, *82*, 189–199.

Zigler, E., & Phillips, L. Social effectiveness and symptomatic behaviors. *Journal of Abnormal and Social Psychology*, 1960, *61*, 231–238.

Zigler, E., & Phillips, L. Social competence and outcome in psychiatric disorder. *Journal of Abnormal and Social Psychology*, 1961, *63*, 264–271.

Zigler, E., & Phillips, L. Social competence and the process-reactive distinction in psychopathology. *Journal of Abnormal and Social Psychology*. 1962, *65*, 215–222.

CHAPTER 7

Assertion Training for Women

Marsha M. Linehan and Kelly J. Egan

INTRODUCTION

In the past several years there has been an increasing amount of both popular and professional attention to the area of assertive behavior. Popular books, articles, and newspaper articles on assertion have proliferated (e.g., Alberti & Emmons, 1978; Bower & Bower, 1976; M. D. Galassi & Galassi, 1977; Gambrill & Richey, 1976; Osborn & Harris, 1975) and reviews of the research literature have been prominent in the professional journals (e.g., Heimberg, Montgomery, Madsen, & Heimberg, 1977; Hersen, Eisler, & Miller, 1973a; Rathus, 1975; Rich & Schroeder, 1976). At least two factors have most likely contributed to this growing interest in assertion. First, the value of assertive behavior as a type of interpersonal interacting has increased. This increased value has undoubtedly been influenced by the growth of the various human rights groups, especially those directed at women's rights. A key emphasis in the feminist movement has been on women asserting themselves by standing up for their rights and expressing their opinions and preferences in ways likely to be attended to by others. In the last several years, almost all national magazines directed toward women have published at least one article on assertion; assertion training groups are heavily attended by women;[1] several popular assertion books have been written especially for women (e.g., Baer, 1976; Bloom, Coburn, & Pearlman, 1975; Butler, 1976; Phelps & Austin, 1975). To a lesser extent,

[1] For example in the assertion program described by Linehan, Goldfried, and Goldfried (1978), over 500 of approximately 700 persons applying to the program were women.

Marsha M. Linehan and Kelly J. Egan • Department of Psychology, University of Washington, Seattle, Washington 98195.

movements to address the rights of other low-power groups also have played a role in increasing attention to assertive behavior. For example, assertion training has been recommended for the elderly (Corby, 1975), blacks (Cheek, 1976), adolescents (McPhail, 1978), and nurses (Herman, 1978). A second factor contributing to the interest in assertive behavior has been the comparatively recent development of effective, brief, assertion-training packages. Although procedures for encouraging assertive behavior have been in existence for some time (e.g., Salter, 1949), it was not until comparatively recently that the effectiveness of these procedures has been rigorously tested (e.g., Linehan, Goldfried, & Goldfried, 1979; McFall & Twentyman, 1973) and systematic, short-term packages have been developed (e.g., J. P. Galassi, Galassi, & Litz, 1974; Wolfe & Fodor, 1977).

The primary emphasis of this chapter is on assertion-training procedures for women. Assertion, however, has been defined in multiple ways and is often confused with other social skills. Therefore, we first will critically review various definitions of assertion and offer a summary of the current meanings of the term. Next we will survey the current status of women and the relationship of the traditional female sex role to assertive behavior. Although we do not deny the need for assertion training for many males, or the applicability of assertion training to many problems irrespective of sex, our primary focus is on the applicability of assertion training for women. In the third section, we will describe methods and procedures which have been found useful in teaching assertive behavior. Since there is little reason to believe that specific procedures are more effective with one sex than the other, we will, for the most part, review those procedures which have demonstrated effectiveness regardless of sex of client. Then we will discuss the content of assertion training programs as it relates to women. Although much attention has been given to *how* to teach assertion, very little research has been done on what to teach in an assertion-training program. Finally, we will suggest future directions for both research and clinical practice.

DEFINITION OF ASSERTION

The construct of assertive behavior is ambiguous and there is little agreement on referents for the term (Goldfried & Linehan, 1977). Given the diversity of definitions, it is extremely difficult to arrive at a single agreed-upon concept of assertion. Assertive behavior, as the term is used in the literature, encompasses particular sets of behavioral *response classes, styles* of expression, levels of social *effectiveness,* and sets of interpersonal *situations.* A careful specification of the domain of assertive behavior is necessary for several reasons. First, definitions which are too broad generally encourage the assumption of generalizability across diverse content areas even when such assumptions are unwarranted. As will be discussed later, evidence suggests that generalization of assertive responding, both within situations across response class, and across situations within response class, is slight. Second, comparability of research results across diverse populations, assessment methods, and intervention procedures is contingent on comparable operational definitions of the construct in question. To date, the complexity and diversity of assertion definitions has made many comparisons in the assertion literature questionable at best. Third, the relationship of behavioral assessment and treatment is such that development of effective behavior-change strategies is contingent on a precise specification of the relevant behavioral responses together with the situations in which they occur. What follows is an attempt to outline the most common components of most, if not all, definitions of assertion.

Assertive Response Classes

Analyses of semantic definitions of assertion, definitions based on comparisons between assertive and nonassertive individuals, factor analyses of assertion self-report inventories, and role-play measures of assertion suggest that the concept of assertion is multidimensional, encompassing a range of behavioral response classes. The most common behavioral referents are *self-expressiveness, standing up for one's rights,* and at times one or more of a variety of other social skills which often reflect general *interpersonal verbal skills*.

Some aspect of self-expressiveness is contained in almost all definitions of assertion. A prevalent theme is that of emotional expression. Wolpe (1973), for example, defines assertion as "expressing emotions other than anxiety" (p. 89); others, however, do not exclude the expression of anxiety (e.g., Lange & Jakubowski, 1976). The expression of preferences (needs, wants) and opinions (beliefs, thoughts) is included in many definitions (M. D. Galassi & Galassi, 1977; Hollandsworth, 1977; Lange & Jakubowski, 1976; MacDonald, 1974; Rich & Schroeder, 1976). Since assertion-training programs often stress the expression of negative emotions and oppositional preferences and opinions (e.g., Linehan, Goldfried, & Goldfried, 1979; McFall & Twentyman, 1973), several authors have argued that the concept of assertion should include an emphasis on verbalizing positive as well as negative feelings, preferences, and opinions (e.g., M. D. Galassi & Galassi, 1977; Hersen, Eisler, & Miller, 1973a; Lazarus, 1973). A counterargument to this emphasis has been offered by Salter (1977) who, although emphasizing the necessity of teaching clients to express positive feelings, suggests the following:

> To define assertion as including warm, friendly feelings is linguistically incorrect. It's like defining "fat" to include its antonym—i.e., *fat* is everybody who is thin, as well as everybody who is overweight. If we were to interview at random a hundred adults who are not psychologists, and if we were to ask them "What does 'assertion' mean?" they would give us a correct dictionary definition. Assertion means when you speak up forcefully. It means that you assert. (p. 34)

This notion of speaking up forcefully is similar to the idea of standing up for one's rights, a second response class included in most definitions of assertion. Lange and Jakubowski (1976) for example, state that "assertion involves standing up for personal rights . . ." (p. 7) and MacDonald included "any act which serves to maintain one's rights" (1974, p. 32). Alberti and Emmons (1978) include the notion of standing up for oneself and the exercise of one's own rights. McFall and his associates (McFall & Lillesand, 1971; McFall & Marston, 1970; McFall & Twentyman, 1973) seem to have limited assertion to the refusal of unjust demands, which is at least one component of maintaining one's rights. Presumably, the activities of "seeking, maintaining, and enhancing reinforcements," which comprise assertive behaviors according to Rich and Schroeder (1976, p. 1083), would include the seeking and maintaining of one's own rights.

A variety of other interpersonal verbal response classes have also been included within the concept of assertion by one author or another. For example, Lazarus (1971), while at one time restricting the term "assertion" to "denote only that aspect of emotional freedom that concerns standing up for one's rights" (p. 116), later expanded the definition to include "four separate and specific response patterns: the ability to say 'no'; the ability to ask for favors or to make requests; the ability to express positive and negative feelings; the ability to initiate, continue, and terminate general conversations" (1973, p. 697). Similar response classes were differentiated by J. P. Galassi and Galassi (1973) and Gambrill and Richey (1975).

A definition of a term is arbitrary. Thus, it is difficult to say *a priori* which response class or classes should be retained as the behavioral referents for the concept of assertion. Two points need to be considered: the common or usual meaning of the term and the dimensionality of the behavioral referents used to describe the term. With respect to the common usage of the term, only MacDonald (1974) has empirically studied the meaning of the term "assertion." She generated a sample of definitions and also consulted several dictionary definitions. Extraction of the common components in the definitions suggested that assertion is most commonly understood to be "the open expression of preferences (by words or actions) which causes others to take them into account; any act which serves to maintain one's rights" (1974, p. 32).

With respect to the dimensionality of the concept of assertion, findings that training on one response class does not necessarily lead to greater ability in other response classes (Lawrence, 1970; McFall & Lillesand, 1971) and the factor analyses of self-report assertion inventories, where anywhere from 4 to 11 independent factors have been found (J. P. Galassi & Galassi, 1973; Gambrill & Richey, 1975; Kipper & Jaffe, 1978), suggest that assertion is a multidimensional construct. Thus, an assumption of generalization across different response classes labeled as assertion is questionable at best. This absence of unidimensionality seriously jeopardizes the utility of the construct. As will be discussed later, this multidimensionality creates special problems for studies claiming differences in male and female assertive behavior.

Styles of Assertion

Besides indicating which response classes are included, most assertion definitions also specify, at least to some extent, how the response is to be made (i.e., the manner or style of the response). The most common stipulation is that assertive self-expressions must be done in a "direct" manner (M. D. Galassi & Galassi, 1977; Hollandsworth, 1977; Lange & Jakubowski, 1976). Similarly, MacDonald's (1974) definition specifies that the manner of expression should be "open"; at least two definitions include the descriptor "honest" (Alberti & Emmons, 1978; Lange & Jakubowski, 1976). A common thread running through many definitions is the suggestion that for a response to be considered assertive, it should be done "comfortably" (Alberti & Emmons, 1978) and without an "undue or excessive amount of anxiety or fear" (M. D. Galassi & Galassi, 1977). The nonverbal responses most often associated with assertion are those likely to communicate confidence or a lack of fear. For example, response latency, fluency, volume, timing, and eye gazing (contact) have been used as outcome measures in assertion-training programs (Kazdin, 1974, 1976; Linehan *et al.*, 1979; Rose, Caynor, & Edleson, 1977). Research by Eisler, Miller, and Hersen (1973b) suggests that some nonverbal behaviors do covary with reports of assertive behavior. They found that subjects rated high in overall assertion on the basis of a questionnaire exhibited shorter response latencies, louder speech, longer speech duration, and greater affect than low-assertion subjects.

With the proliferation of assertion-training programs and the popularization of assertion books, etc., there is an increasing tendency to differentiate "assertive" self-expressions from those that are aggressive, or as Wolpe (1973) put it, to differentiate "expressing legitimate opposition or making demands in socially appropriate ways . . . [from] . . . socially reprehensible oppositional behavior (aggression)" (p. 89). Both Lange and Jakubowski (1976) and Alberti and Emmons (1978) state that the manner of the response should not deny (violate) the rights of others. In general, most authors clearly state that assertion should not include coercive attempts to influence others (e.g., Hollondsworth, 1977).

Unfortunately, there are several problems with defining assertion in contrast to aggression. First, as De Giovanni and Epstein (1978) point out, most self-report inventories measuring assertion confound the concepts. For example, the Rathus Assertiveness Scale (Rathus, 1973) asks respondents to answer the item "Most people seem to be more aggressive and assertive than I am." A second problem has to do with the often judgmental nature of the differentiation which makes the distinction idiosyncratic. Thus, it is not entirely clear how Wolpe would define "socially reprehensible behavior."

This emphasis on social judgments creates special problems for women. Rarely would a culture which is ambivalent about recognizing one's rights (viz., the failure to pass the Equal Rights Amendment) find attempts to assert one's rights "acceptable." Hollandsworth (1977) attempted to reduce the reliance on values in defining assertion by specifying that a response should be labeled aggressive if the use of threats or punishment is made. One can ask, however, when is a statement of consequences in actuality a threat and when not? In addition, as noted by Bower and Bower (1976), the application of social learning principles to assertive encounters would mandate the use of both positive and negative consequenting when necessary. Finally, a third problem in defining assertion in contrast to aggression is the problem of defining any construct in terms of what it is *not*, instead of what it is. Assertion is not the absence of a different response.

Care must also be taken in defining assertion in terms of other stylistic qualities. For example, the emphasis on directness in the assertion literature can result in ineffective and unnecessary bluntness, rudeness, and lack of concern for interpersonal aspects of the encounter. Emphasis on openness and honesty can at times lead to inappropriate and self-destructive self-disclosure. As will be discussed in a following section, these dangers may be especially relevant to women.

Effectiveness

Implicit in almost all definitions of assertion is the assumption that the behaviors described will be effective in producing or maintaining positive consequences while at the same time avoiding negative ones.

MacDonald (1974) includes the phrase "which causes others to take them [expressions] into account" (p. 32); Alberti and Emmons (1978) include acting in one's own best interests; Rich and Schroeder (1976) define assertion as a *skill* of maintaining and enhancing reinforcement; Liberman, King, DeRisi, and McCann (1975) include assertion training as a component in a wider program of personal effectiveness training. Certainly, the objective of the proliferating assertion-training programs is to increase, in one way or another, the interpersonal effectiveness of the trainee. After reviewing a range of assertion measures and definitions, Heimberg *et al.* (1977) suggest that assertive behavior be defined as "effective social problem solving."

Linehan (1977) suggests that the effectiveness of any given assertive response can be evaluated from one of three points of view: the effectiveness in achieving the objectives of the response (objective effectiveness), the effectiveness in maintaining the relationship with the other person in the encounter (relationship effectiveness), and the effectiveness in maintaining the self-respect of the assertive person (self-respect effectiveness). These three types of effectiveness can be related to Goldfried and D'Zurilla's (1969) concept of effective problem solving, which they defined as "a response or pattern of responses to a problematic situation which alters the situation so that it is no longer problematical, and at the same time produces a maximum of other positive consequences and a minimum of negative ones" (p. 158). Objective effectiveness refers to how well the response alters the situation so that it is no longer problematical; relationship and self-respect effectiveness relate to the effect of the behavior on other positive and negative consequences. At times, of course, the objective of the assertion may be to improve a relationship or to enhance self-respect.

Potentially, any response could be evaluated in terms of its effectiveness in each of these three areas. The response a person chooses to make may depend on which type of effectiveness is most important in the given situation. For example, in attempting to reduce an unfair rent increase, objective effectiveness (rent reduction) may be more important than relationship effectiveness (maintaining a positive relationship with the owner). In trying to get a friend to do a favor (the objective), relationship effectiveness may be important enough to restrict the choice of responses to those which will not jeopardize the friendship. Self-respect effectiveness can also vary in importance. To give an extreme example, a woman might be willing to try a "helpless" assertive style to save her own life, whereas she may be unwilling to exhibit the same style to get her spouse to do his share of the housework.

Almost no research has been done to specify empirically which specific behavioral responses are likely to be effective in situations calling for assertion. Thus, the content of most assertion-training programs is a function of the trainer or researcher's judgment of effectiveness, a state of affairs which strikes us as questionable at best. Similarly, scoring manuals for behavioral role-play tests of assertion have been based on arbitrary definitions of assertion rather than empirical scaling of effectiveness. A procedure for developing an empirically based measure of interpersonal effectiveness has been outlined by Goldfried and D'Zurilla (1969) and has been used in other areas of social skill measurement (e.g., Freedman, 1974; Goldsmith & McFall, 1975). To date, however, this approach has not been used to define an assertive response.

Assertive Situations

Behavior cannot be divorced from the situation in which it occurs and, thus, an important determinant of whether or not a particular behavioral response will be defined as assertive is the situational context in which it is observed. The role of situational context is minimized when the assertion definition refers to specific responses, irrespective of behavioral consequences or the preceding events. When this is the case, the recognition that expressiveness, standing up for one's rights, etc., may not be the most acceptable or caring response necessitates the caution that assertion is not always appropriate (Alberti & Emmons, 1978; M. D. Galassi & Galassi, 1977). However, if assertive behavior is not always appropriate, adequate assertion *skill* would require the ability to discriminate when to emit an assertive response (cf. Linehan, 1979). Thus, there is need for an enumeration of those situational contexts where an assertive response would be required. For example, when assertion is defined as standing up for one's rights, one is immediately faced with specifying in which situations a person's rights are jeopardized.

The present tendency to label almost any valued interpersonal behavior or "effective social problem solving" (Heimberg *et al.*, 1977) as assertive further argues against the feasibility of constructing a situation-free definition. Effective and/or valued responses change with the situation. Context is clearly an issue if assertion is limited to "socially acceptable" responses. In what situations is it "acceptable" to express one's emotions? Would a particular style of responding be more appropriate in one situation than in another?

Situational relevance can be determined by either of two methods (Goldfried & Linehan, 1977). The first involves specifying the functional relationship of various situations to the construct in question. Thus, situations in which assertive responses (however defined) are likely to occur across a large sample of individuals would be judged as assertion-relevant situations. A variation on this method would be to determine those sets of situations which are easy to handle for persons otherwise labeled as assertive, and difficult to handle for persons who have an assertion deficit. This method was used by Levenson and Gottman (1978), for example, in determining assertion situations for a social competence inventory subscale.

A second method of determining situational relevance involves asking a representative sample of persons to generate a set of situations which they perceive as relevant to the construct. Both McFall and Marston (1970) and MacDonald (1974) have done this with college populations. However, only MacDonald has analyzed the situations to determine general dimensions which describe assertive-relevant situations. She analyzed a large number of assertion situations generated by college women and found four general dimensions, or types of situations, as follows: (1) a situation whose outcome or resolution is, at present, unclear (either someone is attempting to get the person to give in to a demand contrary to her preferences, or someone could do something the person would like, but only at her initiation); (2) a situation in which the person has been insulted (implicitly or explicitly); (3) a situation in which the person is asked to do someone a favor at her own expense; and (4) a situation in which someone either is or has been inconsiderate to the person. It is unclear whether these same situational parameters would define assertion situations for a college male or noncollege adult male and female population. Examination of the situations generated by Levenson and Gottman (1978) and McFall and Marston (1970) suggests considerable overlap. MacDonald's work represents a sophisticated methodology for defining assertion situations.

In summary, while not all clinicians and researchers would agree, the following are general characteristics of assertion, as the concept is currently described in the literature:

1. Assertion generally involves self-expressiveness, standing up for one's rights, or other more general interpersonal verbal responses to assertive situations.

2. The style of an assertive response is usually direct and open.

3. The person behaving assertively does not exhibit undue or excessive anxiety or fear. Thus, the style is verbally fluent, well timed, with appropriate response latency, speech volume, and eye contact.

4. Assertive responses are socially acceptable and appropriate for a given situation.

5. An assertive response is not coercive or aggressive.

6. Assertive responses are generally chosen to maximize effectiveness, although, according to some, a response could be assertive but not effective.

7. Assertive responses often occur in situations where someone is trying to get a person to give in to a demand or do a favor at the person's own expense, is insulting or inconsiderate, or where someone could do something the person would like, but only at the person's initiation.

ASSERTION AND WOMEN

The traditional feminine sex role dictates that the ideal women act in a basically nonassertive or passive manner. She should *not* be independent, competitive, aggressive, assertive, task oriented, innovative, self-disciplined, objective, analytical, rational, or confident. Instead, she should be dependent, passive, fragile, other oriented, empathetic, sensitive, nurturing, emotional, sympathetic, compassionate, and yielding (Bardwick & Douvan, 1972; Bem, 1974; Douvan & Adelson, 1966; Horner, 1970; Kagan, 1964; Maccoby & Jacklin, 1974; Terman & Tyler, 1954; Witkin, 1954). A woman who fits the feminine stereotype would be primarily oriented toward other people, seeking to nurture others and to receive her satisfactions through the accomplishments of someone else. Such sex-role standards can be conceptualized as the sum of socially designated behaviors that differentiate between women and men (Broverman, Vogel, Broverman, Clarkson, & Rosen-Krantz, 1975). In the past, such consensual standards were uncritically accepted by the public and by mental health workers as essential to healthy personality development and functioning. Rarely were the positive values of standard sex-role behavior questioned. Within the past decade, however, researchers have suggested that close adherence to rigid sex-role standards may actually prove detrimental to the full development of the individual man or woman's capabilities (Blake, 1968; Davis, 1967; Horner, 1969; Maccoby, 1966.).

In a series of studies designed to differentiate behaviors considered typically masculine from behaviors considered typically feminine, Broverman *et al.* (1975) conclude that characteristics ascribed to men are positively valued more often than characteristics ascribed to women; the positively valued masculine traits form a cluster of related behaviors which entail competence, rationality, and assertion, whereas the positively valued feminine traits form a cluster which reflect warmth and interpersonal sensitivity. Not only do sex-role standards assign assertiveness to the male, but the affectionate, dependent, and yielding woman, by definition, is constrained from acting assertively. The traditional model for women is clearly incompatible with the image of a person able to act in her own best interests, to stand up for her rights, or to express her honest feelings, opinions, and preferences comfortably, behaviors commonly attributed to the assertive individual. Despite the apparent shift in sex-role stereotypes in recent years, numerous investigators have noted that stereotypes about the differing characteristics of men and women are widely held (Lunneborg, 1970), persistent (Broverman, *et al.*, 1975), and highly traditional (Komarovsky, 1950; McKee & Sherriffs, 1959).

Are Men More Assertive Than Women?

Do men actually behave more assertively than women? The difficulty in answering this question is due to problems with the way the construct of assertion is defined. Men appear more assertive than women when assertion is defined as the expression of opinions and preferences while women appear more assertive in the expression of feelings. Overall, on self-report measures women tend to see themselves as less assertive than do men. In addition, studies looking at observations of men and women in groups find significant differences in interpersonal behaviors.

On the Wolpe–Lazarus assertion inventory (Wolpe & Lazarus, 1966), Butler (1976) found that women scored as significantly less assertive than men. The expression of positive feelings was the only area of assertion engaged in more frequently than men. The expression of positive feelings is congruent with rigid adherence to the traditional feminine role of being compassionate and sensitive to the needs of others. In other areas of assertion, such as the expression of negative feelings, setting limits on demands made by other people, and self-initiation, or the expression of what a woman wants to do, Butler found that women report themselves as less assertive than men.

A number of studies have assessed the style and amount of communication of men and women in dyads and larger groups. Frequently, the structure of the groups is such that the outcome measure of interest is the degree of influence each individual has within the group setting, specifically in facilitating change in other group members. Typically, women have been found to be less concerned with the task presented, whether problem solving or persuasion of others, and to be more focused on facilitating and supporting the communication of others in the group (Barron, 1971; Hall, 1972). In general, women's behavior in the groups or dyads is entirely consistent with an emphasis on maintaining relationships (relationship effectiveness) almost to the exclusion of obtaining objectives, such as solving the problem or persuading others (objective effectiveness).

Expression of Opinions and Ideas. A major observable difference between the behaviors of men and women in these groups is that men talk more than women, expressing and supporting their own ideas, while women talk less overall, and, in particular, make fewer contributions directly related to the problem to be solved. Hall (1972) found that women talked less than men, tended to support group members in their verbalizations, and did not stand up for their own beliefs and opinions as much as the men did. A number of studies support Hall's finding that men express more ideas and opinions than women (Aries, 1977; Bernard, 1972; Hilpert, Kramer, & Clark, 1975). Swacker (1972) asked men and women to describe a stimulus picture and gave them unlimited time in which to talk. Women spoke an average of 3 min while men talked an average of 13 min per picture. The average time that the men talked is depressed somewhat by the fact that several of the male participants were still talking when the 30 min allotted per subject was up, and they were asked to stop. Contrary to the image of the talkative woman, men clearly seem to express themselves more than women. It is interesting to note how men manage to take the lead in conversations. Several studies indicate that men gain control of the conversation by interrupting women (e.g., Bernard, 1972). When Zimmerman and West (1975) taped conversations of same-sex and mixed-sex pairs in a natural setting, they found that 98% of the interruptions and 100% of the overlaps in conversation which occurred when two people talked at once were made by male speakers.

Expression of Feelings. The results of a study by Barron (1971) support Hall's findings that the content of men and women's verbalizations differs and suggest that women may appear to be more assertive than men when expression of feelings is considered. Barron concluded that men are self-oriented, introducing more references to themselves in the conversation, and are concerned most with action, problem solving, and the projection of themselves into the environment. Barron's women, on the other hand, talked significantly more about their feelings and the feelings of others, and were concerned with bringing more silent members of the group into the conversation.

Implications of Expressing Ideas versus Feelings

It is not surprising that there are relatively few studies measuring expression of feelings as opposed to the numerous studies on the expression of ideas and opinions. Expression of feelings is not as highly valued in our society as expression of ideas and opinions, as is evident by the fact that none of the outcome measures in the above studies centered around facilitating group participation or increasing the expression of feelings by others, skills potentially as valuable to satisfactory decision making as the ability to hold forth on one's own ideas.

Expression of feelings, in contrast to the expression of ideas, is characteristic of the low-power individual (Bales, 1958). Subordinates in work situations are more likely to disclose personal feelings than are superiors (Jacobson, 1972). On the other hand, expression of ideas and opinions is demonstratedly related to level of influence wielded in affecting a group's decisions (Hall, 1972). In general, those who talk more have more visibility within the group. In mixed-sex groups, women rarely are considered to be the leaders. Instead, those people who initiate the most interactions and take up the most time expressing themselves generally are rated higher on leadership (Aries, 1977). Given their behavior when interacting with others, it is hardly surprising that women consistently prove to be less influential than men in affecting decisions. It is clear from the behavior of the women in groups that maintaining and facilitating relationships is their unspoken, perhaps unrecognized, agenda.

The major emphasis in some assertion-training programs is on expression of preferences and opinions, not expression of feelings. Some assertion definitions do not even include expression of feelings as an assertive behavior. By itself, expression of feelings clearly is not related to objective effectiveness or standing up for one's rights, and may even be counterindicated. It seems possible to us that expression of feelings has been included in assertion programs to meet the needs of men, rather than women.

Women in Transition: The Need for Assertion Training

The "Liberated" Woman. The recognition by women themselves that they are in need of assertion training has resulted in the rapid expansion of assertion-training courses for women (Tolor, Kelly, & Stebbins, 1976). Researchers have acknowledged the need for assertion training for populations where chronic feelings of helplessness are common, such as psychiatric patients and persons from a low socioeconomic background. Such environments may punish an individual's attempts to control his or her life and reinforce a passive, helpless posture. The environment of the traditional woman can be compared to that of other "disadvantaged" people. However, the role of women in today's world is in flux; demands and expectations placed on the modern woman are far different from those of the past. The clinical use of assertion training with any population rests on the premise that "people have certain rights which they are fully entitled to exercise, and that proper human adjustment includes exercising them" (Wolpe & Lazarus, 1966, p. 38). But who determines what rights an individual has? Does everyone have the same rights irrespective of the context in which the behavior occurs or the achieved or ascribed "roles" of the participants? Women in transition frequently find themselves in ambiguous social roles where their rights are either ill-defined, in flux, or inadequate to meet the new demands placed on them. The recognition by women of their need for assertion training stems in part from their searching for new and expanded role rights. In addition, as women experiment with different behaviors and are exposed to new situations, it becomes evident to them that they are lacking in particular interpersonal skills that heretofore have been necessary only for men. Skill training is useful with populations which have for some reason been prevented from learning appropriate behavior, especially those who are moving from one milieu to another where different skills are required. Heimberg *et al.* (1977) note that new social skills can be taught to help make the transition from one lifestyle to another go smoothly for such varied groups as the elderly, the recently divorced, or prison inmates preparing for release. Women, like other groups experiencing rapid changes, have become increasingly aware and concerned about personal limitations and have shown by their interest in assertion training a desire to overcome these limitations both in terms of expanded personal rights and more adaptive, broader social skills.

The "Traditional" Woman. Even the woman who is not consciously or deliberately involved with or aware of rapid social shifts is subject to disruption as the structure of her life is inevitably touched by changes outside herself. By their attitudes and behaviors, a large number of women in today's world still conform to the traditional feminine role (Bem, 1974; Broverman *et al.*, 1975). Observation of the role of women in our culture suggests that passive and dependent response patterns may not be useful in terms of overall life satisfaction. Being passive and dependent requires, first and foremost, that a woman have someone on whom to depend, ideally (from a traditional viewpoint) a man. In reality, many women do not have a man on whom to depend; there is a growing trend toward female-headed families, with 14% of American children under 18 being reared by their mother alone (U.S. Department of Commerce, 1974a). In addition, increased longevity of women means the average woman lives alone for at least the last 10 years of her life. According to the March 1974 census, 42% of all the women in the labor force were single, widowed, divorced, or separated, all more or less living on their own. Older women are especially likely to be living without a man on whom to depend; 37% of American women over the age of

40 were living without husbands (U.S. Department of Commerce, 1974b).

How do these women alone fare in a society that has prepared them to live for and through a man? How does the softspoken, gullible, gentle, yielding, childlike, dependent person manage when she is totally responsible for her own life and that of others? Even women who are married and presumably in a position to play out the unassertive feminine role to advantage apparently do not derive general satisfaction from their lives. Foder (1974) found that married women report themselves as more depressed and less satisfied with their lives than single women.

Traditional women have been taught from childhood to gain satisfaction from obedience to authority figures. Despite some cultural variation, girls have been trained to be more compliant than boys, whereas boys have been encouraged to be more assertive than girls (Barry, Child, & Bacon, 1959). What happens when a woman brought up in this milieu discovers that compliance does not yield the rewards she had anticipated? Assertive behaviors that would be more effective and efficient in gaining her goals are likely absent from her repertoire. In fact, she may feel prohibited from learning them by societal pressures to conform to the feminine role. A woman in this position could experience herself as having little effect on her environment and a feeling of helplessness may develop. When intense enough, the experience of being powerless or helpless may lead to clinical depression (Seligman, 1975). At all ages, women suffer from depression at a much higher rate than do men (National Institute of Mental Health, 1965–68). There are data to suggest that depressed women consider themselves more traditionally feminine than the average nondepressed woman (Bart, 1972).

Why is it that both "liberated" and "traditional" women turn to structured training programs in order to learn assertive behavior? It seems reasonable to assume that a woman who becomes aware of both her needs and her right to have them met could quickly learn enough ways of effectively asserting her newly recognized rights. After all, models from whom she could learn new skills account for half the population; consciousness-raising groups exist to support her in redefining her rights and further delineating her needs. Clearly, the learning of assertive behaviors requires a dramatic break with the traditional feminine role, and, in essence, the acquisition of behaviors formerly associated with the traditional masculine role. So pervasive and subtle are role expectations that one of the first functions of assertion-training programs for women is to clarify for participants what the rights of *people* are and that those rights apply to women as well as to men. Skill training in the effective use of assertive behaviors, how to be assertive and when to be assertive, along with informing women of their rights, therefore, constitute the goals of assertion-training programs for women.

METHODS OF ASSERTION TRAINING

Although many assertion books, popular articles, and training programs are specifically aimed at women, to date there have been no empirical studies designed to determine whether the methods and procedures used in assertion training should be varied according to the sex of the client. A number of studies have included both male and female participants. However, with few exceptions (e.g., Friedman, 1968) effectiveness of treatment as a function of client sex has not been examined. Assertion training studies reported in the literature are a mixture of case reports and both analogue and clinical between-group studies. Generally, case reports demonstrate the application of assertion training to novel populations (e.g., Barnard, Flesher, & Steinbook, 1966; Edwards, 1972; Foy, Eisler, & Pinkston, 1975); between-group analogue studies ferret out the effective components of complex treatment packages (e.g., Eisler, Hersen, & Miller, 1973a; Hersen Eisler, & Miller, 1974; Hersen, Eisler, Miller, Johnson, & Pinkston, 1973b, McFall & Lillesand, 1970, 1971; McFall & Marston, 1970; McFall & Twentyman, 1973; Turner & Adams, 1977), and between-group clinical studies demonstrate the effectiveness of a treatment package (e.g., Finch & Wallace, 1977; J. P. Galassi *et al.*, 1974), compare treatment packages (e.g., Linehan *et al.*, 1979; Wolfe & Fodor, 1977), or, on occasion, analyze the relative effectiveness of various treatment components (e.g., Kazdin, 1974, 1976; Nietzel, Martorano, & Melnick, 1977). What follows is a summary of the results of the clinical and analogue treatment studies. Case studies are not included since, although suggestive, their lack of experimental control makes conclusions about treatment efficacy tenuous. In addition, even though treatment studies aimed at modifying social skills other than assertion are relevant for treating the nonassertive individual, we have included only those where the targets are specifically identified as assertive behaviors. We have not, however, excluded the several studies where only males participated. For the most part, these studies have taken place in psychiatric hospitals. For example, four out of the five we found were carried out in Veterans Administration hospitals (Eisler, *et al.*, 1973a; Finch & Wallace, 1977; Hersen *et al.*, 1973b, 1974). Although we are primarily concerned with assertion training for women, the lack of studies examining the relationship of treatment effectiveness to sex of client leaves no grounds for differentiating those procedures, if any, which might be less effective when used with women. Due to space limitations, detailed descriptions of each of the procedures described are not given. The interested reader is referred to any number of texts on behavior therapy (e.g., Goldfried & Davison, 1976; Rimm & Masters, 1979) or clinical articles describing general assertion-training methods (e.g., Linehan, 1979; MacDonald, 1975).

Models of Assertion Training

Most behavioral assertion training packages seem to be based on one or more of three general models (Linehan, 1979). These models are the skill-deficit model, the response-inhibition model, and the faulty-discrimination model. The skill-deficit model assumes that unassertive persons have an assertion skill deficit; that is, they do not have the requisite behavioral responses in their repertoire. In contrast, the response-inhibition model assumes that the nonassertive person has the necessary skills but is inhibited in assertive situations such that the assertive responses are not emitted. There are two major variations of the response-inhibition model. The first suggests that inhibition is due to conditioned anxiety responses in assertive situations; the second hypothesizes that individuals are inhibited due to *maladaptive* beliefs, self-statements, and faulty expectations. The faulty-discrimination model also assumes the presence of the requisite skills, but, in addition, assumes that if unassertive persons knew *how* to apply these skills in particular situations, they would respond in an assertive manner (i.e., the response would not be inhibited by anxiety or faulty beliefs or expectations). Thus, from the vantage of this model, the unassertive person does not discriminate accurately those situations in which an assertive response would be appropriate and effective from those in which it would not. The consequences of these faulty discriminations are, first, inappropriate and ineffective attempts to respond assertively, followed by a consequent extinction of assertive responses in all situations.

A fourth model, the rational choice model, proposed by Linehan (1979) assumes that, at times, the unassertive person may have perfectly valid reasons for choosing not to behave assertively. The person has the skills, is not inhibited, knows when an assertive response is likely to be effective, and, for a wide variety of reasons, chooses not to behave assertively. Religious beliefs, intense caring for the other individual, fatigue, high probability of punishment, or lack of concern about the outcome of the situation are but a few of the variables which may influence a person to choose to behave nonassertively in a given encounter.

Feminist rhetoric suggests that in most, if not all, instances, women are nonassertive because of a high probability of punishing consequences. We have been unable to uncover any well-controlled empirical studies which directly confirm this contention. Individuals of both sexes are likely to be disliked following direct, assertive responses (Ford & Hogan, 1978; Marriott & Foster, 1978). It is possible that, due to her traditional economic dependence on others, maintaining others' goodwill is of greater importance to women than it is to men. If so, this would mean that assertion and the consequent loss of liking would indeed be more punishing for women than for men. A second type of indirect evidence for the feminist contention might be societal attitudes and expectations regarding women's assertion. Linehan and Seifert (1978) found no effects for sex of the subject when male and female community persons were asked to evaluate the appropriateness of assertion. Interestingly, results suggested that females, but not males, expected the *opposite* sex, as compared to the *same* sex, to be more disapproving of their own potential assertive behavior. These results would suggest that a woman's fear of negative consequences from acting assertively may be a function of her erroneous expectations about male disapproval. However, an alternate explanation might be that there is a discrepancy between what men *report* as their attitude about assertive behavior in women and their actual response to an assertive woman. Connor, Serbin, and Ender (in press) found that assertive behavior earned approval for a boy engaging in it, while unassertive, passive behavior earned approval for a girl. Research is needed to study directly whether or not women, in actuality, receive more negative consequences for acting assertively, or if women are inaccurately anticipating negative consequences; such research would clarify whether women's unassertive behavior would best be explained by the rational-choice model or the response-inhibition model (cognitive variation).

Outcome Studies: Complex Treatment Packages

A number of studies have examined the effectiveness of relatively complex treatment packages for increasing assertive behaviors. The typical package is based on the skill-deficit model, is conducted in a group modality, and includes in-session response practice, modeling, coaching and feedback, and *in vivo* response practice assignments. Results of a number of studies taken together suggest that variations of this package are effective with college females (Rathus, 1972), a mixed group of college males and females (J. P. Galassi *et al.*, 1974), community males and females (Schinke & Ross, 1976), and males both with anger problems (Rimm, Hill, Brown, & Stuart, 1974) and diagnosed as schizophrenic (Finch & Wallace, 1977). Linehan, Walker, Bronheim, Haynes, & Yevzeroff (in press) compared a response practice treatment package given individually versus in groups and found that, while both were effective in increasing assertive behaviors and reducing discomfort in assertive situations, there was no difference between the two modalities on any treatment measure.

Other treatment packages have also been developed for treating assertion problems. For example, several studies have tested the effectiveness of systematic desensitization in increasing assertion. The rationale behind this treatment approach is the contention that nonassertive behavior is a function of assertive response inhibition due to conditioned anxiety in assertive situations. Although outcome results are somewhat equivocal, there are some data to suggest that the procedure might be effective (Weinman, Gelbart, Wallace, & Post, 1972), although not as effective as a response-practice package (Thorpe, 1975; Trower, Yardley, Bryant, & Shaw, 1978). To date, there is no controlled outcome study to determine whether the inclusion of systematic desensitization would increase the effectiveness of other assertion-training packages.

A number of investigators have evaluated the effectiveness of a treatment package based on the cognitive-inhibition model of nonassertive behavior. For the most part, cognitive treatments involve teaching clients to ferret out their maladaptive self-statements, beliefs, and expectations about assertive encounters and replace them with more adaptive cognitive contents. Treatment targets, therefore, are the cognitions presumed to mediate assertive verbal responses and nonverbal responses, rather than the actual overt behavioral responses in assertive situations. Both Thorpe (1975) and Linehan *et al.* (1979) compared a cognitive-restructuring package with a response-practice package, and, with few exceptions, they found no difference between the two treatments on measures of both assertive responding and comfort in assertive situations. Wolfe and Fodor (1977) and Linehan *et al.* (1979) examined the effectiveness of a single-treatment package combining cognitive restructuring with response practice. Both conclude that the addition of the cognitive-restructuring component increases the effectiveness of the response-practice package. This conclusion, however, should be viewed with caution. In the Wolfe and Foder study, the combined treatment was superior on self-reports of anxiety in assertive situations but not on measures of assertive behaviors. In the Linehan *et al.* study, the superiority of the combined package, over the response practice package, was concluded on the basis of comparisons of each treatment with the control groups. On no measure was there a significant difference between the two treatments themselves.

Outcome Studies: Component Procedures

Procedures which have received the most attention in the assertion training literature include response practice, modeling, instructions and coaching, feedback, and reinforcement. Although cognitive restructuring and systematic desensitization are sometimes referred to as independent treatment procedures, in actuality they refer to packages of several treatment components and thus are more properly considered as individual complex treatments. Studies have been done to assess the relative effectiveness of various components of both desensitization and cognitive restructuring when applied to other target problems; none, however, has been done in the area of assertion training.

Response Practice and Coaching. As noted above, many complex treatment packages include response practice as well as coaching and instructions. Their relative contribution to the effectiveness of the treatment, however is confounded by the inclusion of other components. McFall and his associates (McFall & Lillesand, 1970, 1971; McFall & Marston, 1970; McFall & Twentyman, 1973) did a series of analogue studies designed to ferret out the contribution of a number of treatment components, including the effects of response practice and coaching. The general methodology of their studies was the same, and treatment included role-play practice with an audio recorder controlled by an experimenter in an adjacent room. Coaching instructions, the same for all participants, were also given via audio recorder. Over a series of six experimental studies, results suggest that combined response practice and coaching is effective in increasing assertive responses both in role-play tests and *in vivo* (McFall & Marston, 1970; McFall & Twentyman, 1973). Comparisons of each component with each other and with a combined treatment suggest that the effects of response practice and coaching are independent and additive (McFall & Twentyman, 1973). Similar findings are reported by Turner and Adams (1977).

Modeling. Results of studies comparing the effectiveness of treatments with and without response modeling are equivocal. McFall and Twentyman (1973) found no increase in effectiveness when either audio- or videotape modeling was added to response practice, coaching, or a combined package. On the other hand, Hersen, Eisler, and Miller found a significant increase in effectiveness when videotape modeling was added to both response practice alone (Eisler *et al.*, 1973a) and to a combined response practice plus coaching (instructions) treatment (Hersen *et al.*, 1973a,b). Friedman (1968) found that a modeling treatment alone was as effective as response practice in increasing assertion, although a combined package was superior to either alone. Given the demonstrated effectiveness of modeling procedures in teaching other social responses (e.g., Bandura, 1969), it is not clear why McFall and Twentyman (1973) did not find an effect when they added modeling to their treatment. One possible explanation might be the simplicity of their target response when compared to the targets in the Hersen *et al.* studies and the Friedman study. McFall and Twentyman confined their study to simple refusal of unreasonable requests, whereas the other investigators required more complex responses. It could be that modeling is relatively unnecessary when the response to be taught is brief and simple.

Covert Treatments. Several studies have reported that both modeling and response practice are just as effective when done covertly, i.e., in the individual's imagination, as when done overtly. McFall and Lillesand (1971), in fact, found that covert response practice was *more* effective than overt response practice when the overt practice individuals also listened to an audio replay of their assertive responses. When playback was deleted, covert, overt, and combined practice were equally effective in increasing assertive responses (McFall & Twentyman, 1973). Kazdin (1974, 1976) conducted two studies examining the effectiveness of covert modeling, where individuals imagined a model person behaving assertively. In both, he found covert modeling beneficial. Imagining the model being reinforced, imagining multiple models, and a combination of both was more effective than covert modeling alone.

Feedback and Response Reinforcement. Feedback refers to the repetition to participants, either verbally or by audio- or videotape, of their previous response(s) in an assertive situation. As a treatment procedure, it is included in almost all complex treatment packages. However, in most clinical studies feedback is confounded with both evaluation of the response and contingent coaching or reinforcement, such that it is unclear whether feedback alone contributes to treatment effectiveness. Studies comparing identical treatments with and without feedback suggest that neither viewing oneself on videotape nor listening to an audiotape replay of one's responses adds appreciably to training effectiveness (Gormally, Hill, Otis, & Rainey, 1975; McFall & Marston, 1971; Melnick & Stocker, 1977).

Surprisingly, response reinforcement and contingent coaching treatment procedures have not been evaluated for treatment effectiveness when applied to assertion training. The exception is the work of Kazdin (1974, 1976) where he compared covert modeling, with and without reinforcement. In both studies, he found that the addition of model reinforcement increased treatment effectiveness. Although there is ample evidence in the behavioral literature to demonstrate the effectiveness of reinforcement as a behavior-change procedure, there is still a need to study its relative effectiveness in assertion training.

EFFECTIVENESS OF ASSERTION-TRAINING CONTENT: A QUESTION OF STYLE

Usefulness of Assertion-Training Programs

While the content of most assertion-training programs is fairly uniform, very little effort has been expended to establish the usefulness of this content from the point of view of the individual participant. Many studies demonstrate that assertion-treatment programs have training effectiveness; that is, participants exhibit a higher rate of the trained behaviors after completion of the program than before (e.g., J. P. Galassi *et al.*, 1974; Linehan *et al.*, 1979; McFall & Lillesand, 1970, 1971; McFall & Marston, 1970; McFall & Twentyman, 1973). However, individuals do not seek out assertion training to become more assertive *per se;* they want to be more effective in interpersonal interactions, to gain important objectives. To date, very few studies have investigated the usefulness of behaviors taught in assertion training as far as accomplishing the goals of the individual is concerned.

Assertion training is a relatively recent venture by behavior therapists into interactional processes. The naïveté of the behavior therapist in the new and complicated sphere of interpersonal interaction becomes clear when one examines the response classes and styles of communication unequivocally promoted by assertion training. Assumptions without benefit of data abound. Fortunately, it is not necessary for the assertion trainers to start from the beginning in establishing the efficacy of their content; a considerable body of knowledge already exists regarding the impact and effectiveness of behaviors clinicians have labeled as "assertive." Social psychology research has long been investigating interactional processes in such areas as communication, bargaining and negotiation, power, leadership, and interpersonal attraction. As will be demonstrated, the data from these sources bring up serious doubts about the efficacy of assertion training program content as it now exists. In particular, we will be examining the utility of such programs for women, as well as suggesting ways to increase the beneficial aspects of assertion training while decreasing potentially adverse effects.

Assertion Training as Typically Conducted

Content. Included in most definitions of assertive behavior is an explicit description of the response classes of expressiveness, standing up for one's rights, and other interpersonal verbal skills, and implicit or explicit information about the style of delivery of these behaviors. The assumption behind the proliferation of training programs designed to increase assertive behaviors is that the specified behaviors, delivered with the specified style, will result in producing or maintaining positive consequences for the individual while at the same time minimizing negative consequences. The traditional content of assertion training programs, in other words, is assumed to be effective in securing for the individual that which he or she desires.

It becomes clear by looking at training manuals and books published to teach assertive skills that the model communication style advocated for use in virtually all interpersonal assertive situations is directness, and directness is usually synonymous with brevity. As McFall and Lillesand (1970) observe, "it is more assertive to use as few words as possible and not go into lengthy explanations" (p. 12). Repeatedly, in the majority of treatment programs designed to increase assertive behaviors, participants are urged to be more direct, brief, and to the point (e.g., Butler, 1976; Richey, 1977). No one seems to have questioned whether indeed there are ways to be assertive other than by using direct and unembellished statements. In particular, assertive-training programs specifically designed for women seem to emphasize brevity almost to the exclusion of any other style of assertion. Butler (1976) admonishes women to send messages "devoid of subtleties, manipulation, or disguise" (p. 85). She points out that in sending a clear message, a woman must remember that she is going against the diplomacy of the stereotypic feminine role. Perhaps it would be useful to think of this preoccupation with simplicity and economy of words as an overreaction to the convoluted style of communication commonly attributed to the traditional woman.

Effectiveness. As described earlier, the effectiveness of any interpersonal response can be evaluated in terms of its objective effectiveness (in achieving the objectives of the response), its relationship effectiveness (in maintaining the relationship with the other person), and self-respect effectiveness (in maintaining the self-respect of the actor). Promotion of the direct style of interaction in assertive situations is based on the assumption that when individuals state what they want in a clear, concise manner, they are more likely to obtain it than if they give an ambiguous, indirect message or, unassertively, say nothing at all. Amazingly, this assumption has no data to support it. Virtually every assertion-training program currently in existence, however, operates as if it were true.

Research in the areas of power, social influence, interpersonal attraction, and bargaining and negotiating effectiveness applies to the usefulness of various communication styles as well as other behaviors related to assertion. Most of the research focuses on two general aspects of the interaction—behaviors associated with influence, persuasion, and power, and behaviors associated with interpersonal attraction, liking, and attribution of positively valued characteristics. Interestingly, these behavioral styles are often antithetical; that is, if one chooses to be liked, a specific set of responses is useful, but if one chooses to be influential, a competing set of responses is most effective.

Examination of research on social power and influence suggests that response classes and styles of delivery typically taught in assertion training are associated with objective effectiveness. Studies investigating the characteristics of the high-power, high-influence person reveal an individual whose behavior is remarkably similar to that of the ideal end product of an assertion training program. The high-power individual freely states opinions and preferences, communicates self-confidence to others, and seeks to have impact on others through direct action (Jacobson, 1972). Thus, if short-term objective effectiveness is of primary importance to the individual interested in assertion training, then the standard, available programs might be useful.

Response classes and styles of delivery typically taught in assertion training are *not* associated in the interpersonal attraction literature with effectiveness in maintaining or enhancing the relationship. Falbo (1977) found that peers evaluated those who used an assertive, direct style of communication as having lower social skills than those who used such embellished styles of communication as compromise, bargaining, and reasoning. Marriott and Foster (1978) discovered that individuals who responded to a request with an ambiguous, indirect refusal were liked significantly more than those who responded directly and unambiguously. Direct-assertive persons and aggressive persons (who engaged in name calling during the refusal) were rated as equally likable (or unlikable). Ratings of respect were highest for the individual making the ambiguous, indirect response. Both men and women participants expressed lowest anger toward the person making the ambiguous refusal. Thus, the direct assertive response might not be the best choice for an individual for whom it is important to be liked and regarded positively by others, at least in the immediate situation.

Three nonverbal behaviors often considered to be components of assertion and included in assertion training programs are fluency, loudness, and eye contact (Eisler *et al.*, 1973b). In a study investigating the effect of these behaviors on the perception of others, Ford and Hogan (1978) found that only fluency led to person perceptions of "assertive" or "competent," while loudness and direct eye contact resulted in the asserter's being perceived as "aggressive" and "authoritarian." A further extension of this line of research indicated that "assertive" behaviors have variable, sometimes negative interpersonal outcomes. Verbal refusals produced some immediate benefits for the asserter, but generated primarily negative affective, attributional, and long-term outcomes (Ford & Hogan, 1978). Again, the direct-assertive response, and, in particular, the nonverbal aspects of that behavior, are detrimental in securing for the individual the goodwill of others.

The assumption of assertion-training programs with respect to self-esteem, or self-respect effectiveness, is that increased assertion is related to an improved self-concept. Little research has been conducted to determine if, indeed, increased self-esteem occurs as a result of assertion training. Tolor *et al.* (1976) found in both sexes a positive relationship between direct assertion and favorableness of self-concept. Thus, direct-assertion training may lead to an increase in self-esteem.

The research suggests, then, that those individuals who use a direct, unembellished style of communication may be enhancing short-term, objective effectiveness at the same time that they are sacrificing relationship effectiveness. Relationship effectiveness is ill-served by the use of direct assertion; the individual using this style is not well liked immediately following the response, nor do others anticipate the continuation of the relationship on a long-term basis. Objective effectiveness, in contrast, seems to be facilitated by the use of the assertive direct style, at least on a short-term basis. However, long-term, objective effectiveness is dependent on maintenance of the relationship. Unless the structure of the relationship is imposed by external sources and, therefore, is not subject to termination (i.e., boss and employee), interpersonal relationship effectiveness will be crucial in maintaining the relationship within which one can pursue objective effectiveness.

Effective Communication Styles Not Included in Assertion-Training Programs

There are communication styles other than unembellished direct which might be more useful at times in gaining both objective and relationship effectiveness, without losing self-effectiveness. In addition to the simple direct style of the traditional assertion-training programs, there are two other general styles of interpersonal communication: complex (or embellished) direct and ambiguous indirect. The complex-direct style consists of the direct message, stated similarly to the message in simple direct ("I don't have the time to help you now"), but also includes embellishments such that a second message is conveyed at the same time ("I can see that you are really having a tough day, but I don't have the time to help you now"). Ambiguous, indirect communication includes the message, but its meaning is not clear to anyone except the sender who hopes it will be understood ("Oh, I was just getting ready to go for lunch, and then I was going to pick up some things at the store").

When we, together with five of our associates, observed our own assertion attempts for a 2-week period, we discovered that at least five embellishments were commonly used, including empathy, helplessness, apology, flattery, and outright lying.[2] Research needs to be done on potential embellishments for differential relationship, objective, and self-respect effectiveness, both from a short-term and a long-term perspective. It could be that refinement of the direct response by the addition of one or more embellishments (i.e., empathy) would result in achieving objective effectiveness without losing self-esteem. We would question the strict application of a simple direct-assertion style to any and all situations.

Despite the emphasis in assertion-training manuals on directness, however, it is unlikely that experienced trainers in the course of conducting assertion training fail to address the issue of when one might reasonably choose to behave unassertively in sensitive situations, or, at least, to soften (or embellish) a refusal or a request with diplomacy. Even if the trainer does not initiate discussion of such situations, a participant most likely will. Thus, styles other than direct are probably already being discussed in assertion-training programs, albeit not systematically. Such inclusion may explain the results of two studies which found that assertion training for women resulted in an *improvement* of their relationships with significant others, as measured by the partner's report of satisfaction with the relationship (Linehan *et al.*, 1979), and by the women's report of increased marital satisfaction (Blau, 1977). If, as seems probable, both clinical wisdom and experience mitigate against a hard-line approach to teaching assertive behaviors, and a rigid adherence to one style of behavior is detrimental for individuals concerned with maintaining interpersonal relationships, the use of inexperienced paraprofessionals as assertion trainers, along with the burgeoning self-help literature for increasing assertion, have grave implications.

FUTURE DIRECTIONS

Linking Assessment to Treatment

To a certain extent the research literature on assertion has been characterized by attempts to construct a model of assertive behavior which would explain all instances of nonassertive behavior. This approach, while useful as a first step, is limited in that it is unlikely that the same variables control assertive or nonassertive behavior in all cases, or

[2] Richard McFall (personal communication) suggests that "lying" might better be labeled "stretching the truth," since it appears to be such when refusing unwanted invitations.

even in the same individual in different situations. For example, it is probable that some individuals do not have assertion skills in their repertoire, whereas other persons may have the skills but are inhibited from using them. Within the inhibited group, some may be inhibited by conditioned anxiety while others may be inhibited by unrealistic beliefs and expectations. Across all people, the absence of assertive behavior in some situations may best be explained by the rational-choice model. Future assertion-training studies need first to differentiate clients in terms of the variables controlling the nonassertive behavior and then construct treatment approaches specific to each client group. Although any experienced clinician will acknowledge that some clients probably need skill training, others may need cognitive restructuring, and still others may need some form of relaxation training, this linkage between assessment and treatment has not occurred to any great extent in the assertion-treatment literature.

Should We Retain the Construct "Assertion"?

Our initial review of assertion definitions, gleaned from semantic definitions, responses taught in assertion-training programs, and measures of assertion, suggest that the construct is variously used to refer to one or more response classes, styles of responding, consequences of responding, and/or situational contexts in which specific responses are made. The terminological obfuscation surrounding the term makes comparisons and generalizations of research results difficult and at times impossible. Usually, it takes close reading of the method section in empirical studies even to determine which response class(es) were studied. When a unidimensional response class is employed (e.g., McFall & Lillesand, 1970, 1971; McFall & Marston, 1970; McFall & Twentyman, 1973), results are not necessarily generalizable to other response classes also defined as assertive. This fact, however, is often obscured by labeling the report an *assertion* study as opposed to a more restrictive title describing the response class(es) studied.

Thus, we believe that the general construct of assertion has outlived its usefulness and should be discarded in favor of more restrictive labels, subsumed under the more general construct of social skills. If this were done, we would be inclined to support a definition closely approximating a correlated set of responses having to do with some aspect of standing up for one's rights. A start in this direction is evident in the increasingly common differentiation of positive assertion from negative assertion. We find it unfortunate, however, that standing up for one's rights and refusing unreasonable requests have been labeled "negative." We suspect that the label is a result of looking at the probable short-term relationship effectiveness of the behavior, to the exclusion of looking at both objective and self-respect effectiveness, long- and short-term, as well as long-range relationship effectiveness. Standing up for one's rights and refusal of unjust demands are likely to have more total positive than negative consequences for women. An additional problem with the differentiation between positive versus negative assertion classes is the absence of data suggesting that either, in fact, form a unidimensional response class. Further research is needed to clarify the behavioral relationships among all responses included under the assertion rubric.

Assertion Training versus Effectiveness Training

There is a surprising paucity of empirical data to support a contention that the behavioral responses taught in most assertion training programs are in fact effective in social situations. As we previously noted, there is reason to believe that, at least for some of the people, some of the time, response content of the typical assertion program may be detrimental rather than helpful. This outcome is not surprising since situational requirements are such that the same response or class of responses is not likely to be effective in all interpersonal situations. Although we would support the narrowing of the assertion concept of a unidimensional response class, we would not support limiting research and training to such a class of behavior. Rather, we believe that there is an increasing need for research to determine what are the effective responses for various situations, together with training programs developed to teach persons how to select and emit such responses.

This emphasis on teaching effective responses is similar to that of Heimberg *et al.* (1977), who suggest that the term "assertion" be redefined to mean effective social problem solving. Thus assertion, or effective social problem solving, would be those responses that change a situation such that it is no longer a problem while at the same time maximizing positive consequences and minimizing negative consequences. Our position differs from Heimberg *et al.* in that we are not suggesting a redefinition of assertion. An additional definition is not going to clarify the field. Instead, we are suggesting that the focus in both research and training be shifted, from one emphasizing specific response classes or behavioral styles to one emphasizing effective responses in predetermined sets of situations. From this point of view, the field of assertion research would be operationally defined by the situations considered relevant. As noted earlier, methods are available for determining which situational parameters are most relevant to assertion.

Analyses of response effectiveness need to be done from more than one point of view. Each response-situation pair must be looked at in terms of both long- and short-range consequences; objective, relationship, and self-respect effectiveness must also be considered. Thus, although objective effectiveness is necessary for adequate problem solving, the effectiveness of the response in terms of maintaining or enhancing the relationship with the other person and the self-respect of the actor is an important side effect which must be addressed. Rather than dictating which type of effectiveness is of greatest value, we believe that the role of the assertion researcher and trainer is to point out to the client the effects, i.e., likely consequences, of various responses. Responsibility for determining the best response in light of this information should be left to the client.

SUMMARY

The focus of this chapter has been on assertion training with women. After reviewing a number of definitions of assertion we concluded that assertion encompasses particular sets of behavioral response classes, styles of expression, levels of social effectiveness, and sets of interpersonal situations. We summarized the various uses of the term in the literature and suggested the following seven characteristics of assertive behavior: (1) assertion involves self-expressiveness, standing up for one's rights, and other more general interpersonal verbal responses, (2) delivered in a direct and open style, (3) without undue anxiety, (4) in a socially acceptable manner, (5) which is not aggressive or coercive. Furthermore, assertive responses (6) are chosen to maximize effectiveness, and (7) most often occur in situations where someone is trying to get a person to give in to a demand or do a favor, is insulting or inconsiderate, or where someone could do something the person would like, but only at the person's initiation.

An analysis of assertion and female sex-role stereotypes indicated that women are typically expected to behave in a nonassertive manner. Indeed, women generally respond in ways antithetical to general assertive requirements. Cultural changes calling for greater assertion on the part of women were explored and the need for assertion-training programs directed at women was discussed.

Four models on which assertion-training programs could be based were outlined: the skills-deficit model, the response-inhibition model, the faulty-discrimination model, and the rational-choice model. For the most part, assertion-training programs have been based on the skills-deficit model, although systematic desensitization and cognitive restructuring, both based on the response-inhibition model, have been evaluated. The treatment components which have received the most extensive attention and have been found to be effective in increasing assertive behavior include response practice and coaching; covert practice and modeling have also been effective. The data on modeling are equivocal but suggest that it increases treatment effectiveness when complex responses are involved. Research on response feedback and reinforcement is limited; reinforcement, at least, is likely to increase assertive responding.

Analysis of the content of the typical assertion-training program suggests that almost no data have been obtained assessing the effectiveness of the behaviors taught in the social interactions of the participants. For the most part, assertion is equated with a direct and brief communication style. Data suggest that this style may not always be the most effective. Several possible other communication styles were reviewed and we suggested that content might be more effective if these styles were included in training packages.

Finally, we offered three suggestions for future directions in assertion-training research. First, future assertion training studies should initially differentiate clients in terms of the variables controlling the nonassertive behavior, and then construct treatment approaches specific to each client group. Second, the general construct of assertion should be discarded in favor of more restrictive labels, subsumed under the more general construct of social skills. Third, the focus, in both research and training, should be shifted, from one emphasizing specific response classes or behavioral styles to one emphasizing effective responses in predetermined sets of situations.

REFERENCES

Alberti, R. E., & Emmons, M. L. *Your perfect right* (3rd Ed.), San Luis Obispo, Calif.: Impact, 1978.

Aries, E. Male–female interpersonal styles in all male, all female and mixed groups. In A. G. Sargent (Ed.), *Beyond sex roles*. St. Paul: West Publishing Co., 1977.

Baer, J. *How to be an assertive (not aggressive) woman in life, in love, and on the job: A total guide to self-assertiveness*. New York: Signet, 1976.

Bales, R. F. Task roles and social roles in problem solving groups. In E. E. Maccoby, T. M. Newcomb, & E. L. Hartley (Eds.), *Readings in social psychology* (3rd ed.), New York: Holt, Rinehart and Winston, 1958.

Bandura, A. *Principles of behavior modification*. New York: Holt, Rinehart and Winston, 1969.

Bardwick, J. M., & Douvan, E. Ambivalence: The socialization of women. In V. Gornick & B. K. Moran (Eds.), *Woman in sexist society*. New York: Basic Books, 1972.

Barnard, G. W., Flesher, C. K., & Steinbook, R. M. The treatment of urinary retention by aversive stimulus cessation and assertive training. *Behaviour Research and Therapy*, 1966, *4*, 232–236.

Barron, N. Sex-typed language: The production of grammatical cases. *Acta Sociologica*, 1971, *14*, 24–72.

Barry, H., III, Child, I. L., & Bacon, M. K. Relation of child training to subsistence economy. *American Anthropologist*, 1959, *61*, 51–63.

Bart, P. B. Depression in middle-aged women. In V. Gornick & B. K. Moran (Eds.), *Woman in sexist society*. New York: Basic Books, 1972.

Bem, S. The measurement of psychological androgyny. *Journal of Consulting and Clinical Psychology*, 1974, *42*, 155–162.

Bernard, J. *The sex game*. New York: Atheneum, 1972.

Blake, J. Are babies consumer durables? *Population Studies*, 1968, *22*, 5–25.

Blau, J. S. *Changes in assertiveness and marital satisfaction after participation in an assertive training group*. Paper presented at the meeting of the Association for the Advancement of Behavior Therapy, Chicago, November 1978.

Bloom, L. Z., Coburn, K., & Pearlman, J. *The new assertive woman*. New York: Dell, 1975.

Bower, S. A., & Bower, G. H. *Asserting yourself: A practical guide for positive change*. Reading, Mass.: Addison-Wesley, 1976.

Broverman, I. K., Vogel, S. R., Broverman, D. M., Clarkson, F. E., & Rosen-Krantz, P. S. Sex-role stereotypes: A current appraisal. In M. Mednick, S. Tangri, & L. Hoffman (Eds.), *Women and achievement*. New York: John Wiley & Sons, 1975.

Butler, P. E. *Self-assertion for women*. San Francisco: Canfield Press, 1976.

Cheek, D. K. *Assertive black . . . puzzled white*. San Luis Obispo, Calif.: Impact, 1976.

Connor, J. M., Serbin, L. A., & Ender, R. A. Responses of boys and girls to aggressive, assertive, and passive behaviors of male and female characters. *Journal of Genetic Psychology*, in press.

Corby, N. Assertion training with aged populations. *The Counseling Psychologist*, 1975, *5*, 69–74.

Davis, K. Population policy: Will current programs succeed? *Science*, 1967, 158, 730–739.

De Giovanni, I., & Epstein, N. Unbinding assertion and aggression in research and clinical practice. *Behavior Modification*, 1978, *2*, 173–192.

Douvan, E., & Adelson, J. *Adolescent experience*. New York: Wiley, 1966.

Edwards, N. Assertive training in a case of homosexual pedophilia. *Journal of Behavior Therapy and Experimental Psychiatry*, 1972, *3*, 55–63.

Eisler, R. M., Hersen, M., & Miller, P. M. Effects of modeling on components of assertive behavior. *Journal of Behavior Therapy and Experimental Psychiatry*, 1973, *4*, 1–6. (a)

Eisler, R. M., Miller, P. M., & Hersen, M. Components of assertive behavior. *Journal of Clinical Psychology*, 1973, *29*, 295–299. (b)

Falbo, T. Multidimensional scaling of power strategies. *Journal of Personality and Social Psychology*, 1977, *8*, 537–547.

Finch, B. E., & Wallace, C. J. Successful interpersonal skills training with schizophrenic inpatients. *Journal of Consulting and Clinical Psychology*, 1977, *45*, 885–890.

Foder, I. The phobic syndrome in women: Implications for treatment. In V. Franks & V. Burtle (Eds.), *Women in therapy*. New York: Brunner/Mazel, 1974.

Ford, J. D., & Hogan, D. R. *Assertiveness and social competence in the eye of the beholder*. Paper presented at the meeting of the Association for the Advancement of Behavior Therapy, Chicago, November 1978.

Foy, D. W., Eisler, R. M., & Pinkston, S. Modeled assertion in a case of explosive rages. *Journal of Behavior Therapy and Experimental Psychiatry*, 1975, *6*, 135–137.

Freedman, B. J. *An analysis of social-behavioral skill deficits in delinquent and nondelinquent adolescent boys*. Unpublished doctoral dissertation, University of Wisconsin, 1974.

Friedman, P. N. *The effects of modeling and role playing on assertive behavior*. Unpublished doctoral dissertation, University of Wisconsin, 1968.

Galassi, J. P., & Galassi, M. D. *A factor analysis of a measure of assertiveness.* Unpublished manuscript, West Virginia University, 1973.

Galassi, J. P., Gallassi, M. D., & Litz, M. C. Assertive training in groups using video feedback. *Journal of Counseling Psychology,* 1974, *21,* 390–394.

Galassi, M. D., & Galassi, J. P. *Assert yourself! How to be your own person.* New York: Human Sciences Press, 1977.

Gambrill, E. D., & Richey, C. A. An assertion inventory for use in assessment and research. *Behavior Therapy,* 1975, *6,* 550–561.

Gambrill, E. D., & Richey, C. A. *It's up to you: Developing assertive social skills.* Millbrae, Calif.: Les Femmes, 1976.

Goldfried, M. R., & Davison, B. C. *Clinical behavior therapy.* New York: Holt, Rinehart and Winston, 1976.

Goldfried, M. R., & D'Zurilla, T. J. A behavioral-analytic model for assessing competence. In C. D. Spielberger (Ed.), *Current topics in clinical and community psychology,* Vol. I. New York: Academic Press, 1969.

Goldfried, M. R., & Linehan, M. M. Basic issues in behavioral assessment. In A. R. Ciminero, K. S. Calhoun, & H. E. Adams (Eds.), *Handbook of behavioral assessment.* New York: Wiley-Interscience, 1977.

Goldsmith, J. B., & McFall, R. M. Development and evaluation of an interpersonal skill-training program for psychiatric inpatients. *Journal of Abnormal Psychology,* 1975, *84,* 51–58.

Gormally, J., Hill, C. E., Otis, M., & Rainey, L. A microtraining approach to assertion training. *Journal of Counseling Psychology,* 1975, *22,* 299–303.

Hall, K. *Sex differences in initiation and influence in decision-making groups of prospective teachers.* Unpublished doctoral dissertation, Stanford University, 1972.

Heimberg, R. G., Montgomery, D., Madsen, C. H., & Heimberg, J. S. Assertion training: A review of the literature. *Behavior Therapy,* 1977, *8,* 953–971.

Herman, S. J. *Becoming assertive: A guide for nurses.* New York: Van Nostrand, 1978.

Hersen, M., Eisler, R. M., & Miller, P. M. Development of assertive responses: Clinical, measurement, and research considerations. *Behaviour Research and Therapy,* 1973, *11,* 505–521. (a)

Hersen, M., Eisler, R., Miller, P., Johnson, M., & Pinkston, S. Effects of practice, instructions, and modeling on components of assertive behavior. *Behaviour Research Therapy,* 1973, *11,* 443–451. (b)

Hersen, M., Eisler, R. M., & Miller, P. M. An experimental analysis of generalization in assertive training. *Behaviour Research and Therapy,* 1974, *12,* 295–310.

Hilpert, F., Kramer, C., & Clark, R. A. Participants' perceptions of self and partner in mixed sex dyads. *Central States Speech Journal,* 1975, *26,* 52–56.

Hollandsworth, J. G. Differentiating assertion and aggression: Some behavioral guidelines. *Behavior Therapy,* 1977, *8,* 347–352.

Horner, M. S. Fail: Bright women. *Psychology Today,* November 1969, 36–38.

Horner, M. S. Femininity and successful achievement: Basic incongruity. In J. Bardwick, E. Douvan, M. S. Horner, & D. Gutman (Eds.), *Feminine personality and conflict.* Belmont, Calif.: Brooks/Cole, 1970.

Jacobson, W. *Power and interpersonal relations.* Belmont Calif.: Wadsworth, 1972.

Kagan, J. Acquisition and significance of sex-typing and sex-role identity. In M. L. Hoffman & L. W. Hoffman (Eds.), *Review of child development research* (Vol. 1). New York: Russell Sage Foundation, 1964.

Kazdin, A. E. Effects of covert modeling and model reinforcement on assertive behavior. *Journal of Abnormal Psychology,* 1974, *83,* 240–252.

Kazdin, A. E. Effects of covert modeling, multiple models, and model reinforcement on assertive behavior. *Behavior Therapy*, 1976, *7*, 211–222.

Kipper, D., & Jaffe, Y. Dimensions of assertiveness: Factors underlying the College Self-Expression Scale. *Perceptual and Motor Skills*, 1978, *46*, 47–52.

Komarovsky, M. Functional analyses of sex roles. *American Sociological Review*, 1950, *15*, 508–516.

Lange, A. J., & Jakubowski, P. *Responsible assertive behavior: Cognitive/behavioral procedures for trainers.* Champaign, Ill: Research Press, 1976.

Lawrence, P. S. *The assessment and modification of assertive behavior.* Unpublished doctoral dissertation, Arizona State University, 1970.

Lazarus, A. A. *Behavior therapy and beyond.* New York: McGraw-Hill, 1971.

Lazarus, A. A. On assertive behavior: A brief note. *Behavior Therapy*, 1973, *4*, 697–699.

Levenson, R. W., & Gottman, J. M. Toward the assessment of social competence. *Journal of Consulting and Clinical Psychology*, 1978, *46*, 453–462.

Liberman, R. P., King, L. W., DeRisi, W. J., & McCann, M. *Personal effectiveness: Guiding people to assert themselves and improve their social skills.* Champaign, Ill.: Research Press, 1975.

Linehan, M. M. Review of *Asserting yourself: A practical guide for positive change* by S. A. Bower & G. H. Bower. *Behavior Modification*, 1977, *1*, 567–570.

Linehan, M. M. Structured cognitive-behavioral treatment of assertion problems. In P. C. Kendall & S. P. Hollon (Eds.), *Cognitive-behavioral interventions: Theory, research, and procedures.* New York: Academic Press, 1979.

Linehan, M. M., Goldfried, M. R., & Goldfried, A. P. Assertion therapy: Skill training or cognitive restructuring. *Behavior Therapy*, 1979, *10*, 372–388.

Linehan, M. M., & Seifert, R. *How appropriate is assertive behavior? Real and perceived sex differences.* Manuscript submitted for publication, 1978.

Linehan, M. M., Walker, R. O., Bronheim, S., Haynes, K. F., & Yevzeroff, N. Group vs. individual assertion training. *Journal of Clinical and Consulting Psychology*, in press.

Lunneborg, P. W. Stereotypic aspects in masculinity–femininity measurement. *Journal of Consulting and Clinical Psychology*, 1970, *34*, 113–118.

Maccoby, E. E. *The development of sex differences.* Stanford, Calif.: Stanford University Press, 1966.

Maccoby, E. E., & Jacklin, C. N. *The psychology of sex differences.* Stanford, Calif.: Stanford University Press, 1974.

MacDonald, M. L. *A behavioral assessment methodology applied to the measurement of assertion.* Unpublished doctoral dissertation, University of Illinois at Urbana, 1974.

MacDonald, M. L. Teaching assertion: A paradigm for therapeutic intervention. *Psychotherapy: Theory, Research and Practice*, 1975, *12*, 60–67.

Marriott, S. & Foster, S. L. *Functional effects of assertive communication styles: Outcome and parameters.* Paper presented at the meeting of the Association for the Advancement of Behavior Therapy, Chicago, November 1978.

McFall, R. M., & Lillesand, D. B. *Behavior rehearsal with modeling and coaching in assertive training: Assessment and training stimuli.* Unpublished manuscript, 1970.

McFall, R. M., & Lillesand, D. B. Behavior rehearsal with modeling and coaching in assertion training. *Journal of Abnormal Psychology*, 1971, *77*, 313–323.

McFall, R. M., & Marston, A. An experimental investigation of behavioral rehearsal in assertive training. *Journal of Abnormal Psychology*, 1970, *76*, 295–303.

McFall, R. M., & Twentyman, C. T. Four experiments on the relative contributors of rehearsal, modeling, and coaching on assertion training. *Journal of Abnormal Psychology*, 1973, *81*, 199–218.

McKee, J. P., & Sherriffs, A. C. Men's and women's beliefs, ideals, and self concepts. *American Journal of Sociology,* 1959, *64,* 356–363.

McPhail, G. W. Developing adolescent assertiveness. In R. S. Alberti (Ed.), *Assertiveness: Innovations, applications, issues.* San Luis Obispo, Calif.: Impact, 1978.

Melnick, J. & Stocker, R. B. An experimental analysis of the behavioral rehearsal with feedback technique in assertiveness training. *Behavior Therapy,* 1977, *8,* 222–228.

National Institute of Mental Health, United States Department of Health, Education and Welfare, *Mental Health Statistics.* Washington, D.C.: Author, 1965–68.

Nietzel, M., Martorano, R. D., & Melnick, J. The effects of covert modeling with and without reply training on the development and generalization of assertive responses. *Behavior Therapy,* 1977, *8,* 183–192.

Osborn, S. M., & Harris, G. G. *Assertive training for women.* Springfield, Ill.: Charles C Thomas, 1975.

Phelps, S., & Austin, N. *The assertive woman.* Fredericksburg, Va.: Book Crafters, 1975.

Rathus, S. A. An experimental investigation of assertion training in a group setting. *Journal of Behavior Therapy and Experimental Psychiatry,* 1972, *3,* 81–86.

Rathus, S. A. A 30 item schedule for assessing assertive behavior. *Behavior Therapy,* 1973, *4,* 398–406.

Rathus, S. A. Principles and practices of assertive training: An eclectic overview. *The Counseling Psychologist,* 1975, *5,* 9–20.

Rich, A. R., & Schroeder, H. W. Research issues in assertiveness training. *Psychological Bulletin,* 1976, *83,* 1081–1096.

Richey, C. A. *How to conduct an assertion training group: A workshop manual for trainees.* Unpublished manuscript, 1977.

Rimm, D., Hill, G., Brown, N., & Stuart, J. Group-assertive training in treatment of expression of inappropriate anger. *Psychological Reports,* 1974, *34,* 791–798.

Rimm, D. C., & Masters, J. C. *Behavior therapy: Techniques and empirical findings* (2nd ed.). New York: Academic Press, 1979.

Rose, S. D., Caynor, J., & Edleson, J. Measuring interpersonal skills. *Social Work,* 1977, *22,* 125–129.

Salter, A. *Conditioned reflex therapy.* New York: Creative Age, 1949.

Salter, A. On assertion. In R. E. Alberti (Ed.), *Assertiveness: Innovations, applications, issues.* San Luis Obispo, Calif.: Impact, 1977.

Schinke, S. P., & Rose, S. D. Interpersonal skill training in groups. *Journal of Counseling Psychology,* 1976, *23,* 442–448.

Seligman, M. *Learned helplessness: On depression, development and death.* San Francisco: W. H. Freeman, 1975.

Swacker, M. The sex of the speaker as a sociolinguistic variable. In B. Thorne & N. Henley (Eds.), *Language and sex: Difference and dominance.* Rowley, Mass.: Newbury House, 1972.

Terman, L. M., & Tyler, L. E. Psychological sex differences. In L. Carmichael (Ed.), *A manual of child psychology* (2nd ed.). New York: John Wiley, 1954.

Thorpe, G. L. Desensitization, behavioral rehearsal, self-instructional training, and placebo effects on assertive-refusal behavior. *European Journal of Behavioural Analysis and Modification,* 1975, *1,* 30–44.

Tolor, A., Kelly, B. R., & Stebbins, C. A. Assertiveness, sex-role stereotyping, and self-concept. *Journal of Psychology,* 1976, *93,* 157–164.

Trower, P., Yardley, K., Bryant, B. M., & Shaw, P. The treatment of social failure: A comparison of anxiety-reduction and skill-acquisition procedures on two social problems. *Behavior Modification,* 1978, *2,* 41–60.

Turner, S. M., & Adams, H. E. Effects of assertive training on three dimensions of assertiveness. *Behavior Research and Therapy*, 1977, *15*, 475–483.

U.S. Department of Commerce, Bureau of the Census. Female family heads (P-23, No. 50). *Current Population Reports*. Washington, D.C.: Author, July 1974. (a)

U.S. Department of Commerce, Bureau of the Census. Marital status and living arrangements: March 1974 (P-20, No. 271). *Current Population Reports*. Washington, D.C.: Author, October 1974. (b)

Weinman, B., Gelbart, P., Wallace, M., & Post, M. Inducing assertive behavior in chronic schizophrenics: A comparison of socioenvironmental, desensitization, and relaxation therapies. *Journal of Consulting and Clinical Psychology*, 1972, *39*, 246–253.

Witkin, H. A. *Personality through perception: An experimental and clinical study*. New York: Harper, 1954.

Wolfe, J. L., & Fodor, I. G. A comparison of three approaches to modifying assertive behavior in women: Modeling-plus-behavior rehearsal, modeling-plus-behavior rehearsal-plus-rational therapy, and conscious-raising. *Behavior Therapy*, 1977, *8*, 567–574.

Wolpe, J. *The practice of behavior therapy* (2nd ed.). New York: Pergamon, 1973.

Wolpe, J., & Lazarus, A. A. *Behavior therapy techniques: A guide to the treatment of neuroses*. New York: Pergamon, 1966.

Zimmerman, D. H., & West, C. Sex roles, interruptions and silences in conversation. In B. Thorne & N. Henley (Eds.), *Language and sex: Difference and dominance*. Rowley, Mass.: Newbury House, 1975.

CHAPTER 8

Communication Skills in Married Couples

Gary R. Birchler

INTRODUCTION

In the past few years there has been a relative explosion in the area of communication skills training. Clinicians, educators, researchers, and workshop leaders alike have increasingly focused on the importance of communication skills in establishing and maintaining the well-being of individuals and their relationships.

This increased attention to communication has evolved over the past two decades. Couples expressing significant dissatisfaction with their relationships have consistently implicated communication as one of the major problems (cf. Brim, Fairchild, & Borgatta, 1961; DuBurger, 1967; Gottman, Notarius, Gonso, & Markman, 1976; McMillan, 1969; Mitchell, Bullard, & Mudd, 1962). More recently, data on couples seeking evaluation and treatment for distressed marriages (52 wives and 47 of their husbands), were collected by Birchler in a psychiatric outpatient setting. The three most frequently mentioned problem areas for husbands and wives were, in descending order: lack of communication, difficulties in sexual expression, and personality factors (e.g., spouse too moody, demanding, critical).

To supplement this finding, data from a more structured assessment instrument, the Areas of Change Questionnaire (ACQ, formally called WC Scale; cf. Birchler, 1973; Weiss, Hops, & Patterson, 1973) were

Gary R. Birchler • Mental Health Clinic, Veterans Administration Medical Center, and University of California School of Medicine at San Diego, La Jolla, California 92161.

obtained from 153 couples seeking marital therapy in the same setting. A comparison sample of 91 "nondistressed" couples having similar demographics was gathered over a 3-year period. In part, the ACQ contains specific behaviors which spouses are asked to rate in terms of the amount of change they seek in their partners. The first five items most frequently endorsed by distressed and nondistressed spouses are listed in Table 1.

Spouses in the distressed group were much more likely to request change than were nondistressed spouses (note percentage of item endorsement). Item content, once again, implicated communication and sexual behaviors as general problem areas, with "arguing" (distressed) and "going out" (nondistressed) group-specific items.

Further analyses of full scale responses suggested that not only do husbands and wives within groups have similar complaints, but that the complaints of distressed and nondistressed couples differ not so much in kind, but rather in perceived severity. Gottman *et al.* (1976) have come to essentially the same conclusion based on very different samples. Considerable marital interaction research over the past few years has supported the belief that it is not the specific problems, but rather the communication skills and specific strategies employed for problem resolution, that differentiate between distressed and nondistressed relationships (Birchler, 1977; Birchler & Webb, 1977; Birchler, Weiss, & Vincent, 1975; Gottman, Markman, & Notarius, 1977; Gottman, Notarius, Markman, Bank, & Yoppi, 1976; Vincent, Weiss, & Birchler, 1975).

Once clinical investigators began to break down marital communication into facilitative and nonfacilitative component behaviors, the emphasis on communication skills training intensified. Witness the number of recent books on this topic (e.g., Gottman, 1976a; Guerney, 1977; Jacobson & Margolin, 1979; Knox, 1975; L'Abate, 1978; Miller, Nunnally, & Wackman, 1975; Strayhorn, 1977; Stuart & Lederer, 1979; Thomas, 1977).

Because of a dramatic increase in the general literature relevant to communication skills training and present space limitations, this chapter will be limited to a review of the literature specifically relevant to communication skills of married couples, which are defined as the observable, trainable, verbal, and nonverbal behaviors concerning the way messages are sent and received between husbands and wives. Most discussion will focus on those communication-training approaches which have demonstrated or are attempting to demonstrate empirical support. A distinction will be made between *enrichment* and *clinical* approaches since they are based on different origins, goals, and target populations.

Table 1. Requests for Behavior Change on the ACQ by Distressed and Nondistressed Husbands and Wives

Rank order	Distressed (N = 153)				Nondistressed (N = 91)			
	Wives about husbands	%	Husbands about wives	%	Wives about husbands	%	Husbands about wives	%
1	"Express emotions more clearly"	86	"Express emotions more clearly"	79	"Give appreciation to spouse"	38	"Express emotions more clearly"	31
2	"Give appreciation to spouse"	78	"Give appreciation to spouse"	63	"Express emotions more clearly"	36	"Initiate having sex"	29
3	"Attend to spouse"	75	"Initiate having sex"	63	"Initiate having sex"	26	"Attend to his sexual needs"	27
4	"Arguing"	73	"Arguing"	61	"Start interesting conversations"	26	"Keep the house clean"	23
5	"Start interesting conversations"	67	"Attend to his sexual needs"	61	"Go out"	22	"Start interesting conversations"	20

MARRIAGE ENRICHMENT AND COMMUNICATION SKILLS TRAINING

Origins and Development

The marriage enrichment movement seems to have had its origins in the work of David and Vera Mace, beginning in 1962 (Mace & Mace, 1976). Following them, a number of religious organizations have established marriage enrichment programs, offered usually as retreats or growth experiences (cf. Bosco, 1972; Mace & Mace, 1976; Regula, 1975; Smith & Smith, 1976). In addition to the programs affiliated with religious groups, a number of more recent marriage enrichment programs have been developed (cf. Miller, Nunnally, & Wackman, 1976a; Rappaport, 1976; Stein, 1975; Travis & Travis, 1975).

Otto (1975) analyzed responses from 30 professionals who conduct marital and family enrichment programs. This analysis and similar reviews (Gurman & Kniskern, 1977; Mace & Mace, 1976) provide a composite view of the typical marital enrichment program. Participants are usually partners to well-functioning, nondistressed relationships, since these programs are designed and best suited to help couples enrich and enhance already committed relationships (Mace & Mace, 1976; Miller *et al.*, 1976a; Otto, 1975). Formats include marathon-type weekends or meeting for shorter sessions over several weeks. The number of couples ranges from 4 to 30; contact time averages about 14 h. Common activities include group discussions, structured small-group exercises, dyadic experiential learning, lectures, and handouts with homework instructions. Content has predominantly focused on aspects of marital communication; however, significant amounts of time are often spent on sexuality, marital roles, conflict resolution, and problem solving. Program evaluation, when conducted, almost always consists of participants' self-report about the program.

Finally, a note about the magnitude of the marital enrichment movement: in 1975, Otto estimated that some 180,000 couples had participated in such programs. The intervention procedures consist largely of communication training. Thus, the remainder of this section will review empirical "enrichment" studies which support the efficacy of communication skills training.

Research and Practice

While an attempt was made to review the literature comprehensively, the sheer volume of new publications and the rate and occasional obscurity of dissertations in this area may have caused some studies to be inadvertently overlooked. However, representative studies demonstrating the efficacy of communication skills training within the marital enrichment field will be discussed. The interested reader is referred to Gurman and Kniskern (1977) for a comprehensive catalogue of studies in this area. A number of articles exist which primarily serve to describe ongoing or developing marriage enrichment programs. In general, specification of training procedures and outcome data are not included (e.g., Clarke, 1970; Nunnally, Miller, & Wackman, 1975; Sauber, 1974). Two programs, Marriage Encounter (Regula, 1975) and the Pairing Enrichment Program (Travis & Travis, 1975), feature group formats in which couples practice communication exercises in private following instruction to the group. Typically, six or more couples meet for a weekend encounter, totaling about 15 contact hours. Another program, entitled Marriage Diagnostic Laboratory (Stein, 1975), consists of a weekly series of five 2-h sessions for six couples. An interesting exercise involves participants practicing effective communication with opposite-sex nonspouses prior to working with their own spouses. In light of dyadic interaction research which suggests that individuals communicate more positively with nonspouses than spouses (Birchler *et al.*, 1975; Vincent *et al.*, 1975), this structured exercise may help elicit positive methods of communication which may then be generalized to the couple.

Outcome Studies. There are relatively few marital enrichment outcome studies which test the efficacy of communication skills training *per se*. However, several studies have investigated related outcome variables. A review of the literature to date revealed only one significant study (Zarle & Boyd, 1977) in addition to those already catalogued by Gurman and Kniskern (1977, Table 1, p. 4) in their comprehensive review and analysis of outcome research on marital enrichment programs. In addition to studies selected from this group, the evaluation research described by Miller *et al.* (1976a, b) on the Minnesota Couples Communication Program (MCCP) will be considered in the discussion which follows. These studies can be grouped into four program categories: those associated with MCCP, Conjugal Relationship Enhancement (CRE), Behavioral Exchange (BE), and those incorporating miscellaneous enrichment procedures.

Let us first consider the MCCP, as described by Miller *et al.* (1976a). The two types of skills taught pertain to *awareness* and *communication*. The former relates to partners' ability, individually and together, to perceive dyadic interaction processes; the latter refers to the couples' ability to engage in open, clear, and direct communication. In awareness training, participants are taught six skills relating to "self-disclosures": speaking for self, making sense statements, making interpretive statements, making feeling statements, making intention statements, and making action statements. Next, partners learn four skills relating to "awareness of others." One of them is *checking out,* i.e., learning to ask for clarification regarding spouse's thoughts, feelings, or actions. The other skills comprise the process called *sharing a meaning,* one component of which is "active listening" (Gordon, 1970). This process, which is now common to most communication-training programs, involves teaching partners to communicate so that the message intended by the sender equals the message heard by the listener. To accomplish this, a three-step process is taught: (a) send short, clear, direct messages, and ask for acknowledgment; (b) the receiver learns to make statements which acknowledge, summarize, and reflect back the messages received; and (c) the sender is taught to clarify or confirm the original message.

MCCP groups are composed of five to seven couples who meet weekly for 3 h for four weeks. Training procedures include didactic minilectures, experiential exercises, simulations, group discussion and feedback, and supplementary reading of *Alive and Aware: Improving Communication in Relationships* (Miller, Nunnally, & Wackman, 1975), given to encourage transfer of learning.

Miller *et al.* (1976a) cite four studies as providing evidence for the efficacy of the MCCP (Campbell, 1974; Miller, 1971; Nunnally, 1971; Schwager & Conrad, 1974). Since the MCCP has reportedly undergone six revisions since its conception, one can only presume that the skill variables and training procedures described above were approximated in the studies offered as empirical support for the program. In the Miller and Nunnally dissertations 32 volunteer, premarried couples were randomly assigned to an experimental ($N = 17$) and a time-matched waiting list control group ($N = 15$). As a function of training, experimental relative to control couples were expected to recall more accurately partners' communication behaviors following a discussion of "plan something the two of you can do together" (Nunnally, 1971). Comparisons were made against objective data obtained from audiotaped recordings of the same discussion using an interaction scoring procedure. Experimental couples were also expected to engage in significantly more work statements, comprised of facilitative, communication skill behaviors taught during training (Miller, 1971).

Results indicated significant differences between the experimental (E) and control (C) groups on both accuracy of recall ($p < .02$) and work-pattern communication ($p < .05$). Although none of the C couples increased on both dimensions, only four E couples did so. Overall, 14 E compared to 8 C couples increased on at least one of the variables. Thus, over half the waiting-list control couples demonstrated positive change on one of the variables. Conceivably, the dependent measures had reactive or unreliable properties.

For Campbell's dissertation (1974), 60 volunteer married couples were evenly divided into the MCCP and waiting-list control groups. A posttest-only control-group design was used to detect group differences in communication effectiveness and self-disclosure, which were measured by means of two paper-and-pencil instruments and by objectively coded audiotape recordings of marital interaction. Results indicated a significant difference between the two groups on the evaluator-rated measures of self-disclosure and work-pattern communication. However, the self-report measures not only failed to discriminate between the groups but correlated negatively with evaluator-rated effectiveness scores. Thus, it appears that the communication skills training component of the program was relatively effective while the awareness training component was not.

A fourth study (Schwager & Conrad, 1974, cited by Miller *et al.* 1976a) was designed to test the effectiveness of MCCP on measures of self-acceptance and acceptance of others. Reportedly, large increases on both measures obtained for both spouse-together and spouse-apart treatment conditions, although no control groups were used.

In summary, the general effectiveness of communication training within the MCCP has received some empirical support. Samples of premarried and married couples have generally improved in their abilities to engage in verbal behaviors associated with "work-pattern" effectiveness. Improvements in self-disclosure and interaction awareness skills have been less demonstrable across studies. Unfortunately, lack of published details regarding training procedures and variations in populations and criterion measures employed make it difficult to draw definitive conclusions about the effectiveness of the program.

A *second* enrichment program with some empirical support is called Conjugal Relationship Enhancement (CRE). Developed originally by Bernard Guerney, this approach teaches couples client-centered communication skills relating to direct expression of feelings and empathic listening in an atmosphere of nonpossessive warmth. Four CRE studies have been reported which focus on communication skills in nondistressed married couples (Collins, 1971; Ely, Guerney, & Stover, 1973; Rappaport, 1976; Wieman, 1973). Two have been selected for detailed discussion in this section.

Following Collins's (1971) demonstration that a 6-month weekly meeting format was effective in improving couples' communication skills and marital satisfaction scores, Rappaport (1976) sought to determine if an intensive, revised CRE program would do as well.

Twenty volunteer married couples met in groups of four couples each for two 8-h and two 4-h sessions over a 2-month period (24 contact hours). The basic skills taught included "speaker" skills (i.e., expressing thoughts and feelings in a clear, direct, situation-specific fashion) and "listener" skills (i.e., empathizing and helping the speaker verbalize and clarify thoughts and feelings in a nonjudgmental fashion). Training methods included instruction and feedback, modeling, behavior rehearsal with spouses and nonspouses, homework, and reading assignments from a programmed text. The revised CRE program also included rather intensive work on couple-determined areas of marital conflict. Although couples were asked to apply CRE skills in the acts of compromising, bargaining or suggesting alternative behaviors, it is not clear whether couples actually received problem-solving skill training.

Couples served as their own controls over a 2-month waiting period; no other control groups were used. Data were collected during prewaiting period, postwaiting period, and posttreatment period interviews. Dependent measures included several standard paper-and-pencil instruments designed to assess changes in such variables as marital adjustment and satisfaction, marital communication, trust and intimacy, and problem solving. In addition, two behavioral rating scales were used to score couple interactions taped during the interviews.

Spouses alternately discussed self-identified areas of change for 4 min each. The coding scales, which measured speaker effectiveness and listener empathy, were employed by reliable, trained judges who were blind as to which interaction sequence was being scored.

Results indicated that, for all nine dependent variables, the posttreatment means were significantly higher than means obtained during the prewaiting and postwaiting period assessments (which did not differ significantly). Thus, relationship enhancement and increased speaker and listener skills were attributed to the intensive CRE program. The limitations of this study relate to design, population, and selection of outcome criteria. First, the own-control design, as discussed by the author (Rappaport, 1976), does not control for passage of time, it increases the potential for reactivity to the assessment measures, and it does not rule out change due to nonspecific effects of treatment. Second, couples were primarily college student volunteers and thus were not representative of marriages at large. The high attrition rate of 33% following the prewaiting period interview also affects the representativeness of this sample. Finally, seven of the nine dependent measures were highly subjective indices, and the two behavioral coding scales closely matched the specific target behaviors intensively practiced during the program. Nevertheless, this study offers significant evidence that the CRE program is effective in a short-term intensive group format led by relatively inexperienced graduate students.

Further evidence supporting the CRE approach comes from Ely *et al.* (1973), who investigated the efficacy of the *training* phase of conjugal therapy as one part of a larger program. Twenty-three volunteer student couples, averaging 10 years of marriage, were randomly assigned to experimental or waiting-list control groups. Couples on the waiting list later participated in the treatment program, which allowed for a replication study.

Treatment groups met for 8–10 weeks, 2 h each week. The basic skills and training procedures were similar to those described above (Rappaport, 1976), with the exception that much less time was devoted to conflict-resolution activities. The dependent measures, obtained before and after training (or waiting) included: (a) the Ely Feeling Questionnaire (EFQ), which elicits responses that can be scored as direct expressions of one's own feelings, or restatements of spouse's feelings; (b) audiotaped accounts of a couple's responses to 12 role-play situations which were also scored for feeling expression or feeling clarification; (c) the Primary Communication Inventory (PCI), a self-report measure of verbal and nonverbal communication (Navran, 1967); and (d) the Conjugal Life Questionnaire (CLQ), selected to assess degree of harmony between spouses.

Findings indicated that relative to the waiting-list controls, the trained couples made significant gains on all measures ($p < .05$ or better) except for the CLQ and the feeling expression part of the EFQ. Similarly, after the control couples completed training, all measures except the CLQ showed significant increases ($p < .05$ or better). These results suggest that training couples to engage in direct expressions of feelings and to be empathic, reflective listeners can be accomplished effectively. Once again, however, some rather common methodological procedures constrain the conclusiveness of this study: the investigator served as the sole group trainer, couples were volunteer students, self-report measures or measures closely related to the skills being taught predominated, and a training group controlling for nonspecific effects of treatment was lacking.

Nevertheless, based on this study and those by Collins (1971) and Rappaport (1971), CRE would appear to have significant empirical support. Participation in CRE programs relative to nonparticipation has demonstrably increased couples' basic speaker and listener skills, and their general relationship satisfaction. Findings, however, are rather limited to fairly young, well-educated, nondistressed couples.

A third enrichment approach to communication training features behavioral exchange procedures. Representative outcome studies with nondistressed couples have dealt with conflict-negotiation skills (Harrell & Guerney, 1976), reciprocity counseling (Dixon & Sciara, 1977), and behavioral exchange, negotiation, and contracting (Roberts, 1975). Treatment rationale is based generally on the following premises: (a) that marital satisfaction is determined in part by a couple's ability to manage and resolve marital conflicts; (b) that such conflicts often elicit negative and coercive behavior-change strategies; and therefore (c) if partners could learn positive behavior-change strategies to resolve marital conflicts, there would be an increase in marital satisfaction. Communication skills are central because partners must learn to express their feelings, to clearly specify target behaviors to be changed, and to engage in the negotiation and contracting process efficiently and positively.

Roberts (1975) examined the effects on marital satisfaction of 3 h of training in behavior-exchange negotiation. Eighteen couples received behavioral-contract training, six couples each served as attention-placebo and waiting-list controls. The attention-placebo couples engaged in "dialogue" for equal amounts of time. Five self-report measures of marriage adjustment and relationship change administered before and after training indicated that the contract-trained couples did significantly better on these general measures than either of the control groups. While this analogue study was limited by heavy reliance on self-report measures, it provided initial encouragement for the application of the behavior-exchange model to nonclinical couples.

Additional support derives from research conducted by Dixon and Sciara (1977), who examined the efficacy of group reciprocity counseling for seven married couples who signed up for a university extension marriage enrichment workshop. Couples met 2 h weekly for 8 weeks. Training and assignment procedures were integrated throughout the program using a modified version of the Marital Happiness Scale (MHS) (Azrin, Naster, & Jones, 1973), and partners made daily satisfaction ratings on nine 10-point scales covering such content areas as finances, affectionate interaction, and household responsibilities. A multiple-baseline procedure allowed couples to monitor changes in all MHS content areas while sequentially implementing behavior exchanges in some. Group training procedures focused on communication and negotiation skills. The Marital Pre-Counseling Inventory (MPI; Stuart & Stuart, 1972), was administered before and after training.

Although no control groups were employed and data were completely derived from self-report measures, results were fairly encouraging. In particular, mean weekly group satisfaction ratings for three areas from the MHS increased significantly from the third through the eighth week (affection interaction, communication with spouse, child and/or family interaction). Means for the other six satisfaction areas increased, but not significantly. The multiple-baseline analysis of weekly MHS ratings suggested that changes in content area ratings followed the application of reciprocity exchange procedures to those areas.

Another behavior-exchange study was reported by Harrell and Guerney (1976). The intent of developing the program was to go beyond the basic training of speaker and listener skills demonstrated to be effective in the Conjugal Relationship Enhancement Program (cf. Rappaport, 1976), and to teach couples negotiation skills better to resolve relationship conflicts. Nine behavior-exchange steps comprise the program: (1) listening carefully, (2) locating a relationship issue, (3) identifying (one's own) contributions to the issue, (4) identifying alternative solutions, (5) evaluating alternative solutions, (6) making an exchange, (7) determining conditions of the exchange, (8) implementation of the behavior-exchange contract, and (9) renegotiation of the behavior-exchange contract. In small-group training sessions couples learn and practice the negotiation process while receiving feedback from the trainer and other couples. Homework is assigned throughout the 8-week, 16-h program.

For the evaluation study, 30 fairly young, well-educated couples were recruited and randomly assigned to a behavior-exchange experimental group or a no-treatment control group. Behavior-exchange couples, relative to the controls, were expected to demonstrate improvement in conflict-management skills, perceptions of marital satisfaction, and perceived rate of relationship change. A variety of nine self-report and behavioral measures were employed to detect the changes. Of particular interest, the authors designed the Marital Conflict Negotiation Task (MCNT). Before and after training (or waiting), each couple role-played for 10-min their attempts to resolve three standardized marital conflicts: running a household, affection and sex, and mutual leisure. Audiotaped MCNT interactions were later scored yielding three categories of behavioral measures: behavior-exchange skill scores related to accomplishment of steps 3–7 of the program; positive verbal behavior (frequency counts of "agree" and "approve" statements); and negative verbal behavior (frequency counts of "interrupt" and "put down" statements).

Significant changes between experimental (*E*) and control (*C*) groups were detected only on the behavioral interaction measures. Analysis of variance indicated that *E* relative to *C* couples made significant gains on two of the five behavior-exchange skills measures (steps 4, $p < .03$, and 5, $p < .02$) with a near-significant difference for a third (step 3, $p < .09$). In addition, *E* versus *C* couples evidenced significant decreases in frequency of both positive and negative verbal behaviors.

Apparently, as couples learned to make more "summarizing statements" (step 1), they emitted fewer "agree" statements, which comprised 97% of the positives recorded during pretesting. Surprisingly, none of the self-report measures designed to assess changes in marital satisfaction or perceived rate of relationship change resulted in significant changes for either group.

Of the plausible explanations offered by the authors to account for this finding, two are of particular interest. It may be that for nondistressed, enrichment-seeking couples, a training program which primarily emphasizes behavior-change skills will have little impact on subjective measures of marital satisfaction. First, there is evidence that nondistressed couples, relative to distressed, engage in far fewer arguments (Birchler, 1973) and far fewer displeasing behaviors in daily interaction (Birchler *et al.*, 1975). As a group, they also express significantly less desire for change in spouses' behavior (Birchler & Webb, 1977) and report less of a tendency to respond maladaptively to marital conflict (Birchler, 1977).

Accordingly, nondistressed couples would seem to have relatively little need and few opportunities to practice conflict-resolution skills. A second possibility is that negotiation training, problem solving, and overtly reciprocal exchange procedures are inherently not very appealing or romantic, to nondistressed couples in particular. Other investigators using such procedures have also commented that certain couples are less than enchanted with such a rational, mechanistic, *quid pro quo* approach to relationship counseling (Jacobson, 1978; Tsoi-Hoshmand, 1975). Related to these points, it should be noted that the investigation of behavioral, problem-solving approaches is relatively new in the area of marriage enrichment. However, these early applications discussed above should encourage further program development and research along these lines.

A number of unrelated studies contribute by methodology or outcome to our understanding of the effects of communication skills training (CST) for nondistressed couples. Hines (1976) investigated the effects of CST compared to insight-oriented group therapy and no treatment. Procedures used in the CST included instruction, practice, videotape feedback, and group discussion. Judges rated "helper's degree of helpfulness" as spouses interviewed one another before and after treatment or waiting. Results indicated CST had a greater positive effect on husbands' helpfulness than either of two control conditions. In a treatment comparison study, Wieman (1973) solicited 36 couples through the newspaper and assigned them to one of three groups: Conjugal Relationship Modification (CRM), Reciprocal Reinforcement therapy (RR), or waiting-list control (WL). Basically, CRM couples were taught to be effective speakers and listeners, while RR couples learned behavior-exchange procedures. Both self-report and objective behavioral measures were obtained before, during, and after treatment. Relative to the WL group, both the CRM and RR treatment groups increased significantly in outcome measures of marital functioning, but they did not significantly differ from each other. These relationships were maintained at the 10-week follow-up. A similar dissertation study (Venema, 1976) sought to compare the effects of communication training, behavior-exchange negotiation training, and the two combined on several self-report measures. Results suggested the superiority of the combined treatment, although all groups improved minimally. These studies indicate that both basic communication training and behavior-exchange procedures incorporate effective treatment components.

Finally, a more recent study (Zarle & Boyd, 1977) sought to compare videotaped modeling versus experiential practice in facilitating couples' self-disclosure. Self-disclosure, whether called "leveling" (Gottman *et al.*, 1976), "feeling talk" (Strayhorn, 1977), or "emotional expression" versus problem solving (Weiss & Birchler, 1978), is considered to be a critical skill for increasing relationship satisfaction. Twenty-seven young couples were randomly assigned to one of three conditions: CST, videotaped modeling, or no contact. Treatment groups received 7 1/2 h of training, the last half spent either practicing self-disclosure statements to videotaped stimulus materials or viewing a wide range of self-disclosure behavior in videotaped models. Pre- and posttesting for all couples consisted of 15-min interactions around standard topics relevant to marriage. Time samples were analyzed for self-disclosure statements using the Hill Interaction Matrix (Hill, 1965) and the Facilitative Self Disclosure Scale (Carkhuff, 1969). Results indicated that, relative to the control group, the CST group improved in four of five self-disclosure measures, while the modeling group improved on three of five. Thus, both procedures seemed to employ facilitative elements. Actually, a careful review of the training procedures indicates much overlap concerning the use of videotaped stimulus materials. More discrimination may have occurred between the two training procedures if videotaped stimuli had been used in only one treatment.

Current Status and Future Directions

It is clear that CST is the most prevalent training procedure applied annually to thousands of couples seeking marital enrichment (Otto, 1975). Unfortunately, outcome studies designed to establish the efficacy of programs and CST procedures are lagging far behind practice. Studies were reviewed which, taken together, document substantially the efficacy of CST for the enrichment population (e.g., Ely *et al.*, 1973; Harrell & Guerney, 1976; Miller *et al.*, 1976a; Rappaport, 1976; Wieman, 1973). Nevertheless, so few studies have incorporated comprehensive behavioral measures designed to validate CST that definite conclusions cannot be drawn. Gurman and Kniskern (1977) calculated that 84% of the criterion measures employed in enrichment studies relied on couples' subjective responses, and of these, significant change was reported on only 57%.

While Miller, Corrales, and Wackman (1975) have discussed recent *conceptual* progress in the area of facilitating marital communication, others have called for more *empirical* progress (Gurman & Kniskern, 1977; L'Abate, 1978; Mace, 1975; Olson, 1976; Otto, 1975). Relevant to CST, these research issues include: effects of differential formats (e.g., weekend vs. weekly sessions); effects of trainer variables (e.g., professionals vs. paraprofessionals); multimethod assessment of program outcome; analyses of skill components; analyses of methods of training (e.g., modeling, videotape, behavioral practice, group process); follow-up to assess durability of treatment; development of CST programs to meet the needs of nonwhite, non-middle-class consumers; and the effectiveness of enrichment programs for clinical populations. Each of these issues is worthy of well-controlled research, and this will require considerably more research. However, the marital enrichment (communication skills training) movement is yet in its infancy and both the tendency and technology to evaluate its effectiveness are growing.

MARRIAGE THERAPY AND COMMUNICATION SKILLS TRAINING

Origins and Development

Most of the research on the relationship between marital communication and marital satisfaction has occurred in the past decade. Early investigations of unhappy relative to happy marriages sought to demonstrate a positive relationship between marital communication and adjustment or satisfaction, while validating a variety of self-report inventories designed to measure marital communication. In an initial study by Navran (1967), distressed and nondistressed couples were administered an early version of the now widely used Locke–Wallace (LW) Marital Adjustment Scale (Locke & Wallace, 1959). Results indicated not only that the two groups were discriminated by the Primary Communication Inventory (PCI), but that the high intercorrelation of PCI and LW scores suggested a positive relationship between communication and marriage adjustment.

Kahn (1970) demonstrated a positive relationship between accurate nonverbal communication and marital satisfaction. He found that well-adjusted couples performed significantly better on a nonverbal interaction task than poorly adjusted couples. This study also replicated Navran's earlier findings (1967) mentioned above.

Further evidence was gathered on the positive relationship between marital communication and marital satisfaction after Bienvenu (1970) developed the Marital Communication Inventory (MCI) to measure the *process* of communication as an element of marital interaction. Based on MCI scores, spouses' self-reports of how they communicated significantly discriminated between groups of distressed and nondistressed couples. Further validation was provided for the MCI in a study by Murphy and Mendelson (1973a). Based on responses to the Locke–Wallace Marital Adjustment Scale, groups of high- compared to low-scoring couples also scored significantly higher on the MCI. Thus, on the basis of comparative group studies relying on self-report instruments, there appears to be a significant relationship between marital communication and marital satisfaction.

Beginning with Stuart's pioneering paper (1969), behaviorally oriented clinicians started to explore the relationship between marital dysfunction and a couple's conflict-resolution skills. Liberman (1970) advocated the application of behavioral approaches to marital therapy. A group of investigators in Oregon presented a social learning formulation of marital distress based on principles of reciprocity and coercion (Patterson & Hops, 1972), behavior-exchange, and conflict-negotiation skill deficits (Weiss *et al.*, 1973). Certain aspects of this approach to understanding marital distress were validated in early studies comparing the interaction of distressed and nondistressed, married and stranger dyads (Birchler, 1973; Birchler, Weiss, & Wampler, 1972; Vincent, 1972). Following these early studies, behavioral marriage therapy (BMT) gained rather considerable momentum (cf. Gurman, 1973). Much of the remainder of this chapter will be taken up in an examination of studies related to BMT and communication skills training involving clinical couples.

Research and Practice

Observational Methods for Communication Skills Analysis. Not surprisingly, behaviorally oriented investigators have given relatively little attention to designing self-report measures of marital communication skills. Much more effort has gone into the development of more objective behavioral observation methods for analyzing *in vivo* samples of couples' marital communication. These methods and data regarding their reliability and validity have been reviewed in detail elsewhere (Weiss & Margolin, 1977); however, an overview here will acquaint the reader with the methods used and their typical applications. Observational and objective assessment procedures have included audiotape recording, videotape recording, electromechanical recording devices, and the use of live observers. Most of these procedures have been applied in both laboratory and home settings, and some investigators have used more than one method.

Carter, Thomas, and associates combined the use of audiotape recordings and an innovative electromechanical device to assess and modify couples' marital communication and verbal problem-solving behavior (Carter & Thomas, 1973a, b; Thomas, Carter, & Gambrill, 1971). McLean, Ogston, and Grauer (1973) had couples tape-record problem-oriented discussions at home and deliver the samples to the investigator for analysis. They were also given electrical "cue boxes" to take home in order to signal positive and negative responses during practice. Other investigators have used audiotapes and typescripts to study problem-related marital communication (Miller & Wackman, 1974; Pierce, 1973; Wackman & Miller, 1975).

Two other applications of electromechanical devices to assess marital communication are notable. First, the "talk table," described by Gottman *et al.* (1976b), allowed the investigators to determine the degree to which the *intent* of positive and negative messages sent by a speaker were equal to the *impact* of those received by the listener. Investigations of distressed and nondistressed couples' problem discussions using the talk table indicated that the two groups did not differ on the *intended* impact of messages sent, but differed significantly in measures of the *actual* impact on the receiver. Distressed partners perceived messages sent by spouses as more negative and less positive than their nondistressed counterparts.

Second, Margolin and Weiss (1978b) describe a button-press procedure which can be used in both the assessment and treatment of marital communication skill deficits. In the case study presented, spouses independently rated their "helpfulness" as senders and receivers while coding videotaped samples of their own previous problem-solving negotiations. Additional training procedures utilized electromechanical cueing devices.

With videotape-recording systems came the opportunity to analyze both the verbal and nonverbal aspects of marital communication. Early investigations sought to determine the reliability and validity of videotaping compared to existing alternatives. Such studies indicated that (a) percentages of interrater agreement were significantly higher using videotaped communication samples (81.54% agreement) than typescripts prepared from audiotapes (60.34%; Murphy & Mendelson, 1973b); (b) a positive relationship was found between coders' observations from videotapes and couples' self-reports of their communication process (Murphy & Mendelson, 1973b); and (c) reliabilities in scoring discrete marital interaction behaviors such as "looking" and "smiling" were found to be as high for videotape analysis as for live observation and scoring (Eisler, Hersen, & Agras, 1973b). Thus, videotaping is advantageous in that it provides for delayed analysis of both verbal and nonverbal interaction behaviors.

Accordingly, contemporary marital researchers have developed assessment technologies, observational coding systems, and treatment interventions based on the use of videotaping procedures (cf. Weiss & Margolin, 1977). In particular, Weiss, Patterson, and their associates have employed videotape technology in the development of the Marital Interaction Coding System (MICS; Hops, Wills, Patterson, & Weiss, 1972); in assessing discriminating interaction behaviors among distressed and nondistressed couples (e.g., Birchler *et al.*, 1975; Royce & Weiss, 1975; Vincent *et al.*, 1975); in case analyses of distressed marital communication (Margolin, 1977; Weiss *et al.*, 1973); and in the development of videotape feedback procedures for therapy intervention (Margolin & Weiss, 1978b; Patterson, Hops, & Weiss, 1975).

Other researchers also have used videotape observation procedures and have either adapted the MICS for their own evaluative purposes (cf. Jacobson, 1978c; Liberman, Levine, Wheeler, Sanders, & Wallace, 1976) or they have developed a similar behavior coding system to facilitate videotape analysis (Gottman *et al.*, 1977). In a very innovative assessment study, Gottman *et al.* (1977) investigated interaction patterns of distressed and nondistressed couples as they engaged in conflict resolution. These interactions were videotaped and coded behaviors were later sequentially analyzed according to conditional probabilities. Results were quite revealing and demonstrated objectively what marital therapists have often subjectively experienced, i.e., during conflict resolution, distressed partners quickly lose focus on the topic and begin to blame one another for any number of tangential problems (cross complaining). This process often escalates into an aversive, negatively charged interaction marked by mindreading, insults, and increased recriminations (negative exchange loop), which typically ends either in cold, hostile silence or violence. By contrast, nondistressed couples are much more likely to maintain focus on the issue, to listen to and support one another's feelings, to avoid negative exchanges, and to end their discussion with proposals for solution and agreement. Clearly, when replicated, these findings will help validate many of the basic assumptions which have guided marital therapists who offer communication skills training to their distressed clients.

In concluding this section, one final observational method should be mentioned that involves *in vivo* home observation. Relatively little home observation of marital interaction *per se* has been reported. In an early case study, Patterson and Hops (1972) used home observation procedures to assess family interaction prior to and following marital therapy. Trends indicated overall improvement in family interaction as a function of marital intervention, but results were not statistically significant. In a more recent application, Haynes, Follingstad, and Sullivan (1978) observed the home interactions of distressed and nondistressed couples on three separate evenings, 2 weeks after subjects completed marital satisfaction inventories. Time-sampled behaviors, adapted from the MICS, included: eye contact, positive physical contact, agreement, criticism, interruption, disagreement, smile, compliment, suggestion, and no response. Results indicated that the first 6 of the 10 behavior codes listed were reliable and significantly discriminated between the distressed and nondistressed couples. These are encouraging results since such code-by-code discriminative power has generally not been found when the MICS has been used to analyze similar interactions in the laboratory.

In sum, one can see that the technology developed to assess communication skills and the effects of communication training have become increasingly sophisticated. To self-report questionnaires, scale scores, and hand analysis, we have since added audiotaping, typescript frequency counts, and desk calculators, and gone on to videotaping, elaborate behavioral coding systems, and sequential probability computer analysis. But before we get carried away, let us return to a discussion of the practicalities of communication skills training with distressed couples.

Program Descriptions and Uncontrolled Studies. A review of the literature revealed few treatment program descriptions which did not include both basic communication and problem-solving components. Programs which seemed to omit an emphasis on problem solving (e.g., Pierce, 1973; Schauble & Hill, 1976; Wells, Figuerel, & McNamee, 1975) were primarily based on the communication training model of Carkhuff (1972). In the remainder of this chapter, communication skills training refers to both components: general communication training (e.g., sender and receiver skills), and training in conflict resolution (e.g., techniques of problem solving, negotiation training, and contingency contracting). Case studies will be mentioned for the reader's reference, but programmatic efforts will be emphasized. Finally, an attempt will be made to focus on communication skills training *per se,* and therefore certain otherwise major aspects of some programs will not be covered here.

Publications have appeared during the past 10 years which have described a therapeutic approach to distressed marital communication and the authors have presented one or more case studies to illustrate their programs and representative findings. As mentioned earlier, an important stimulus was the 1969 paper by Stuart, which was a kind of theoretical, behaviorally oriented beginning for the field. Subsequently, several case studies appeared describing the application of the behavioral model to marriage and family counseling (Liberman, 1970), negotiation training and multimethod assessment (Patterson & Hops, 1972), and behavior-exchange procedures (Hickok & Komechak, 1974; Knox, 1973; Rappaport & Harrel, 1972). Others suggested that distressed marital communication could be improved through assertive training procedures (Alberti & Emmons, 1976; Eisler, Miller, Hersen, & Alford, 1974), corrective feedback and instruction (Carter & Thomas, 1973a), and by combining communication and sexual skills training to the behavior-exchange model (Wieman, Shoulders, & Farr, 1974).

Two relatively recent papers describe the application of systematic communication skills training to reduce spouse abuse (Margolin, 1978) and to develop behavioral competencies in seriously distressed couples (Peterson & Frederiksen, 1976). Margolin describes a cognitive-behavioral treatment program which incorporates five components: (1) identification of anger cues—within oneself and emitted by spouse, using behavioral diaries, anger checklists, and in-session role-plays; (2) learning alternative responses to the anger cues once identified—particularly avoiding exchanges of aversive stimuli; (3) modification of faulty cognitions about the relationship—e.g., "If I don't express negative feelings things will get better"; (4) learning problem-solving skills—altering the aversive and coercive behavior-change strategies often employed; (5) relationship enhancement—learning to increase positive exchanges and mutually rewarding activities. A case example suggests that the approach is promising for anger-prone couples who are still committed to the relationship and who can gain some immediate control over resorting to violence.

The second report (Peterson & Frederiksen, 1976) describes a rather elaborate modular skills training approach for distressed couples which systematically assesses and teaches couples how to negotiate problems and/or express positive and negative emotions, whichever skills are deficient. The training program includes five sequentially applied components: rationale and instructions about the skill to be learned; videotaped modeling stimuli; behavioral rehearsal which is videotaped;

videotape feedback and coaching regarding the rehearsal; and homework assignments to practice the skills being learned. Results from two cases of severely distressed couples were presented. The program seemed effective in altering behavioral rates of targeted behaviors, although self-report measures of satisfaction were not particularly affected. Follow-up data were not presented. While the program looks promising, together with all the above case studies, more definitive research is required before conclusions can be drawn.

At this point let us review three of the well-established clinical treatment programs which have varying degrees of empirical support. All three programs have given increasing attention to couples' basic communication skills, in addition to their long-term emphasis on strategies for behavior change and marital conflict resolution. Available outcome studies will be summarized in a later section.

First to be considered will be the program commonly understood to be a product of the Oregon Marital Research Project, initiated in the late 1960s and co-directed for years by Gerald Patterson and Robert Weiss. More recently, under the direction of Weiss, clinical research has continued as part of the Oregon Marital Studies Program. Weiss, together with his current students in Oregon (e.g., Weiss & Aved, 1978), former students (e.g., Birchler, 1977; Margolin, 1978), and young colleagues elsewhere (e.g., Jacobson, 1977a), has continued to develop and evaluate components of the so-called Oregon approach. Figure 1 is a flow chart representation of the treatment process and intervention modules. Detailed descriptions of each phase are available elsewhere and thus only brief summaries will be presented here (cf. Margolin, Christensen, & Weiss, 1975; Weiss, 1978; Weiss & Birchler, 1978).

Naturally, steps in the flow chart are idealized. In practice some variation in procedure or ordering of modules may occur, depending on

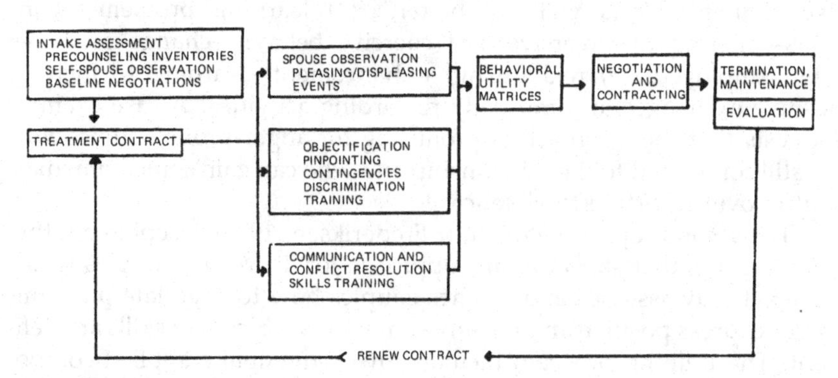

Figure 1. Flow chart of behavioral intervention modules: The "Oregon" approach.

the needs, skills, and deficiencies of each couple. Prior to the first person-to-person contact, couples are mailed a set of precounseling marital inventories. This package (cf. Margolin *et al.*, 1975) is designed to assess global marital satisfaction (Locke–Wallace Marital Adjustment Scale; Locke & Wallace, 1959); marital problem areas (Areas of Change Questionnaire, ACQ); who does what with whom (Inventory of Rewarding Activities, IRA); and steps toward divorce (Marital Status Inventory, MSI; Weiss & Cerreto, 1975).

Once the inventories are returned, couples begin a 10-week assessment–treatment unit. The first 2 weeks are devoted to baseline assessments, including spouse observation and self-recording of pleasing and displeasing events in the natural environment. Also, during assessment, couples provide four 10-min videotaped samples of their marital communication and problem-solving skills, which are then analyzed using the Marital Interaction Coding System (MICS). It features 30 behavior codes which can be analyzed sequentially (e.g., Margolin, 1977) or summarized into categories representing problem-solving skills, positive and negative verbal behaviors, and positive and negative nonverbal behaviors.

Following assessment, the couple is given a behavioral formulation of strengths and deficits in their relationship and a *treatment contract* is formalized. At this point, the first three treatment modules become operational rather concurrently. Couples are asked to continue tracking *P*'s (pleases) and *D*'s (displeases) at home on a daily basis, and they are phoned every few days for the data. A second module consists of increasing *objectification skills,* i.e., helping couples translate vague complaints (e.g., "My husband doesn't love me") into specific behaviors which can be increased (e.g., "I would like my husband to talk with me after he comes home from work"). Depending on couples' baseline abilities, more or less time is spent on basic communication skills training. Therapists employ videotape feedback, modeling, behavior rehearsal, and homework assignments to increase partners' general sending and listening skills, and skills related to conflict resolution.

Once couples are aware of pleasing and displeasing interactions in their relationship, have learned to pinpoint problems, and have acquired constructive communication skills, they are then ready for the next intervention module: developing *behavioral utility matrices.* This consists of generating lists of potential reinforcements (*P*'s) and penalties (*D*'s) which can be integrated into behavioral contracts (cf. Weiss, Birchler, & Vincent, 1974).

The succeeding treatment module is *negotiation and contracting*. During this phase couples are taught how to negotiate actively and to design rather formal behavior-change contracts, typically of the "good faith" variety (cf. Weiss *et al.*, 1974). Areas of change for husband and wife are balanced for value, and the positive and negative contingencies are applied to each target behavior (spouse) independently. Couples typically implement two or three contracts under the therapists' supervision before termination, maintenance, and evaluation. During this final step, the therapists' direction is increasingly faded out as couples take over complete responsibility for their relationship. All of the initial assessment procedures are repeated. Typically, couples are invited to return to the clinic for "booster" sessions if they feel the need.

The BMT program described by Liberman, Wheeler, and Sanders (1976a) is very similar to the Oregon program described above. One major difference is that couples are seen individually for intake and assessment (1–3 meetings), and then are placed in conjoint couples groups for the treatment which lasts for 8–10 weekly sessions (cf. Figure 2). Three to five couples comprise a group.

During intake and orientation, couples are allowed to ventilate and recriminate as therapists shape positive interactions and help set goals for group intervention. The major elements of treatment are analogous to those in the Oregon model. Couples are taught to pinpoint and discriminate pleases using large "please" checklists and a diary format for spouse observation. Couples are encouraged to identify or develop and nurture core symbols, i.e., activities, places, or objects which signify special meaning to the relationship. In addition, couples are encouraged to develop mutually rewarding recreational activities. This strategy is

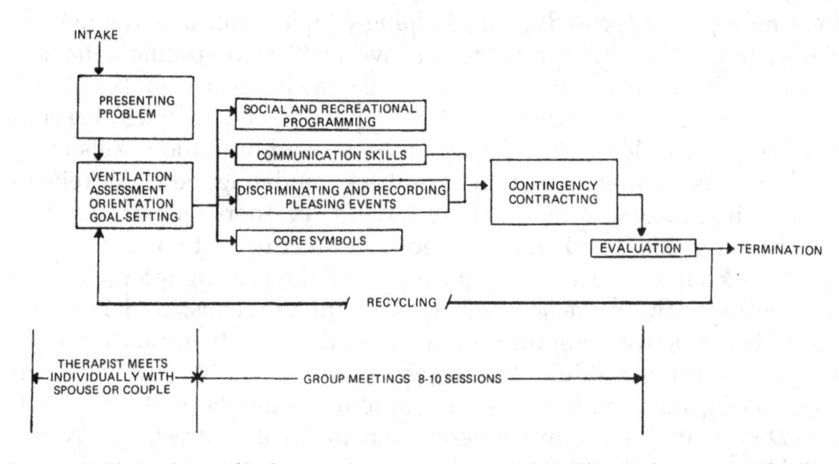

Figure 2. Flow chart of behavioral treatment modules: The "Liberman, Wheeler, and Sanders" approach. Reprinted from Liberman, Wheeler, and Sanders (1976a), Figure 1, copyrighted by and used with the permission of the American Association for Marriage and Family Therapy.

advisable since distressed, relative to nondistressed, couples have been found to engage in significantly smaller proportions of spouse-only elective activities, and often larger proportions of alone activities (Birchler, 1977; Birchler & Webb, 1977; Birchler *et al.*, 1975). Accordingly, helping couples place a higher priority on mutually reinforcing couple activities has been a common thrust in the remediation of marital distress.

According to Liberman *et al.* (1976a), the most important treatment element has been communication training, which includes structured exercises, therapist modeling, behavior rehearsal, coaching, and homework assignments. Couples practice positive and negative emotional expressiveness, general assertiveness, and communication about sex and affection.

Finally, contingency contracting is used on a selective basis for appropriate couples. The contracting procedures, carried out in the group, feature written contracts, "good faith" parallel contracting, emphasize positive contingencies, and include home observation and record keeping. Contracts are seen as temporary aids which help certain couples to structure positive exchanges rapidly. Moreover, the negotiation and contracting process itself is often therapeutic.

Evaluation procedures typically consist of the LW and ACQ, a consumer-satisfaction questionnaire, and data related to exchanges of "pleases" and contract performances. A comparative group evaluation study will be reviewed (cf. Liberman, Levine, Wheeler, Sanders, & Wallace, 1976b).

The third well-established BMT program which features communication skills training has been developed by Richard Stuart (Stuart, 1976). As indicated in Figure 3, Stuart describes eight steps. Briefly summarized these are:

Step I—*Completion of the Marital Pre-Counseling Inventory* (Stuart & Stuart, 1972), which is given pre- and posttreatment as the only assessment instrument.

Step II—*Formalizing the treatment contract.* This step is typically accomplished over the phone with both partners, and the agreement is made to see the couple conjointly for an initial six sessions.

Step III—The first conjoint session and the primary goal is to discuss the *rationale for the operant interpersonal approach to treatment*. Emphasis is placed on positive strategies for behavior change and focusing on the here and now.

Step IV—*Initiating caring days,* consists of helping partners to identify simple, positive caring behaviors which can be exchanged frequently (e.g., backrubs, a special meal, a cup of coffee, a flower, a compliment, a hug or kiss). Partners are assigned to do 8–20 of these per day, indepen-

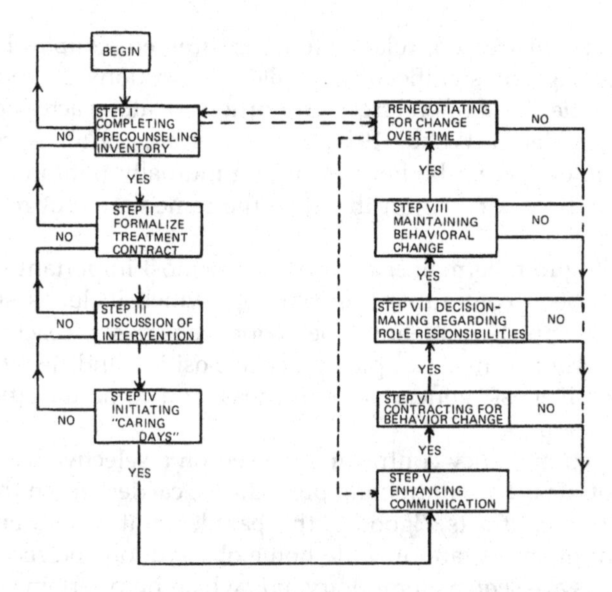

Figure 3. Flow chart of behavioral treatment process: The "Stuart" approach. Reprinted from Stuart (1976), Figure 1, by permission of the author.

dent of the number received. The idea is to increase dramatically positive behaviors while deemphasizing the negatives.

Step V—*enhancing communication* really occurs throughout treatment. However, in this phase couples are taught a number of specific "rules" which help them communicate honest and constructive timely messages, while they learn to modify messages which may have a negative impact. Once these rules and basic communication skills are learned,

Step VI—takes couples on to *contracting for behavior change.* Such contracts and procedures for implementation are virtually identical to those described above for the Oregon and Liberman programs.

Step VII—Consists of *decision making regarding role responsibilities.* Here Stuart helps couples to analyze, and if necessary modify, their current allocation of responsibilities for major decision making.

Step VIII—*Maintaining behavioral changes,* is the final step. Couples review the new "relationship rules" they have formed regarding communication, behavior change, and decision making. They are also alerted to the potential for spontaneous remission of the old patterns and offered a series of exercises to help them anticipate and deal with this possibility.

Finally, couples are asked to complete the Marital Pre-Counseling

Inventory at 4-month intervals. This procedure provides ongoing periodic assessments for both the couple and the therapist.

In sum, it can be seen that three of the major BMT programs have numerous similarities. Obviously, they have borrowed significantly from one another over the years, and yet they each have some unique features. Let us now turn to a brief examination of three preliminary single-group studies designed to evaluate the basic effectiveness of the Oregon model.

Weiss *et al.* (1973) presented data from two small pilot groups who participated in a program similar to the one outlined in Figure 1. Basically, expectations were that intervention would: (a) improve problem solving in the laboratory (MICS); (b) generalize to daily interactions in the home (*P*'s and *D*'s); (c) increase marital satisfaction (LW); (d) decrease requests for change (ACQ); and (e) increase the proportion of spouse-together activities (MAI-TAO, now IRA). In Study I (Weiss *et al.*, 1973), the focus was on five couples' laboratory problem solving (MICS) and daily interactions at home (*P*'s and *D*'s). Results indicated a signficant increase in compromise statements and decreases in accept responsibility, assent, disagree, put down, and problem description. With the exception of the assent code (positive nonverbal head nods), the other changes were to be expected after intervention. An analysis of *P*'s and *D*'s indicated a significant increase from baseline in *P*'s, although levels of *D*'s did not change. Thus, at first glance the intervention package seemed effective in producing certain expected changes.

Study II (Weiss *et al.*, 1973) also included five distressed couples treated for 8–13 weeks. The 30 MICS codes were collapsed into six summary categories to analyze negotiation sessions. As predicted, analysis of variance for repeated measures indicated that "problem solving" increased ($p < .001$); "problem description" decreased ($p < .001$); "negative verbal" ($p < .05$) and "nonverbal behaviors" ($p < .005$) decreased; and "positive verbal" ($p < .005$) and "nonverbal behaviors" ($p < .025$) increased. In addition, couples' LW scores increased ($p < .01$); their WC (now ACQ) scores decreased ($p < .05$); *P*'s increased ($p < .01$); *D*'s decreased ($p < .02$); but there was no significant group change on the MAI-TAO (couple activities).

A third single-group study added more support for the Oregon BMT approach (Patterson *et al.*, 1975). Ten couples obtained 6–9 h of training. MICS codes were collapsed into two overall summary categories: facilitating and disrupting behaviors. Results indicated that pre- to posttraining, six of eight couples significantly increased their positive conflict-resolution skills and seven of eight decreased disrupting communication behaviors. Regarding *P*'s and *D*'s at home, six of nine couples significantly increased *P*'s, while six of eight decreased *D*'s. Overall, laboratory and self-report data suggested that six of eight couples had made significant gains as a result of training. These early studies, though uncontrolled, set the stage for more rigorous studies.

Analogue Studies and Components Analysis. To date, very little well-controlled clinical research exists which demonstrates the efficacy of widely used treatment components (e.g., conjoint vs. concurrent vs. group treatment, positive target behaviors vs. positive and negative target behaviors, verbal instruction and feedback, therapist modeling, behavior rehearsal, videotape feedback, problem solving and decision making, contingency contracting, and the use of homework assignments). The preliminary research which does exist for some of these variables will now be discussed.

One intervention procedure which has received considerable attention has been *videotape feedback* (VTF) (cf. Griffith, 1974). However, mixed results have been obtained when VTF has been applied to married couples. While one study has suggested caution in its use for severely disturbed couples (Alkire & Brunse, 1974), others have found VTF to have a relatively small incremental effect when compared to focused verbal instructions (Eisler, Hersen, & Agras, 1973a). A similar unpublished analogue study by Birchler, Hoehne, and Martorelli (1977) also failed to demonstrate the incremental effectiveness of VTF over alternative conditions, including focused instructions and an "irrelevant television" control group.

In contrast, Edelson and Seidman (1975) compared the effects of VTF plus verbal feedback, relative to verbal feedback alone or no feedback, in changing student married partners' interpersonal perceptions about self and spouse. Results indicated that VTF combined with verbal feedback had superior effects. Mayadas and Duehn (1977) developed a stimulus-modeling videotape which depicts dysfunctional and effective segments of marital problem solving. Thirty distressed couples were equally assigned to 8 h of verbal counseling (VC), stimulus-modeling tape (SM) only, or SM plus VTF. Regarding couples' five most frequently checked problematic verbal behaviors (e.g., countercomplaining), SM plus VTF significantly reduced all five behaviors; SM alone reduced three of five; while VC decreased only one of five target behaviors.

In summary, it appears that VTF in combination with verbal instruction or other discriminative cues (e.g., stimulus modeling) can be effective in improving couples' marital communication skills. However, the effect of VTF alone has not yet been reliably demonstrated.

Concerning therapy format considerations, a thorough review of the literature by Gurman and Kniskern (1978c) indicated that conjoint and conjoint group marital therapy result in the best improvement rates and least deterioration due to intervention (cf. Cookerly, 1976).

A comprehensive outcome study initiated by Turner (1972) was designed to investigate the relative effectiveness of: conjoint vs. concurrent format; intervention on positive vs. positive and negative behaviors; and single vs. co-therapists. Eight groups of 10 couples each met for six 2-h weekly sessions. Training included pinpointing, home observation and data collection, modeling, behavior rehearsal, and contingency contracting for problem behaviors. Preliminary results (Turner, 1978) suggest that the best therapeutic approach was conjoint therapy by co-therapists, using contingency contracting designed to decrease negatives and increase positives. Details of the study await the publication of Turner's results.

Integrated within behavioral approaches to marital dysfunction, communication skills training (including problem solving) has often been seen as a prerequisite treatment component to behavioral contracting (cf. Figures 1, 2, 3). While contingency contracting *per se* is not the focus of this chapter, the two treatment components are almost always offered in combination. Some initial research has sought to determine the relative efficacy of the communication/problem-solving and the contracting components. For example, clinical studies by Jacobson have indicated that problem-solving training alone (Jacobson, 1977b) and in combination with contingency contracting (Jacobson, 1977a) are significantly effective components in intervention programs designed to remediate dysfunctional marital communication and marital dissatisfaction. Another study (Jacobson, 1978c) compared two forms of contingency contracting with attention-placebo and waiting-list control groups. While both the *quid pro quo* and "good faith" forms of contracting proved more effective than the control groups on multimethod dependent variables, results were not significantly different for the two forms of contracting.

While contingency contracting seems instrumental for some couples to effect behavior changes efficiently, it is nevertheless a cumbersome and sometimes unpalatable procedure for consumers (Jacobson, 1979; Tsoi-Hoshmand, 1975). Therefore, research demonstrating more effective behavior-change strategies would be very important.

One final analogue study to be considered sought to evaluate and compare three therapeutic components commonly associated with BMT programs (Margolin & Weiss, 1978a). Twenty-seven couples recruited for distress were randomly assigned to one of three treatment conditions; nonspecific effects (NS), NS plus behavioral therapy (BT), and NS plus BT plus attitudinal restructuring (AB). During intervention all groups engaged in a number of 10-min videotaped conflict negotiation sessions, after which 10-min feedback discussions were held. In the BT condition, training procedures included: pinpointing, practicing "sender" and "receiver" skills, and behavior rehearsal of "helpful" responses. At home, BT partners were encouraged to increase baseline rates of "pleasing" behaviors by 100%. AB couples followed a pattern similar to that of BT couples except that during the above procedures AB couples were continually given a prorelationship cognitive set. For example, problem areas, target communication skills, "pleases" to be increased, etc., were all mutually defined for the good of the relationship. The NS couples followed a similar format but were neither trained in using behavioral procedures nor given instructions to change at home.

The outcome criteria included measures of marital satisfaction (e.g., LW, ACQ, adjective checklists), daily reported *P*'s and *D*'s, and laboratory observations of marital conflict resolution analyzed using the MICS. Results largely substantiated expectations. The AB condition, combining all three treatment components, consistently demonstrated superiority in bringing about positive changes. Out of the nine multimethod outcome criteria, significant changes were noted for the AB, BT, and NS groups on eight, three, and three measures, respectively. Interestingly, all three conditions produced significant reductions in negative behaviors. Unfortunately, since a no-treatment control group was not employed, it was not possible to determine whether these latter findings were due to treatment effects, demand characteristics, or measures of reactivity.

In addition, the implications of this study were also limited by subject recruitment procedures, short duration of treatment (4 intervention hours), inexperienced therapists, and highly standardized training procedures. Nevertheless, support was clearly gained for the inclusion of cognitive restructuring and attitude change features into BMT treatment programs.

In summary, we must note that currently available treatment component investigations have consisted of short-term analogue studies often limited by short duration of treatment, small sample size, couples not specifically seeking marital therapy, poor controls, and other methodological problems. Overall, while clinical biases certainly exist concerning the efficacy of certain procedures, there is relatively little empirical consistency or replication concerning their effects.

Uncontrolled Comparative Studies. Let us now consider a number of studies which involved CST and compared the relative effectiveness of one or more treatments. "Uncontrolled" refers to the fact that no-treatment and/or attention-placebo control groups were not included in the experimental designs. Some of these studies have already been discussed above (e.g., Margolin & Weiss, 1978a; Mayadas & Duehn, 1977; Turner, 1972, 1978), while others were not available to this writer (Crowe, 1973) or were published as a dissertation abstract which lacked sufficient detail for careful review (Becking, 1973; Swan, 1972). Thus, there are three remaining comparative treatment studies to be described in this section.

One study involved the application of BMT procedures in the treatment of depression (McLean *et al.,* 1973). Twenty severely depressed patients between the ages of 20 and 55 were randomly assigned to an experimental or comparison group. The behavioral treatment group met for eight conjoint therapy sessions, studied principles of social learning, practiced communication skills daily at home facilitated by an electromechanical cue box, and implemented contingency contracts to modify marital problems. The comparison group received whatever alternative treatment was normally administered (e.g., medication, follow-up by family physician, group therapy). Pre- and postintervention and 3-month follow-up measures included the Depression Adjective Check List (DACL; Lubin, 1965), half-hour audiotapes of marital problem solving at home, and ratings of change in five problematic target behaviors related to the patients' depression (e.g., sleep disturbance, suicidal preoccupation, withdrawal).

Results suggested that, by all measures, the experimental group showed the greatest response to treatment and the differences were maintained at follow-up. While this research neither compared alternative approaches to marital dysfunction nor involved couples seeking marital therapy, it does provide support for the effectiveness of systems (conjoint) intervention in general and BMT in particular.

A second study which sought to compare the effectiveness of a behavioral versus a more traditional marital therapy approach was conducted by Liberman *et al.* (1976b). Four couples in the experimental group received rather standard BMT (Figure 2), including coaching, modeling, behavior rehearsal, and homework procedures (all designed to improve marital communication), and contingency contracting to facilitate behavior change. In contrast, the "interactional-insight" comparison group (5 couples) focused on expression of feelings, discussion of problems, group support and feedback. Groups met for approximately 16 h.

Multiple pre- and posttreatment outcome measures were obtained, some also at follow-up. These included self-report, spouse observation, nonverbal interactional measures during sessions, and MICS analysis of conflict-resolution discussions. Data obtained did not conclusively demonstrate the superiority of one treatment over the other. Both groups improved on LW and ACQ satisfaction measures. Regarding communication behaviors, BMT was significantly better in increasing couples' looking and smiling behaviors during sessions, in decreasing MICS rates of negative verbal and nonverbal behaviors, and in increasing MICS positive nonverbal behaviors. Somewhat surprisingly, on almost all of the other outcome measures (at least eight) there were no significant within- or between-group differences found.

The significance of the Liberman *et al.* (1976b) study, as suggested by the authors, is that while both approaches reflected improvement in couples' subjective marital satisfaction, only the combined procedures of systematic communication skills training and contingency contracting produced objective behavioral changes. Despite a number of methodological limitations to the study (e.g., potentially biased therapists, lack of random assignment to groups, very small samples, and no untreated control group), the efficacy of communication training using behavioral procedures gained significant empirical support.

Finally, Valle and Marinelli (1975) compared the effects of a systematic didactic and experiential communication training program (based on the work of Carkhuff, 1972), to a more traditional therapy group. The training group learned "helper" and "listener" skills, and applied these to the resolution of marital problems. The comparison group emphasized cathartic release and problem-solving discussion, led by a nondirective therapist. Two groups of five distressed couples obtained over 50 h of treatment. Dependent variables were three forms of self-report measures relating to the target skills and relationship functioning. Results indicated that the training group improved significantly more than the traditional-insight therapy group on all measures.

In summary, the uncontrolled comparative group studies reviewed, though methodologically imperfect, suggest fairly convincingly that systematic skill training approaches are generally superior to several more traditional forms of intervention for improving distressed couples' communication and marital satisfaction. Let us turn now to a review of the better controlled outcome studies which incorporate communication skills training.

Controlled Outcome Studies. If one were seeking methodological perfection in the pursuit of outcome studies on communication skills training with therapy-seeking distressed couples (cf. O'Leary & Turkewitz, 1978), one would be forced to conclude that none exist. Indeed, there are few studies which even approximate this goal. However, both the number and quality have increased in just the past couple of years.

Pierce (1973) reported a study similar to that by Valle and Marinelli (1975). However, the Pierce study included a no-treatment control group in addition to the "skill training" and "traditional-insight therapy" treatment groups. Also, the two outcome measures (interviewers' "helper" skills and interviewees' self-exploration statements) were based solely on trained coder ratings of spouses' pre- and post-treatment 15-min interviews of one another. Results indicated superior within-group and between-group differences on both measures for the experimental group relative to both comparison groups.

In passing, we should mention an outcome study by Tsoi-Hoshmand (1976). It suggested that a behavioral-learning treatment program was superior to two no-treatment control groups in increasing couples' scores on self-report measures of caring and satisfaction, and on a measure testing amount of learning in therapy. Unfortunately, there were so many methodological limitations to this study that one can only cautiously accept the results.

A better example of the "state of the art" in outcome studies of BMT communication skill training is represented by the recent research efforts of Jacobson and Turkewitz. Jacobson has been quite active in the past 5 years, and in the process of evaluating primarily the Oregon BMT model described earlier he has conducted several empirical investigations. Thus far, two have included conventional control groups (Jacobson, 1977a, 1978c). In the first study (Jacobson, 1977a), 10 dissatisfied couples were randomly assigned to experimental or waiting-list control groups. The primary focus of nine intervention sessions was training in problem-solving communication skills. Through coaching, modeling, videotape feedback, behavior rehearsal, and structured homework assignments, couples first learned communication strategies for problem solving, followed by the implementation of "good faith" contingency contracts (cf. Weiss *et al.*, 1974).

Pre- and posttherapy (or waiting period) dependent measures included the LW and two behavior measures derived from two 5- to 10-min videotaped negotiation samples analyzed using the Marital Interaction Coding System (MICS). MICS behavior codes were collapsed into overall positive and negative categories, each expressed in terms of a ratio, with total responses in the denominator. Couples in treatment also monitored and recorded multiple target behaviors at home 2 weeks before and throughout intervention.

MICS results indicated that treated couples nearly doubled their proportions of positive communication behaviors (pre- to posttest means were .204 and .393, respectively); no significant change was noted for the controls (.159 to .139). For negative communication behaviors, similar scores for the treated couples decreased from .514 to .269; control group means were .574 and .612. The final differences between the treatment and control groups were significant for both positive ($p < .002$) and negative ($p < .009$) interactional behaviors. LW scores increased an average of 32.3 points and 2.2 points for the treatment and control groups, respectively ($p < .003$). For the treated couples, the gains in marital satisfaction scores were maintained at 1-year follow-up.

To reinforce and extend the above findings, Jacobson (1978c) investigated the relative efficacy of two behavioral treatments, a nonspecific treatment, and no treatment. The behavioral treatments consisted of training in problem-solving skills and one of two forms of contingency contracting (which did not produce differential effects on the outcome measures).

Procedures employed for the nonspecific control groups (NS) successfully controlled for a host of competing explanations for change. In addition to the NS variables, the behavioral treatment groups received explicit problem-solving skill training and contingency contracting. The waiting-list control group completed pretest and posttest measures.

Thirty-two volunteer, rather young couples were recruited for the study, primarily through advertisements. Final group sizes were fairly even. The dependent measures taken pre- and postintervention were similar to those described above. All treatment groups met weekly for 1 to 1 1/2 h over 8 weeks. Three therapists saw couples in each condition.

Results indicated that for both observational measures (MICS positive and negative communication behaviors) and the LW scale, only the behavioral groups changed significantly in the desired direction from pre- to posttest, and both behavioral treatments were significantly more effective than either of the control conditions. On one marital happiness self-report measure, all three treatment groups registered significant changes. Follow-up LW scores obtained at 6 months indicated that the posttreatment group differences were maintained.

The conclusiveness of these studies is limited because the couples were not representative of help-seeking, severely distressed marriages; the sample sizes were quite small; and the investigator served as a therapist in each study. However, the results of these investigations strongly suggest that one component of communication training (problem solving) in combination with contingency contracting is demonstrably more effective than no treatment or attention-placebo factors in producing desirable behavioral and cognitive changes in distressed couples.

One final communication training outcome study to be reviewed was conducted by Turkewitz (1977). Thirty distressed couples were assigned equally to one of three conditions: behavior-exchange and negotiation training (BMT), general communication training (CT), or waiting-list control (WL). The treated couples were seen in conjoint sessions by one of five therapists, each of whom saw two couples in each condition. Couples were seen weekly for 10 weeks. BMT focused on techniques to increase the spouses' desired behaviors: communication training relevant to problem solving, modeling, coaching, contingency contracting, and homework assignments. CT included techniques to increase clear and decrease destructive interchanges, and to increase support to one another: modeling, feedback, behavior rehearsal, empathy training, homework assignments.

Pre–post dependent measures assessed marital satisfaction (LW), communication patterns (PCI), change in client-targeted problems, and eight communication behaviors observed during problem-solving discussions.

Surprisingly, there were no significant group differences in change scores for the following planned comparisons: Locke–Wallace, a Positive Feelings Questionnaire, and all eight observed communication behaviors. Both treatment groups were found to improve significantly more than the control group on a number of subjective self-report measures, including the PCI. Interestingly, *post hoc* analyses discovered that young couples in BMT (mean ages 29.4 years, married an average of 6.8 years) changed significantly more in marital satisfaction and communication patterns (PCI) than couples in CT or WL. In contrast, older couples (mean ages 40.9 years, mean length of marriage 18 years) seemed to respond better to CT.

The pattern of results in the Turkewitz study is complex and difficult to interpret. For example, six of eight observed behaviors showed no within-group changes; two changed in all three groups. Based on experimental results using similar coding systems, one might question the sensitivity and reliability of the system adopted for the present study.

In sum, the lack of differential treatment effects could be accounted for by uncontrolled, nonspecific effects, by too much overlap in communication training, or by the fact that both treatments, though different, were equally effective. The discovery that age was a mediating variable is important and definitely merits further research.

Current Status and Future Directions for Clinical Approaches

There is no doubt that during the past decade significant advances have been made in the application of communication skills training to marital distress. The empirical work with clinical populations has largely been forthcoming from behaviorally oriented investigators. From a series of uncontrolled, though highly suggestive case and single-group studies (e.g., Azrin *et al.*, 1973; Margolin *et al.*, 1975; Patterson & Hops, 1972; Weiss *et al.*, 1973) to a number of more recent and better controlled studies (e.g., Jacobson, 1977a, 1978c; Liberman *et al.*, 1976b; Margolin &

Weiss, 1978a; Turkewitz, 1977), communication skills training in general and the problem-solving component in particular have been shown to be superior to a variety of comparative and control groups. Nevertheless, as reviewed above, all of these studies were limited by methodological considerations.

Let us briefly outline, as have others, directions for future research and development (Greer & D'Zurilla, 1975; Gurman & Kniskern, 1978c; Jacobson & Martin, 1976; Weiss & Margolin, 1977).

1. In the matter of assessment, work is progressing but more is needed to pinpoint the communication deficits which distressed couples possess. Recent sequential analyses of marital conflict-resolution interactions have helped us to delineate distressed couples' maladaptive communication patterns (cf. Gottman *et al.*, 1977; Margolin, 1977), but we need more understanding of the stimulus conditions and antecedents to marital conflict.

Certainly, the multimethod assessment methodology employed by Weiss at the Oregon Marital Studies Program (Weiss & Margolin, 1977) and his colleagues (e.g., Birchler, 1977; Jacobson, 1977a) is the most sophisticated to date. However, the degree of complexity is both an asset and a liability. The liability derives from the fact that few investigators and even fewer clinicians command the resources necessary to implement the assessment procedures (e.g., the MICS, which requires videotape equipment, trained coders, and computer analysis; or the Spouse Observation Checklist procedure, which requires near daily telephone calls, scorers, and computer analysis). Eventually, more practical and less cumbersome procedures will have to be developed if this technology is to be widely adopted in the field.

2. Initial or replication studies are required to determine the relative effectiveness of: (a) conjoint versus conjoint group formats; (b) nonspecific effects associated with BMT (Jacobson, 1978c; Margolin & Weiss, 1978a); (c) single versus co-therapists (Turner, 1972); (d) videotape feedback (e.g., Eisler *et al.*, 1973a; Mayada & Duehn, 1977); (e) modeling; (f) feedback and instruction (Carter & Thomas 1973a); (g) behavior rehearsal; (h) structured homework; (i) cognitive restructuring (Margolin & Weiss, 1978a; Strayhorn, 1978); (j) contingency contracting (Jacobson, 1978c); (k) non-problem-solving communication skills training (Pierce, 1973; Turkewitz, 1977); and (l) communication skills training relevant to problem solving (Jacobson, 1977a). All of the above procedures have received at least indirect support as part of overall treatment packages, but substantial knowledge of their unique effectiveness is lacking.

3. We definitely need more well-controlled studies to document the

effectiveness of communication skills training for significantly distressed couples. Few clinical treatment studies exist which describe the exclusive use of treatment-seeking couples, including data attesting to their levels of significant distress (e.g., Jacobson, 1977b; Liberman *et al.*, 1976b). More work is required with typical clinic populations, i.e., older, less educated couples who manifest both individual and relationship malfunction.

4. Until recently, most of the outcome studies in the marriage literature have suffered greatly from a lack of specification of treatment procedures. It is often difficult to determine exactly what was done, much less to extract enough detail to attempt replication. The recent trend toward detailed specification of treatment procedures should be encouraged (cf. Jacobson, 1977c, d; Margolin *et al.*, 1975; Turkewitz, 1977).

CONCLUSION

Integration of the Enrichment and Clinical Models

In reviewing the *empirical* literature concerned with marriage enrichment and marriage therapy, one is struck by the lack of reference by members of one group to those in the other. Indeed, there have been historical and developmental differences between these two groups. The enrichers, relative to the therapists, have generally (a) adopted a preventive versus a remedial approach, (b) focused on functional versus dysfunctional marriages, (c) used group and weekend formats versus individualized weekly meetings, (d) emphasized general "sender" and "listener" skills versus problem-solving and conflict-resolution skills, (e) avoided versus embraced systematic program evaluation, and (f) been influenced more by communications theory and client-centered principles compared to the clinicians' alignment with principles of social learning theory.

However, despite these considerable differences and the lack of acknowledgment, there has been an impressive convergence of the two approaches. For example, originators of relationship enhancement programs based primarily on the Rogerian model of expressing feelings and empathic listening (cf. Guerney, 1964; Rappaport, 1976) have described extended programs designed to increase couples' conflict-negotiation skills (Harrell & Guerney, 1976; Rappaport & Harrell, 1972). Similarly, the behavior therapists, who originally focused on negotiation training and behavior exchange (e.g., Liberman, 1970; Patterson & Hops, 1972; Stuart, 1969; Weiss *et al.*, 1973), have given increased emphasis to communication training modules designed to facilitate partners' expression of support and understanding (e.g., Liberman *et al.*, 1976a; Margolin & Weiss, 1978b; Stuart, 1976; Weiss & Birchler, 1978).

Another comparison which illustrates the integration of the two approaches concerns the increasing similarity in training procedures. Most CST programs now use feedback and instruction, modeling, behavior rehearsal, and structured practice at home to facilitate spouses' ability to send and receive verbal and nonverbal messages openly, honestly, and constructively. Based on these procedures, our review has indicated that general communication training is effective in enhancing marital satisfaction, particularly for nondistressed couples. Moreover, the clinical literature suggests that intervention packages featuring problem solving and contingency contracting are effective in enhancing marital interaction, behaviorally and subjectively.

The work which remains to be done is considerable, but fairly clear-cut (cf. Jacobson, 1978). We need more well-designed research (cf. O'Leary & Turkewitz, 1978) to help us prescribe selective intervention packages which are effective for couples varying in levels of adjustment. Concomitantly, we need to determine which treatment components are important and which are unnecessary, depending on the goals of intervention. For example, are behavior-exchange contracts appropriate for enrichment programs? Are they really necessary for therapy programs? Based on the increasing quantity and quality of research, together with the current levels of critical appraisal and vitality in the field (cf. Gurman & Kniskern, 1978a, b; Gurman, Knudson, & Kniskern, 1978; Jacobson & Weiss, 1978), one would expect that answers to these questions will be forthcoming in a relatively short period of time.

SUMMARY

This chapter has been concerned with contemporary research and practice in communication skills training for married couples. Two distinct bodies of literature were found which have focused on this topic. Therefore, considered separately, research and practice in the fields of marriage enrichment and marriage therapy were thoroughly reviewed. General descriptions were provided regarding several contemporary intervention programs, followed by reviews of the currently available empirical research relevant to these approaches. Included at the end of the marriage enrichment and marriage therapy sections were discussions of the current status of and future directions for each specialty area. Finally, a case was made for the increasing integration of communication training procedures developed in both fields, and a call was issued for continued research in order to develop, document, and refine the effectiveness of communication skills training.

REFERENCES

Alberti, R. E., & Emmons, M. L. Assertion training in marital counseling. *Journal of Marriage and Family Counseling*, 1976, *2*, 49–54.

Alkire, A. A., & Brunse, A. J. Impact and possible causality from videotape feedback in marital therapy. *Journal of Consulting and Clinical Psychology*, 1974, *42*, 203–210.

Azrin, N. H., Naster, B. J., & Jones, R. Reciprocity counseling: A rapid learning-based procedure for marital counseling. *Behaviour Research and Therapy*, 1973, *11*, 365–382.

Becking, E. P. Pretraining effects on maladaptive marital behavior within a behavior modification approach. (Doctoral dissertation, California School of Professional Psychology, San Francisco, 1972.) *Dissertation Abstracts International*, 1973, *33*, 5007B.

Bienvenu, M. J. Measurement of marital communication. *The Family Coordinator*, 1970, *19*, 26–31.

Birchler, G. R. Differential patterns of instrumental affiliative behavior as a function of degree of marital distress and level of intimacy (Doctoral dissertation, University of Oregon, 1972.) *Dissertation Abstracts International*, 1973, *33*, 14499B–4500B.

Birchler, G. R. *A multimethod analysis of distressed and nondistressed marital interaction: A social learning approach*. Paper presented at Western Psychological Association Meeting, Seattle, April 1977.

Birchler, G. R., & Webb, L. J. Discriminating interaction behaviors in happy and unhappy marriages. *Journal of Consulting and Clinical Psychology*, 1977, *45*, 494–495.

Birchler, G. R., Hoehne, U., & Martorelli, J. *Effects of videotape and instructional feedback on nonverbal and verbal marital interaction*. Unpublished manuscript, University of California at San Diego, 1977.

Birchler, G. R., Weiss, R. L., & Wampler, L. D. *Differential patterns of social reinforcement as a function of marital distress and level of intimacy*. Paper presented to the Western Psychological Association meetings, Portland, Oregon, April 1972.

Birchler, G. R., Weiss, R. L., & Vincent, J. P. Multimethod analysis of social reinforcement exchange between maritally distressed and nondistressed spouse and stranger dyads. *Journal of Personality and Social Psychology*, 1975, *31*, 349–360.

Bosco, A. *Marriage encounter: The rediscovery of love*. St. Meinrad, Ind.: Abbey Press, 1972.

Brim, O. G., Fairchild, R. W., & Borgatta, E. F. Relations between family problems. *Marriage and Family Living*, 1961, *23*, 219–226.

Campbell, E. E. The effects of couple communication training of married couples in the child rearing years: A field experiment. (Doctoral dissertation, Arizona State University, 1974.) *Dissertation Abstracts International*, 1974, *35*, 1942–1943.

Carkhuff, R. R. *Helping and human relations* (Vols. 1 & 2). New York: Holt, Rinehart and Winston, 1969.

Carkhuff, R. R. *The art of helping*. Amherst, Mass.: Human Resource Development Press, 1972.

Carter, R. D., & Thomas, E. J. Modification of problematic marital communication using corrective feedback and instruction. *Behavior Therapy*, 1973, *4*, 100–109. (a)

Carter, R. D., & Thomas, E. J. A case application of a signaling system (SAM) to the assessment and modification of selected problems of marital communication. *Behavior Therapy*, 1973, *4*, 629–645. (b)

Clarke, C. Group procedures for increasing positive feedback between married partners. *The Family Coordinator*, 1970, *19*, 324–328.

Collins, J. D. *The effects of the conjugal relationship modification method on marital communication and adjustment*. Unpublished doctoral dissertation, Pennsylvania State University, 1971.

Cookerly, J. R. Evaluating different approaches to marriage counseling. In D. H. Olson (Ed.), *Treating relationships*. Lake Mills, Iowa: Graphic Publishing Co., 1976.

Crowe, M. J. Conjoint marital therapy: Advice or interpretation? *Psychosomatic Research*, 1973, *17*, 309–315.

Dixon, D. N., & Sciara, A. D. Effectiveness of group reciprocity counseling with married couples. *Journal of Marriage and Family Counseling*, 1977, *3*, 77–83.

DuBurger, J. E. Marital problems, help-seeking, and emotional orientation as revealed in help-request letters. *Journal of Marriage and Family Living*, 1967, *29*, 712–721.

Edelson, R. I., & Seidman, E. Use of videotaped feedback in altering interpersonal perceptions of married couples: A therapy analogue. *Journal of Consulting and Clinical Psychology*, 1975, *43*, 244–250.

Eisler, R. M., Hersen, M., & Agras, W. S. Effects of videotape and instructional feedback on nonverbal marital interactions: An analogue study. *Behavior Therapy*, 1973, *4*, 551–558. (a)

Eisler, R. M., Hersen, M., & Agras, W. S. Videotape: A method for the controlled observation of nonverbal interpersonal behavior. *Behavior Therapy*, 1973, *4*, 420–425. (b)

Eisler, R. M., Miller, P. M., Hersen, M., & Alford, H. Effects of assertiveness training on marital interaction. *Archives of General Psychiatry*, 1974, *30*, 643–649.

Ely, A. L., Guerney, B. G., & Stover, L. Efficacy of the training phase of conjugal therapy. *Psychotherapy: Theory, Research and Practice*, 1973, *10*, 201–207.

Gordon, T. *Parent effectiveness training*, New York: Peter H. Wyden, 1970.

Gottman, J., Markman, H., & Notarius, C. The topography of marital conflict: A sequential analysis of verbal and nonverbal behavior. *Journal of Marriage and Family*, 1977, *39*, 461–477.

Gottman, J., Notarius, C., Gonso, J., & Markman, H. *A couples' guide to communication*, Champaign, Ill.: Research Press, 1976. (a)

Gottman, J., Notarius, C., Markman, H., Bank, S., & Yoppi, B. Behavior exchange theory and marital decision making. *Journal of Personality and Social Psychology*, 1976, *34*, 14–23. (b)

Greer, S. E. & D'Zurilla, T. J. Behavioral approaches to marital discord and conflict. *Journal of Marriage and Family Counseling*, 1975, *1*, 299–315.

Griffith, P. D. Videotape feedback as a therapeutic technique: Retrospect and prospect. *Behavior Research and Therapy*, 1974, *12*, 1–8.

Guerney, B. G., Jr. Filial therapy: Description and rationale. *Journal of Consulting Psychology*, 1964, *28*, 304–310.

Guerney, B. G., Jr. *Relationship enhancement*. San Francisco: Jossey-Bass, 1977.

Gurman, A. S. Marital therapy: Emerging trends in research and practice. *Family Process*, 1973, *12*, 45–54.

Gurman, A. S., & Kniskern, D. P. Enriching research on marital enrichment programs. *Journal of Marriage and Family Counseling*, 1977, *3*, 3–11.

Gurman, A. S., & Kniskern, D. P. Behavioral marriage therapy: I. A psychodynamic systems analysis and critique. *Family Process*, 1978, *17*, 121–128. (a)

Gurman, A. S., & Kniskern, D. P. Behavioral marriage therapy: II. Empirical perspective. *Family Process*, 1978, *17*, 139–148. (b)

Gurman, A. S., & Kniskern, D. P. Research on marital and family therapy: Progress, perspective and prospect. In S. L. Garfield & A. E. Bergin (Eds.), *Handbook of psychotherapy and behavior change,* New York: Wiley, 1978. (c)

Gurman, A. S., Knudson, R. M., & Kniskern, D. P. Behavioral marriage therapy: IV. Take two aspirin and call us in the morning. *Family Process,* 1978, *17,* 165–180.

Harrell, J., & Guerney, B. Training married couples in conflict negotiation skills. In D. H. Olson (Ed.), *Treating relationships.* Lake Mills, Iowa: Graphic Publishing Co., 1976.

Haynes, S. N., Follingstad, D. R., & Sullivan, J. C. *Assessment of marital satisfaction and interaction.* Unpublished manuscript, Southern Illinois University, 1978.

Hickok, J. E., & Komechak, M. G. Behavior modification in marital conflict: A case report. *Family Process,* 1974, *13,* 111–119.

Hill, W. *Hill interaction matrix monograph.* Los Angeles: University of Southern California, Youth Studies Center, 1965.

Hines, G. A. Efficacy of communication skills training with married partners where no marital counseling has been sought. (Doctoral dissertation, University of South Dakota, 1975.) *Dissertation Abstracts International,* 1976, *36,* 5045–5046A.

Hops, H., Wills, T. A., Patterson, G. R., & Weiss, R. L. *Marital interaction coding system.* Unpublished manuscript, University of Oregon, 1972.

Jacobson, N. S. Problem solving and contingency contracting in the treatment of marital discord. *Journal of Consulting and Clinical Psychology,* 1977, *45,* 92–100. (a)

Jacobson, N. S. *The role of problem-solving training in behavioral marital therapy.* Paper presented at the annual meeting of the Association for the Advancement of Behavior Therapy, Atlanta, December 1977. (b)

Jacobson, N. S. Training couples to solve their marital problems: A behavioral approach to relationship discord: I. Problem-solving skills. *International Journal of Family Counseling,* 1977, *5,* 22–31. (c)

Jacobson, N. S. Training couples to solve their marital problems: A behavioral approach to relationship discord: II. Intervention strategies. *International Journal of Family Counseling,* 1977, *5* (2), 22–30. (d)

Jacobson, N. S. A stimulus control model of change in behavioral couples' therapy: Implications for contingency contracting. *Journal of Marriage and Family Counseling,* 1978, *4,* 39–35. (a)

Jacobson, N. S. A review of the research on the effectiveness of marital therapy. In T. J. Paolino & B. S. McCrady (Eds.), *Marriage and the treatment of marital disorders from three perspectives: Psychoanalytic, behavioral, and systems theory.* New York: Brunner/Mazel, 1978. (b)

Jacobson, N. S. Specific and nonspecific factors in the effectiveness of a behavioral approach to marital discord. *Journal of Consulting and Clinical Psychology,* 1978, *46,* 442–452. (c)

Jacobson, N. S. Contingency contracting with couples: Redundancy and caution. *Behavior Therapy,* 1979.

Jacobson, N. S., & Margolin, G. *Marital therapy: Strategies based on social learning and behavior exchange principles.* New York: Brunner/Mazel, 1979.

Jacobson, N. S., & Martin, B. Behavioral marriage therapy: Current status. *Psychological Bulletin,* 1976, *83,* 540–566.

Jacobson, N. S., & Weiss, R. L. Behavioral marriage therapy: III. The contents of Gurman, et al. may be hazardous to our health. *Family Process,* 1978, *17,* 149–163.

Kahn, M. Nonverbal communication and marital satisfaction. *Family Process,* 1970, *9,* 449–456.

Knox, D. Behavior contracts in marriage counseling. *Journal of Family Counseling,* 1973, *1,* 22–28.

Knox, D. *Dr. Knox's marital exercise book.* New York: David McKay, 1975.

L'Abate, L. *Enrichment: Structured interventions with couples, families, and groups.* Washington, D.C.: University Press of America, 1978.

Liberman, R. P. Behavioral approaches to family and couple therapy. *American Journal of Orthopsychiatry,* 1970, *40,* 106–118.

Liberman, R. P., Wheeler, E., & Sanders, N. Behavioral therapy for marital disharmony: An educational approach. *Journal of Marriage and Family Counseling,* 1976, 2, 383–396. (a)

Liberman, R. P., Levine, J., Wheeler, E., Sanders, N., & Wallace, C. Experimental evaluation of marital group therapy: Behavioral vs. interactional-insight formats. *Acta Psychiatrica Scandinavia,* 1976, Supplement. (b)

Locke, H. J., & Wallace, K. M. Short-term marital adjustment and prediction tests: Their reliability and validity. *Journal of Marriage and Family Living,* 1959, *21,* 251–255.

Lubin, G. Adjective checklists for the measurement of depression. *Archives of General Psychiatry,* 1965, *12,* 57–62.

Mace, D. R. Marriage enrichment concepts for research. *The Family Coordinator,* 1975, *24,* 171–173.

Mace, D. R., & Mace, V. Marriage enrichment—a preventive group approach for couples. In D. H. Olson (Ed.), *Treating relationships.* Lake Mills, Iowa: Graphic Publishing Co., 1976.

McLean, P., Ogston, K., & Grauer, L. A behavioral approach to the treatment of depression. *Journal of Behavior Therapy and Experimental Psychiatry,* 1973, *4,* 323–330.

McMillan, E. L. Problem build-ups: A description of couples in marriage counseling. *The Family Coordinator,* 1969, *18,* 260–267.

Margolin, G. *A sequential analysis of dyadic communication.* Paper presented at the annual meeting of the Association for the Advancement of Behavior Therapy, Atlanta, December 1977.

Margolin, G. *Training couples to enhance anger control and reduce spouse abuse.* Unpublished manuscript, University of California at Santa Barbara, 1978.

Margolin, G., & Weiss, R. L. A comparative evaluation of therapeutic components associated with behavioral marital treatments. *Journal of Consulting and Clinical Psychology,* 1978, *46,* 1476–1486. (a)

Margolin, G., & Weiss, R. L. Communication training and assessment: A case of behavioral marital enrichment. *Behavior Therapy,* 1978, *9,* 508–520. (b)

Margolin, G., Christensen, A., & Weiss, R. L. Contracts, cognition and change: A behavioral approach to marriage therapy. *Counseling Psychologist,* 1975, *5,* 15–26.

Mayadas, N. S., & Duehn, W. D. Stimulus-modeling (SM) videotape for marital counseling: Method and application. *Journal of Marriage and Family Counseling,* 1977, *3,* 35–43.

Miller, S. L. *The effects of communication training in small groups upon self-disclosure and openness in engaged couples' systems of interaction: A field experiment.* Unpublished doctoral dissertation, University of Pennsylvania, 1971.

Miller, S., Corrales, R., & Wackman, D. B. Recent progress in understanding and facilitating marital communication. *The Family Coordinator,* 1975, *24,* 143–152.

Miller, S., & Wackman, D. B. *Couple communication patterns and marital satisfaction.* Unpublished manuscript, University of North Carolina at Greensboro, 1974.

Miller, S., Nunnally, E. W., & Wackman, D. B. *Alive and aware: Improving communication in relationships.* Minneapolis: Interpersonal Communication Programs, 1975.

Miller, S., Nunnally, E. W., & Wackman, D. B. Minnesota couples communication program (MCCP): Premarital and marital groups. In D. H. Olson (Ed.), *Treating relationships.* Lake Mills, Iowa: Graphic Publishing Co., 1976. (a)

Miller, S., Nunnally, E. W., & Wackman, D. B. A communication training program for couples. *Social Casework,* 1976, *57,* 9–18. (b)

Mitchell, H. E., Bullard, J. W., & Mudd, E. H. Areas of marital conflict in successfully and unsuccessfully functioning families. *Journal of Health and Human Behavior,* 1962, *3,* 88–93.

Murphy, D. C., & Mendelson, L. A. Communication and adjustment in marriage: Investigating the relationship. *Family Process,* 1973, *12,* 317–326. (a)

Murphy, D. C., & Mendelson, L. A. Use of the observational method in the study of live marital communication. *Journal of Marriage and the Family,* 1973, *35,* 256–263. (b)

Navran, L. Communication and adjustment in marriage. *Family Process,* 1967, *6,* 173–184.

Nunnally, E. W. *Effects of communication training upon interaction awareness and empathic accuracy of engaged couples: A field experiment.* Unpublished doctoral dissertation, University of Minnesota, 1971.

Nunnally, E. W., Miller, S., & Wackman, D. B. The Minnesota couples communication program. *Small Group Behavior,* 1975, *6,* 57–71.

O'Leary, K. D., & Turkewitz, H. Methodological errors in marital and child treatment research. *Journal of Consulting and Clinical Psychology,* 1978, *46,* 747–758.

Olson, D. H. Bridging research, theory, and application: The triple threat in science. In D. H. Olson (Ed.), *Treating relationships.* Lake Mills, Iowa: Graphic Publishing Co., 1976.

Otto, H. A. Marriage and family enrichment programs in North America—report and analysis. *The Family Coordinator,* 1975, *24,* 137–142.

Patterson, G. R., & Hops, H. Coercion: A game for two: Intervention techniques for marital conflict. In R. Ulrich & P. Mountjoy (Eds.), *The experimental analysis of social behavior.* New York: Appleton-Century-Crofts, 1972.

Patterson, G. R., Hops, H., & Weiss, R. L. Interpersonal skills training for couples in early stages of conflict. *Journal of Marriage and the Family,* 1975, *37,* 295–303.

Peterson, G. L., & Frederiksen, L. W. *Developing behavioral competencies in distressed marital couples.* Paper presented at the 10th annual meeting of the Association for the Advancement of Behavior Therapy, New York, December 1976.

Pierce, R. M. Training in interpersonal communication skills with the partners of deteriorated marriages. *The Family Coordinator,* 1973, *22,* 223–227.

Rappaport, A. F. *The effects of an intensive conjugal relationship modification program.* Unpublished doctoral dissertation, Pennsylvania State University, 1971.

Rappaport, A. F. Conjugal relationship enhancement program. In D. H. Olson (Ed.), *Treating relationships.* Lake Mills, Iowa: Graphic Publishing Co., 1976.

Rappaport, A. F., & Harrell, J. A behavioral-exchange model for marital counseling. *The Family Coordinator,* 1972, *21,* 203–212.

Regula, R. R. Marriage encounter: What makes it work? *The Family Coordinator,* 1975, *24,* 153–159.

Roberts, P. V. The effects on marital satisfaction of brief training in behavioral exchange negotiation mediated by differentially experienced trainers. (Doctoral dissertation, Fuller Theological Seminary 1974.) *Dissertation Abstracts International,* 1975, *36,* 457B.

Royce, W. S., & Weiss, R. L. Behavioral cues in the judgement of marital satisfaction: A linear regression analysis. *Journal of Consulting and Clinical Psychology,* 1975, *43,* 816–824.

Sauber, S. R. Primary prevention and the marital enrichment group. *Journal of Family Counseling,* 1974, *2,* 39–44.

Schauble, P. G., & Hill, C. G. A laboratory approach to treatment in marriage counseling: Training in communication skills. *The Family Coordinator,* 1976, *25,* 277–284.

Schwager, H. A., & Conrad, R. W. *Impact of group counseling on self and other acceptance and persistence with rural disadvantaged student families.* Counseling services report No. 15, Washington, D.C.: National Institute of Education, 1974.

Smith, L., & Smith, A. Developing a nationwide marriage communication labs program. In H. A. Otto (Ed.), *Marriage and family enrichment: New perspectives and programs.* Nashville: Abingdon, 1976.

Stein, E. V. MARDILAB: An experiment in marriage enrichment. *The Family Coordinator,* 1975, *24,* 167-170.

Strayhorn, J. *Talking it out.* Champaign, Ill.: Research Press, 1977.

Strayhorn, J. Social exchange theory: Cognitive restructuring in marital therapy. *Family Process,* 1978, *17,* 437-448.

Stuart, R. B. Operant interpersonal treatment for marital discord. *Journal of Consulting and Clinical Psychology,* 1969, *33,* 675-682.

Stuart, R. B. An operant interpersonal program for couples. In D. H. Olson (Ed.), *Treating relationships.* Lake Mills, Iowa: Graphic Publishing Co., 1976.

Stuart, R. B., & Lederer, W. J. *How to make a bad marriage good and a good marriage better.* New York: W. W. Norton, 1979.

Stuart, R. B., & Stuart, R. *Marital pre-counseling inventory.* Champaign, Ill.: Research Press, 1972.

Swan, R. W. Differential counseling approaches to conflict reduction in the marital dyad. (Doctoral dissertation, Arizona State University, 1972.) *Dissertation Abstracts International,* 1972, *32,* 6629B.

Thomas, E. J. *Marital communication and decision making: Analysis, assessment, and change.* New York: The Free Press, 1977.

Thomas, E. J., Carter, R. D., & Gambrill, E. D. Some possibilities of behavior modification with marital problems using "SAM" (signal system for the assessment and modification of behavior). In R. D. Rubin, H. Fensterheim, A. A. Lazarus, & C. M. Franks (Eds.), *Advances in behavior therapy.* New York: Academic, 1971.

Travis, R. P., & Travis, P. Y. The pairing enrichment program: Actualizing the marriage. *The Family Coordinator,* 1975, *24,* 161-165.

Tsoi-Hoshmand, L. The limits of *quid pro quo* in couple therapy. *The Family Coordinator,* 1975, *24,* 51-54.

Tsoi-Hoshmand, L. Marital therapy: An integrative behavioral-learning model. *Journal of Marriage and Family Counseling,* 1976, *2,* 179-191.

Turkewitz, H. *A comparative outcome study of behavioral marital therapy and communication therapy.* Unpublished doctoral dissertation, State University of New York at Stony Brook, 1977.

Turner, A. J. *Couple and group treatment of marital discord: An experiment.* Paper presented at the annual meeting of the Association for the Advancement of Behavior Therapy, New York, October 1972.

Turner, A. J. Personal communication, July 26, 1978.

Valle, S. K., & Marinelli, R. P. Training in human relations skills as a preferred mode of treatment for married couples. *Journal of Marriage and Family Counseling,* 1975, *1,* 359-365.

Venema, H. B. Marriage enrichment: A comparison of the behavioral exchange negotiation and communication models. (Doctoral dissertation, Fuller Theological Seminary 1975.) *Dissertation Abstracts International,* 1976, *36,* 4184-4185B.

Vincent, J. P. *The relationship of sex, degree of intimacy, and degree of marital distress to problem solving behavior and exchange of social reinforcement.* Unpublished doctoral dissertation, University of Oregon, 1972.

Vincent, J. P., Weiss, R. L., & Birchler, G. R. A behavioral analysis of problem solving in distressed and nondistressed married and stranger dyads. *Behavior Therapy*, 1975, *6*, 475–487.

Wackman, D. B., & Miller, S. *Analyzing sequential interaction data: Two empirical studies.* Paper presented to the Interpersonal Communication Association, Chicago, April 1975.

Weiss, R. L. The conceptualization of marriage and marriage disorders from a behavioral perspective. In T. J. Paolino & B. S. McCrady (Eds.), *Marriage and the treatment of marital disorders from three perspectives: Psychoanalytic, behavioral, and systems theory.* New York: Brunner/Mazel, 1978.

Weiss, R. L., & Aved, B. M. Marital satisfaction and depression as predictors of physical health status. *Journal of Consulting and Clinical Psychology*, 1978, *46*, 1379–1384.

Weiss, R. L., & Birchler, G. R. Adults with marital dysfunction. In M. Hersen & A. J. Bellack (Eds.), *Behavior therapy in the psychiatric setting*. Baltimore: Williams & Wilkins, 1978.

Weiss, R. L., & Cerreto, M. *Marital status inventory: Steps to divorce*. Unpublished manuscript, University of Oregon, 1975.

Weiss, R. L., & Margolin, G. Marital conflict and accord. In A. R. Ciminero, M. S. Calhoun, & H. E. Adams (Eds.), *Handbook for behavioral assessment*. New York: John Wiley & Sons, 1977.

Weiss, R. L., Hops, H., & Patterson, G. R. A framework for conceptualizing marital conflict, a technology for altering it, some data for evaluating it. In F. W. Clark & L. A. Hamerlynck (Eds.), *Critical issues in research and practice: Proceedings of the fourth Banff international conference on behavior modification*. Champaign, Ill.: Research Press, 1973.

Weiss, R. L., Birchler, G. R., & Vincent, J. P. Contractual models for negotiation training, *Journal of Marriage and the Family*, 1974, *36*, 321–331.

Wells, R. A., Figurel, J. A., & McNamee, P. Group facilitative training with conflicted marital couples. In A. S. Gurman & D. G. Rice (Eds.), *Couples in conflict: New directions in marriage therapy*. New York: Aronson, 1975.

Wieman, R. J. *Conjugal relationship modification and reciprocal reinforcement: A comparison of treatments for marital discord*. Unpublished doctoral dissertation, Pennsylvania State University, 1973.

Wieman, R. J., Shoulders, D. I., & Farr, J. H. Reciprocal reinforcement in marital therapy. *Journal of Behavior Therapy and Experimental Psychiatry*, 1974, *5*, 291–295.

Zarle, T. H., & Boyd, R. C. An evaluation of modeling and experimental procedures for self-disclosure training. *Journal of Counseling Psychology*, 1977, *24*, 118–124.

PART THREE

METHODOLOGICAL ISSUES

CHAPTER 9

Social Skills: Methodological Issues and Future Directions

James P. Curran

INTRODUCTION

The basic premise of social skills training is that there are individual differences with respect to the degree of social competency exhibited by individuals in social situations and that for some of these individuals this lack of adequate performance is problematic. The presumed etiological bases for poor social skill performance are multiple and not mutually exclusive (Curran, 1977). The skills deficit hypothesis for poor social performance is that an individual does not possess in his behavioral repertoire the requisite skills needed for a competent performance. A variation of the skills-deficit interpretation is that while an individual at some point in time may have possessed the requisite skills, these behaviors are no longer readily available to him because of some decaying process (e.g., the decrement in skill performance often seen in individuals who have been institutionalized for long periods of time). Another interpretation of inadequate social performance is that the individual does possess the requisite skills, but due to his previous learning history a less adequate or socially inappropriate response has a higher probability of occurrence. For example, an individual who has experienced a learning history where aggressive behavior "paid off" may become oriented to this form of responding even though it is now inappropriate and even though the individual possesses the requisite skills for a more socially appropriate response. Other interpretations of poor social skill

James P. Curran • Mental Hygiene Clinic, Veterans Administration Medical Center, and Brown Medical School, Providence, Rhode Island 02908.

performance revolve around an interference mechanism, either of an emotional or cognitive nature, which inhibits and/or disrupts the effective application of existing skills. Emotional states may interfere with effective skills application, as in the case where an individual has experienced previous aversive conditioning episodes in social situations, and, consequently, these situations now elicit major anxiety reactions. There are numerous ways in which cognitive processes can interfere with skill performance. An individual may read social cues poorly and, consequently, may perform inadequately, not because of a lack of adequate skills, but because of inadequate perceptual ability. Faulty cognitive assumptions and/or illogical reasoning may lead to faulty interpretations of perceived cues and, ultimately, to inadequate social performance (Goldfried & Sobocinski, 1975). Internal standards with respect to what is regarded as acceptable performance may be so unduly stringent that individuals become overly critical of their performance, producing anxiety that interferes with subsequent performances. Parenthetically, the cognitive factors just described are regarded by some investigators as a subset of social skills and not as separate processes.

On an intuitive level, the course of treatment for individuals demonstrating poor social skill performance would vary depending on the various etiological factors associated with such poor performance. Unfortunately, the lack of sophistication of our assessment procedures does not permit us to specify these etiological factors with any degree of certitude. Improvements in our ability to assess these etiological factors should lead to innovations in treatment so that programs could be altered to address the specific factors responsible for an individual's inadequate social performance. In the meantime, however, social skill training appears to be a multimodal treatment regime robust enough to improve the social functioning of individuals differing in etiological considerations (Trower, Yardley, Bryant, & Shaw, 1978).

Regardless of the etiology of poor social skill performance, it is evident that social incompetency is associated with psychopathology. A number of investigators (Argyle & Kendon, 1967; Lentz, Paul, & Calhoun, 1971; Libet & Lewinsohn, 1973; Sylph, Ross, & Kedward, 1978; Zigler & Levine, 1973) have presented data relating the level of social competence to psychiatric disorders. The work of Zigler and Phillips (1961, 1962) indicated that premorbid level of social functioning was the single best prognostic indicator of posthospital functioning. Lentz, Paul, and Calhoun (1971) demonstrated that level of social functioning was related to discharge from a hospital and to recidivism rate. Not only is the lack of social skills related to major forms of psychopathology, but it is also related to numerous other problematic behaviors, such as alcohol abuse (Kraft, 1971; Miller, Hersen, Eisler, & Hilsman, 1974), homosexuality (Feldman & MacCulloch, 1971), sexual dysfunctioning (Barlow, 1973), drug addiction (Callner & Ross, 1976), dating anxiety (Curran, 1977), and marital problems (Eisler, Miller, Hersen, & Alford, 1974).

Although the relationship between social skill level and psychopathology has been noted for some time, until the development of a social skills training program no systematic approach to the alleviation of social skills deficits existed. As attested to in this book, social skills training appears to be an effective treatment modality for many types of problematic behaviors. However, the social skills training area is also beset by severe methodological problems. We have witnessed a disproportionate number of studies applying social skills training to a variety of populations and very few studies addressing methodological issues. I am afraid that, unless we improve our methodological sophistication, social skills training will join the long list of treatment procedures which achieved popularity after their introduction but were quickly abandoned as soon as the next treatment fad was introduced. As Hersen noted in Chapter 6, "Now that the easy studies have been done, it is probably time to begin doing the difficult ones." In this chapter, I would like to address what I regard as some of the fundamental issues which must be resolved in order for social skills training to remain a viable treatment procedure. I will first discuss the definitional problem which greatly affects our research methodology. Next, the question of what constitutes social skills training will be explored. Selected assessment issues will be addressed, followed by a discussion of methodological problems such as subject selection, collection of follow-up data, and data analyses. The last section of this chapter contains speculations on future directions, in which some suggestions are made that may help to clarify definitional issues and measurement problems.

DEFINITIONAL PROBLEM

The definitional problem most succinctly stated is that everyone seems to know what good and poor social skills are but no one can define them adequately. The most widely used definition of social skill was proposed by Libet and Lewinsohn (1973). They defined social skills as "the complex ability to maximize the rate of positive reinforcement and to minimize the strength of punishment from others" (p. 311). What is the utility of such a definition? If a boxer ducks when an opponent throws him an overhand right, then the boxer is minimizing the strength of punishment from others, but is that an example of social skill? Why, on the other hand, do we not consider ducking as a social skill but regard eye contact as a component of social skill? Also, how do we differentiate the response of ducking from a cleverly timed repartee which could be regarded as a social skill? Let us consider another example. If a psychiatric patient exhibits "crazy talk" in the presence of a nurse, which is followed by attention from the nurse, then the patient may be said to be maximizing the rate of positive reinforcement in this situation. But does that mean the patient's "crazy talk" is an example of social skill? A short-term gain seems apparent (i.e., increased attention); however, the continued delivery of "crazy talk" could interfere with the patient's discharge, which may or may not be a reinforcing event.

Liberman, Vaughn, Aitchison, and Falloon (1977b) list three dimensions of social skills:

> One includes discreet nonverbal behaviors that constitute person-to-person communications: eye contact, facial expression, posture, gestures, loudness, tone of voice, pacing or speed of speech, latency, duration of responding, and fluency of speech. A second dimension includes the content of speech or conversation: requesting something of another person, praising, thanking, or complimenting other persons, saying "no" to an unreasonable request, going through a job interview, reacting appropriately to criticism, and managing other daily instrumental and affectional encounters. A third dimension involves reciprocity in communicating: such as giving reinforcement to another to maintain conversation, initiating conversations, terminating conversations, and timing one's entry and exit from social groups. (p. 35)

I do not believe Liberman *et al.* (1977b) intended to be all-inclusive in listing these components. Certainly, more components could be listed, but, nevertheless, a listing of components does not constitute a definition.

To further complicate matters, Liberman *et al.* (1977b) have expanded the definition of social skill beyond just motoric responses. They state "that a patient's consistent sending of skillful messages would require that they actually receive or comprehend incoming interpersonal messages and be able to generate response alternatives and evaluate the adequacy of these alternatives in terms of his rights, responsibilities, and goals prescribed by the situation" (p. 30). These authors further state that they "have stressed the importance of a conceptual model of social skill that incorporates receiving, processing, and sending functions" (p. 30).

Other examples can be found where investigators appeared to have incorporated cognitive processes under the rubric of social skill. For example, in Chapter 6 of this book, Hersen viewed social perception training as a component of social skills training. Hersen states, "for effective interpersonal functioning to occur, the ability accurately to perceive feelings and intentions of others as well as the ability to make appropriate responses are required."

Frankly, I am somewhat ambivalent about the expansion of the construct of social skill to include cognitive processes. I think, on one hand, that such an expansion will have some use because it will push the model to the limit of its utility and it should provide innovative approaches to treatment. On the other hand, however, I am troubled by the expansion of the construct when we are still far from a definition of social skill with respect to motoric behaviors. If we do not restrain ourselves and put some limits on the construct of social skill, it will expand to include all human behavior, and social skills training will soon come to mean any process which is capable of producing changes in human behavior. If such an expansion does occur, it would obviously render these terms meaningless. My bias, at this time, is to limit the construct of social skill to motoric behavior. We should measure cognitive processes because these processes are important with respect to both theory and treatment. However, let us agree not to call these cognitive processes components of social skill. It is, of course, important to assess whether an individual knows the correct response in a particular social situation, but I prefer to limit the construct of social skill to his actual performance in that situation. I also do not think that ducking one's head when a punch is thrown is an example of social skill, and would like to exclude from the definition of the construct those behaviors that are not primarily social in nature. While the consequences of social performance are important determinants in evaluating the social skill level, the idiosyncratic nature of these consequences and their situational specificity provide difficult obstacles to coming to any definition of social skill.

As a researcher who has been actively engaged during the last decade in training raters to evaluate the social skill level of various subject populations, I feel it important to share with you some of my frustrations. In training our raters, we give them some indicators of what we believe are good social skills. For example, we inform our raters that appropriate cye contact is an indicator of good social skill. However, we are unable to specify how much eye contact is "good." We feel that both lack of eye contact and continuous eye contact over long durations are inappropriate. Another indicator that we give to our raters as an example of good social skill is the delivery of "genuine-sounding compliments" when an individual has performed a favor. Again, we regard sarcastic-sounding compliments or lack of compliments as poor skill in this particular situation. One can see from just these two examples the complexity involved in the judgmental process. In the case of eye contact, we regard both the duration and the timing of the contact as important. In the case of a compliment, not only is the content of the verbalization crucial, but so may be the tone and style of its delivery. In training our raters, we do not give them the impression that we have an exhaustive list of all possible indicators of social skill, nor do we tell them how to weigh each of these indicators in making their judgment. We do not do so because we prefer to be "sloppy," but rather because we feel we have not reached the degree of sophistication needed before we can specify all the components of social skill and operationalize them. Moreover, nor do we feel we can specify how these components should be weighed.

In addition to giving our raters indicators which we feel are *relatively* generalizable across different types of situations, we also give our raters situation-specific indicators for the various scenes which they may be rating. For example, in a scene in which a subject has to respond to criticism of his work by his boss, we may give instructions such as "In this scene a good skill response would involve the subject's maintaining control and not getting angry, asking his supervisor for specific feedback with respect to his dissatisfaction with his performance, a possible mention of his previous satisfactory work record, and the presentation of a workable solution to the problem incurred by the faulty workmanship." Our raters are extensively trained on many practice scenes and given feedback with respect to both their reliability and their accuracy. Despite all our attempts at training, the average interrater reliability coefficients reported in most of our studies range in the mid 0.70s, indicating a moderate degree of agreement (but far from consensus).

There is no question that we could improve our interrater reliability if we specified a limited set of behaviors to be rated and operationalized these terms. In fact, many of our colleagues in the area of social skill research have pursued this alternative. However, as I have mentioned in a previous paper (Curran, 1978), I feel we are not able, given the "state of the art," to specify exactly what behaviors constitute social skill. For example, in one study (Fischetti, Curran, & Wessberg, 1977), we demonstrated that high and low socially skilled subjects did not differ in their frequency of a particular response but did differ in the timing of that response. Questions such as the appropriate level of response (too much or too little may be regarded as inappropriate) need to be resolved, as well as questions of situational or cultural specificity. Most investigators who specify and operationalize behaviors to be rated also include global ratings. But very often there is a confound because the raters are led to believe that the specific behaviors they are rating are the crucial behaviors and constitute, in some sort of summary fashion, a definition of social skills. There exist very few consistent empirical data regarding the specification of the components of social skill, how these terms should be operationalized, or how these components should be weighed in evaluating an individual's social skill level. I still do not know what to say to my raters if they ask me if four 10-sec periods of eye contact during a 2-min role-play is better than three 15-sec periods and how much they should weigh eye contact in relationship to what the subjects said during the role-play.

While I have not proposed a definition of social skill, I have begun to exclude some forms of behavior which I do not feel justifiably fall under the rubric of social skill. I encourage other interested individuals to do the same so that we may gradually move toward some consensus concerning the construct that we are addressing. It is an extremely complicated problem and is the most fundamental issue in the whole social skill paradigm. Unless we can come to some consensus about the construct we are addressing, we are likely to be building a Tower of Babel. The definitional problem affects all the other issues in the field. Our definition of the construct not only affects the content of what we train but also how we develop our training programs. It affects what we decide to measure and how we decide to measure it. It affects the manner in which we select subjects, the types of control groups we employ, and how we choose to analyze our data. The definitional problem indeed even affects the types of questions we ask in our study. In the last section of this chapter (on future directions) I will propose two types of experimental strategies which I feel will help clarify the definitional problem. I would now like to address an issue that has seldom been discussed, that is, the wide divergences in procedures which have been labled social skills training. Our independent variable (social skills training) in our treatment outcome studies varies to such an extent that comparing results across studies is akin to comparing apples to oranges.

SOCIAL SKILLS TRAINING: WHAT IS IT?

Goldsmith and McFall (1975) described social skills training as

> a general therapy approach aimed at increasing performance competence in critical life situations. In contrast to the therapies aimed primarily at the elimination of maladaptive behaviors, skills training emphasizes the positive educational aspects of treatment. It assumes that each individual always does the best he can, given his physical limitations and unique learning history in every situation. Thus, when an individual's best effort is judged to be maladaptive, this indicates the presence of a situation specific skill deficit in the individual's repertoire. . . . Whatever the origin of this deficit (e.g., lack of experience, faulty learning, biological dysfunction), it often may be overcome or partially compensated through appropriate training in more skillful response alternatives. (p. 51)

Hersen and Bellack (1976, p. 564) listed what they regarded as chief components of the social skills training approach with psychiatric patients:

> (1) Concerns itself with critical social interactions with psychiatric patients; (2) Views the psychiatric patient as needing education or reeducation; (3) Attempts to provide that education or reeducation for dysfunctional psychiatric patients with specific objectives using specific techniques; and (4) Introduces a positive expectancy in the treatment of psychiatric patients irrespective of diagnostic labeling and chronicity of individual psychopathology. (p. 564)

They further state that social skills training differs from traditional psychotherapy in two ways:

> (a) A focus on the acquisition of specific behavioral skills such as appropriate eye contact, short response latency; (b) a one-to-one relationship between diagnosis and treatment such that the specific behaviors measured during assessment are the very ones targeted for modification. (p. 564)

Not one of these characteristics of social skills training definitively distinguishes social skills training from other forms of therapy. Most forms of therapy can be characterized as attempting to increase the patient's competency in critical life situations. Many forms of therapy also emphasize the positive educational aspects of treatment. One of the characteristics of behavior therapy itself is the close relationship between diagnosis and treatment. Many forms of treatment attempt to teach specific behavioral skills in order to reach treatment goals such as improved communication between spouses. These stated characteristics are more reflective of a philosophy of treatment than a definition of treatment. Taken in total, they do tend to describe, but not in a definitive sense, a generic form of treatment that has become known as social skills training.

Social skills training is, in essence, a variety of techniques including behavioral rehearsal, the use of prompts, modeling, instructions, feedback, reinforcement, self-monitoring procedures, and *in vivo* practice. The manner in which these techniques are used can be characterized as a response-acquisition approach to treatment (Bandura, 1969). Although these techniques are the most widely employed components of the social skills training "package," most studies use only a subset of these components. In addition, other therapeutic techniques which are not generally regarded as a part of the social skills training are often incorporated into a treatment "package" and labeled as social skills training. For example, in a study by Monti, Fink, Norman, Curran, Hayes, and Caldwell (1979), several sessions were devoted to relaxation training, which is generally not regarded as a component of the social skills training "package." In this section, we will examine variations in the social skills training "package." While it is commendable that investigators modify a particular treatment "package" to a particular patient population, the lack of consistency in training programs across studies leaves serious doubts about the comparability of the independent manipulation across studies. The variations employed by different investigators should also be examined in their own right in order to compare their relative effectiveness and efficiency.

Social skills training has been conducted on both an individual basis and in groups, and variations in between (Eisler *et al.*, 1974; treatment of married couples). It would appear on an intuitive basis that individual treatment would allow for greater effectiveness because the program could be "tailor-made" for the particular deficiencies of the individual patient. However, there may be some advantages to group treatment, such as peer pressure, increased feedback sources, etc., which might make group treatment as effective as individual treatment. Of course, group treatment may be more efficient because of the greater number of individuals involved.

In one study, Liberman, Lillie, Falloon, Vaughn, Harper, Leff, Hutchison, Ryan, and Stoate (1977a) had members of the patient's immediate family participate in a family group social skills training program in addition to the patient's receiving individual treatment. In treating our patients at the Veterans Administration Medical Center, it appears to us that while patients have many deficits across many different types of social situations, their deficiency in relating to one or two "significant others" (such as a spouse or son) is especially problematic. In order to promote transfer to the natural environment and to maintain treatment gain, it might prove effective to include significant others in the actual treatment. In many cases, it appears that the "significant other" also needs to learn alternate ways of handling difficult situations.

The number of individuals responsible for implementing the social skills training program also varies from study to study. In most cases, an individual therapist is responsible for implementation of the program, but often two or more therapists are used (especially in group-based programs). In our own social skills training program, we feel it is important to have two therapists implement the program. For any one particular session, one of the therapists is responsible for task-oriented behavior (i.e., presentation of the instruction material, organizing the role-play, etc.), while the other therapist is mainly responsible for attending to the process of the group and to use that process in order to implement changes. In addition to helping facilitate the group process, multiple therapists also increase the number of appropriate role models, increase the number of feedback sources, and increase the number of stimulus persons involved in role-plays.

In the behavioral rehearsal component of social skills training, the other person, in addition to the targeted subjects, generally has either been the therapist or other subjects. The effect of varying the number of rehearsal participants and the type of participants has not been systematically investigated. One recent study by Wood (1978) compared the effects of different types of participants and the effects of such variations on subsequent performances. Targeted subjects were low-assertive elementary school pupils. Various groups of individuals conducted treatment and rehearsals, including peers, parents, and teachers. While the effects of treatment were somewhat minimal, it did appear that on some measures the group of subjects who were trained to be assertive by teachers increased their assertive behavior in teacher–pupil relations more than other groups who were trained by nonteachers. In our own work with VA patients, we have begun to use adjunct role-players in our training sessions. These adjunct role-players are in addition to our co-therapists, and they mainly participate in the behavioral rehearsal sections of our program. Our hope is that, by having our patients interact with numerous individuals, generalization will increase. One could vary the characteristics of these adjunct role-players along different dimensions in order to promote generalization. As mentioned previously, "significant others" in the patient's life could be utilized during behavior rehearsal in order to promote transfer of learning through the natural environment. Of course, including adjunct role-players does increase the therapist–patient ratio, and hence becomes more costly.

Another dimension on which skills training programs have varied greatly is the number of sessions included in the program. These have varied from as few as 4 sessions in therapy analogue studies such as Eisler, Hersen, and Miller (1973), to 30 or more sessions (e.g., Lamont, Gilner, Spector, & Skinner, 1969). In a study by Liberman *et al.* (1977a), patients received 30 h a week of social skills training for 8 weeks, in addition to 5 h a week of social skills training in a family group. In a most ambitious project currently being conducted by Liberman *et al.* (1977a), patients will receive 5 h of social skills training per day, 7 days a week, for 9 weeks. The length and number of treatment sessions, of course, will vary depending on a number of factors, including the scope and complexity of the behaviors to be taught, subject variables, etc. Consequently, one should remain cautious in trying to compare results from different studies which very greatly in the actual amount of treatment time.

Social skills training programs have also varied greatly with respect to their emphasis on the modality of behaviors to be taught. Most of the previous studies have focused on actual behavioral performances or, as Liberman *et al.* (1977a) have labeled it, the sending of interpersonal messages. In addition to training individuals to send interpersonal messages, Liberman's program also emphasizes both the receiving and processing of interpersonal messages. Part of the program is geared to test whether a patient can receive or comprehend incoming interpersonal messages. Another component of the program deals with processing these messages in which emphasis is placed on the generation of response alternatives and the evaluation of the adequacy of these response alternatives. Liberman's training program is consistent with his expanded definition of social skills training which subsumes some cognitive functions. Other social skills training programs do not ignore receiving and processing aspects of human functioning, but there is less emphasis on systematic training of these functions.

In addition to differential emphasis on the modalities to be trained, social skills training programs also differ with respect to the level of behavior (molecular vs. molar) to be changed. The training program at the University of Pittsburgh Medical School (Hersen & Bellack, 1976) appears to emphasize the teaching of very specific molecular behaviors such as eye contact. Other programs, for example Goldsmith and McFall (1975), appear to attempt to teach more molar types of behaviors. Some of the topics covered in the Goldsmith and McFall (1975) study involved initiating and terminating conversations, dealing with rejection, becoming more self-disclosing, etc. In our own program at the VA, we are also teaching on a more molar level (e.g., training patients to deal with criticism and approaches to intimate situations). The level of behavior being taught can interact with many different variables including patient characteristics and treatment components. The degree to which targeted behaviors generalize to real-life situations may also be a function of the level of behavior being taught. I think it is fair to say that neither the molecular nor the molar approach ignores the other; in a molar approach program, feedback is given with respect to molecular behaviors, such as eye contact, and in a molecular program, training is given in various situational contexts and feedback is given with respect to molar behavior. However, I do feel that the differential emphasis in these programs makes these approaches significantly different.

The behaviors to be taught in a skills-training program as well as the situational context in which they are presented should be chosen after careful assessment of each individual subject's situationally specific skill deficit in order to achieve maximum effectiveness and efficiency. In many studies, especially those applying group designs, this is not the case. Individualization of a treatment program in group designs is sacrificed in order to insure comparability of replication. However, even in the single-case approach, where there seems to be a high degree of individualization of the treatment program, this individualization is somewhat more superficial than real. For example, let us consider the single-case approach conducted by Hersen and Bellack (1976). In their studies, patients are assessed on their Behavioral Assertiveness Test–

Revised (BAT-R). The original BAT was based on situations which were similar to those developed by McFall and Marston (1970) for a college population. The behaviors rated from the BAT-R, such as eye contact, loudness of voice, etc., were derived by asking a number of clinicians to nominate what behaviors they thought comprised assertive behavior. From the ratings of these behaviors in these situations, individual deficits are specified. Consequently, identification of individual patients' deficits are restricted both by the behaviors to be rated as well as by the situations in which they are rated. These BAT-R situations and the behavioral ratings derived from them comprise the major content of the treatment program of the Pittsburgh group. That is, treatment consists of behavioral rehearsal in BAT-R-type situations and feedback is directed at correcting those behaviors which were deemed inadequate in the pretreatment ratings. I am not in any way urging the abandonment of consistent assessment procedures; however, I do wish to alert the reader to the fact that our treatment programs may be glossing over major individual differences with respect to situationally specific deficits.

There are examples in the literature where investigators have tried to develop situations and modify behaviors which are specifically (at least for the group) problematic for their subject population. Cox, Gunn, and Cox (1976) first attempted to determine critical life situations for their population of children. Extensive behavior observations were conducted and problematic situations were listed. These situations were then discussed with the children with respect to their difficulty level. Cox *et al.* (1976) then attempted to develop a social skills training program to provide remediation in these particular areas of interpersonal conflict. However, their enthusiasm was dampened by the realization that their training program failed to account for developmental components which interacted with the attainment of skills. The authors stated that:

> We became increasingly more concerned that we were trying to teach adult-level responses to children who have not attained the maturity to allow for maximum benefit from the training. In addition, we realize that many of the social responses that we were so rigorously teaching were not based on any empirical data. (p. 2)

They then utilized a contrasted-group design employing problematic and nonproblematic school-age children. These contrasted groups interacted in the chosen situations. Videotapes of these interactions were scored for a number of verbal and nonverbal behaviors, chosen on the basis of a literature search. In analyzing the data from these contrasted-group designs, it was determined that some of the behaviors chosen to be scored differentiated the groups, but many did not. The behaviors which did differentiate the groups became focal parts of the training program. Here, then, is an example where both behaviors to be trained and the situations used in the training were developed on an empirical basis for a particular subject population.

Other examples can be found where the situations used in training and in assessment were based on empirical data. Goldsmith and McFall (1975) asked a number of psychiatric patients similar to those subjects used in their study to nominate problematic social situations which were used for both assessment and training. Finch and Wallace (1977) administered a list of situations to a patient population and chose those particular situations in which the patients claimed that they were most anxious. Of course, in this example, a major limitation is that the situations presented may not have corresponded to a listing which could have been generated by the patients themselves. In many social skills training programs, the situations chosen for rehearsal and the behaviors to be trained have been selected mainly on the basis of face validity rather than on any empirical data source.

In our own work with psychiatric patients, our self-report and observational assessments are based on the factor-analytic work of Richardson and Tasto (1976). Richardson and Tasto obtained numerous items from investigators in the social skills area and administered these items to a large undergraduate population. They factor-analyzed these items and came out with seven factors which they felt corresponded to seven different types of social situations. We (Curran, Corriveau, Monti, & Hagerman, in press) have replicated this factor structure in our own psychiatric population. Our observational situations are constructed to resemble these seven types of social situations. We feel that we have a broad-base assessment procedure which samples different types of problematic situations (Curran, Monti, Corriveau, Hagerman, Hay, & Zwick, 1978). The behaviors taught in our social skills training program were selected after an extensive literature search. We attempt to teach our patients categories of behaviors (e.g., giving compliments, strategies to deal with criticism, etc.) which we feel will generalize over a broad range of social situations. The program is group based but the training is somewhat individualistic. That is, the role rehearsals used during the actual training are designed to tap those situations which are most problematic for any individual patient. A determination of which situations are problematic for any particular patient is based on reading the patient's records, talking to his or her primary therapist, talking to the patient, and on data from our assessment procedures. We feel that we have a treatment program which teaches behaviors that generalize across situations in a format which individualizes training for each particular patient.

Another way in which various training programs differ is the manner in which the behavior rehearsals are conducted. Some investigators chose an arbitrary number of times a particular rehearsal can be repeated. Others use some sort of performance-based criterion while others appeared to conduct as many rehearsals as possible during a particular time frame.

Social skills training programs also differ with respect to the procedures used to promote transfer of training to the natural environment. Liberman *et al.* (1977a) conducted family group therapy in addition to conducting individual social skills training with a schizophrenic population. Within the family group, social skills training for patients and the parents of these patients was conducted for a total of 5 h a week for 8 weeks. Monti *et al.* (1979) encouraged the use of *in vivo* practice by assigning homework tasks which the patients recorded and reported on during subsequent sessions. From the description of their training program, other investigators (e.g., Goldsmith & McFall, 1975) appear to have put little effort into promoting transfer of training.

The point emphasized in this section is that our independent variable (social skills training) is not a homogeneous variable but differs considerably across investigators and across investigations. The structure of the treatment itself may be individually based or group based. The number of therapists employed in the training varies and the therapists are sometimes assisted by adjunct role-players. The types of modalities emphasized in the treatment range from strictly motoric to the inclusion of perceptual and process factors. There appears to be differential emphasis with respect to the molarity or molecularity of the behaviors taught. The situations used in rehearsal as well as the behaviors taught are not, in general, based on any empirical data and vary across investigators. There are also differences with respect to the emphasis placed on promoting transfer of the training to the natural environment. To illustrate my point even further, let us examine several authors' descriptions of their social skills training program.

The first description is from Lomont *et al.* (1969). Treatment was group based and conducted by a single therapist. Treatment sessions were 90 min in duration and were conducted 5 days a week for 6 weeks:

> The assertion therapy essentially consisted of systematic practice by the patients of numerous behaviors for coping effectively with a variety of social situations. . . . Most of the training in assertiveness involved the practice of assertive behaviors in roleplaying. . . . Except for the last interactions, practice work on a given interpersonal situation began with a script. The script specified all the verbal responses between two or more people in an imaginary but common sort of situation. One of the people in the interaction was intended to be a model for the patients of effectively assertive behavior in the

> interaction. The patients were divided into groups of two or three which independently practiced the interaction specified by the script. The patients in each subgroup took turns playing the part intended as a model for the patients and also the other parts involved in the interaction. As the subgroups rehearsed in this way, the therapist circulated among them coaching their patients in their practice of the patient-model role. After the patients had all had the opportunity to familiarize themselves with the model role through repeated practice of it, they were encouraged in their practice of it to express the specified responses of the role in their own words as much as possible. . . . Toward the end of the therapy, the patients engaged in discussion of various topics of general interest in order to gain practice of assertion in group discussion. In some of the discussions the therapist challenged statements made by the patients so they could practice effectively but appropriately defending their positions when strongly challenged. The patients were repeatedly encouraged to practice in real life situations the behaviors they practiced in the assertion therapy. (pp. 464–465)

The next description of a social skills training program comes from Finch and Wallace (1977). Here, the skills training was group based and conducted by two therapists in hourly sessions three times per week for 4 consecutive weeks:

> Generally, the two therapists modeled appropriate and inappropriate behavior in a given situation requiring socially skilled behavior. Frequently, they exaggerated the inappropriate behavior. Patients then engage in response practice, rehearsing the behavior in a situation with either one of the therapists or with each other, and reported their SUDS level after each rehearsal. The patients were often asked to practice their behavior two or three times to achieve a more appropriate performance.
>
> However, to keep the sessions as positive as possible, the patients were not required to rehearse more than three times even if their performance was not appropriate or their SUDS level had not decreased. Rather, they were given an opportunity to practice the behaviors in later sessions, during which they could utilize an expanded repertoire of skill behaviors. The patients received specific instructions and were coached to improve their interpersonal skills by focusing on the six components of skill behavior. The patients were trained to give each other feedback on these components.
>
> Between-session assignments were given at the end of each session. The patients were paired according to their own preference or, if no preference was expressed, at the discretion of the therapist. The assignments proceeded in a graded manner from less to more difficult. During each session, the patients discussed their successes and failures in completing the previous assignment. When appropriate, the patients roleplayed the assignments and received feedback from the group. As the assignments became more complex and patients were required to engage in social activities with each other, the patients suggested the assignments and extended invitations to other members with similar interests. (p. 887)

The next description comes from Weinman, Gelbart, Wallace, and Post (1972). This was a ward-based program called by the authors "social environmental therapy" but labeled by Hersen in Chapter 6 as social skills training. The patients in this study received 3 months of this form of treatment:

> Essentially it consists of the following three major components: (a) a core of five weekly group activities which require social interaction; (b) informal social activities which encourage social contact; (c) a staff trained to stimulate patient participation in the formal and informal aspects of the program. (p. 247)

ASSESSMENT OF SOCIAL SKILLS

There are several recent chapters (Curran & Mariotto, in press; Hersen & Bellack, 1977) including one in this volume by Bellack (Chapter 3) which address the problems and issues involved in the assessment of social skill. However, since social skill assessment is a major methodological problem, I would feel remiss if I did not briefly address this subject. Before proceeding, I would like to issue a warning; that is, unless we pay attention to and utilize the principles and strategies of psychological assessment, then the assessment instruments used in the social skill area will remain in a primitive state. Behaviorists, in their fervor to reject certain personality models (e.g., trait-dispositional models), have either incorrectly rejected or ignored the basic principles and strategies of assessment. If we do not reacquaint ourselves with these principles and strategies and become more sophisticated with respect to their use, our assessment procedures will remain dubious and the conclusions drawn from our studies may lack substance. I will address three assessment issues in this section: (1) the lack of correspondence of data derived from different assessment procedures; (2) the ecological validity of our assessment procedures; and (3) the problem of assessing clinically significant change. I will also make several suggestions which I feel will be useful in developing more sophisticated assessment procedures.

Lack of Correspondence across Assessment Procedures

In this section, I will review several studies which examine the correspondence between data collected utilizing different modes of assessment (self-report and observational) and different methods within the same mode of assessment. The first two studies discussed are in the heterosexual-social-skill/anxiety paradigm. The first study, by Wallander, Conger, Mariotto, Curran, and Farrell (1978), examined the comparability of five heterosexual/social-anxiety questionnaires, dating experience information, and self- and judges' ratings of anxiety and skill in simulated social situations. Three of these questionnaires—the Social Avoidance and Distress Scale (Watson & Friend, 1969), the Situation Questionnaire (Rehm & Marston, 1968), and the Survey of Heterosexual Interaction (Twentyman & McFall, 1975)—have been commonly used in the heterosexual-social-skill/anxiety paradigm (Curran, 1977). The other two questionnaires used by Wallander *et al.* were the Social Anxiety Inventory (Richardson & Tasto, 1976) and the Social Anxiety and Skill Questionnaire, which was developed for this particular study. The Social Anxiety Inventory was designed to be a broad-band measure of social anxiety tapping anxiety reactions in a variety of social situations. The Social Anxiety and Skill Questionnaire consisted of a detailed description of two heterosexual-social simulated situations. Subjects were asked to rate how anxious and skillful they would imagine themselves to be in the situations described. After completing the Social Anxiety and Skill Questionnaire, subjects were asked to role-play the situations previously described with a confederate. After the role-plays, subjects rated the degree of anxiety and skill they experienced in these role-plays. The subjects' performance in these role-plays was also rated by trained observers for anxiety and skill. The results from this study indicated moderate correlations among the questionnaires commonly used in this paradigm (the Social Avoidance and Distress Scale, Situation Questionnaire, and the Survey of Heterosexual Interactions). Correspondence was generally low between these questionnaires and the Social Anxiety Inventory and the Social Anxiety and Skill Questionnaires. The questionnaire scores, in general, failed to predict significantly either the self- or observer ratings based on the simulated role-plays. This was especially distressing in the case of the Social Anxiety and Skill Questionnaire because this questionnaire consisted of a description of these simulated situations. In other words, there was poor correspondence between a subject's own self-report when given a description of a situation and asked to rate his imagined skill and anxiety level, and his self-report immediately following participation in such a situation. The dating experience measure which dealt with the quantity and quality of the subject's dating experience also correlated poorly with the observational and self-ratings from the simulated situations and with the questionnaire scores.

The second study in the heterosexual-social-skill/anxiety paradigm was conducted by Farrell, Mariotto, Conger, Curran, and Wallander (1979). This study dealt with the correspondence between several types of self- and observer ratings in two heterosexual-social simulated situations. The subject's self-ratings consisted of three types. First, subjects were given a description of simulated social situations and asked to rate their imagined anxiety and skill in these situations (Social Anxiety and Skill Questionnaire). Then the subjects were asked to role-play these situations, after which they rated their anxiety and skill. Last, at a week's time interval, subjects were permitted to view the videotapes of their role-plays and were asked to rate the degree of anxiety and skill they exhibited. Judges ratings consisted of two types (confederate and videotape observers). Two confederates who interacted with the subjects during the role-plays and two video judges who observed the videotaped performance rated the subjects on anxiety and skill. The data were analyzed in terms of generalizability theory (Cronbach, Glaser, Nanda, & Rajaratnam, 1972). Correspondences among the various types of self-ratings were in the moderate range. Correspondence between the two confederates' ratings and between the two videotape judges' ratings, as well as the correspondence among all four judges, were in the moderate range. However, the correspondence between self- and judges' ratings was low.

In another generalizability study, Curran, Monti, Corriveau, Hay, Hagerman, and Zwick (1978) obtained self-report and observer ratings on psychiatric patients in a number of simulated social situations. Two different types of self-report ratings were obtained. First, the subjects completed the Social Anxiety Inventory developed by Richardson and Tasto (1976). This inventory, consisting of 105 items, has been factor-analyzed (Curran *et al.*, in press; Richardson & Tasto, 1976) with seven factors extracted. After completing the inventory, subjects were asked to role-play simulated situations which corresponded to the seven factors on the inventory. After each of the role-plays, subjects rated themselves with respect to their anxiety and skill in the role-plays. These role-plays were videotaped and shown to two trained judges who rated subjects' performances on anxiety and skill. The degree of correspondence between the two self-report measures (i.e., the inventory and the ratings after the role-play) was low for the skill ratings and only in the moderate range for anxiety ratings. There was a decided lack of relationship between observer ratings and either type of self-rating.

These three studies taken together indicate low to moderate correspondence between scores obtained by different methods within the same mode and poor correspondence across assessment modes. While a certain lack of correspondence between self-report and observational ratings is expected on theoretical grounds (Curran, 1977), the poor correspondence obtained in these studies would indicate that we need to improve both types of assessment procedures. I would now like to offer some suggestions which I feel would improve the quality of both types of procedures.

In the Wallander *et al.* (1978) study, there was a lack of correspondence between the subjects' own self-report on how they imagine themselves feeling and acting in a situation and how they rated themselves after such an interaction. How a subject thinks he will behave and how he behaves may differ for several different reasons. A study by Eisler, Frederiksen, and Peterson (1978) illustrates this point. Eisler *et al.* (1978) had subjects view a videotape of a particular scene followed by various response alternatives. Subjects were then asked which of the alternatives they would choose if placed in such a situation. Then, subjects were placed in these simulated situations and asked to role-play. There was less than perfect correspondence between how the subjects said they would react in these situations and how they actually acted. A number of studies (e.g., Fishbein & Azzen, 1975) have demonstrated that correspondence between self-report and actual behavior in a variety of situations can be increased if the subject's behavioral intent is assessed. Fiedler and Beach (1978) demonstrated that the difference between participants who chose an assertive response and those who did not could be predicted from an assessment of the subject's estimation of probabilities that good or bad consequences would or would not occur contingent upon their response. In other words, if one can determine a subject's perception of the risks involved in being assertive and derive the subject's behavioral intent, one can increase the correspondence between self-report and assertive behavior.

Another manner in which to increase the correspondence between self-report and observational data is to reduce differences in content. Cone (1977) has indicated that often, when comparing data across modes of assessment, there is a confound because of differences in content. For example, the self-report question may be "Are you afraid of snakes?" while the observational procedure consists of observing of how close a subject will approach a snake. Lick and Unger (1977) have demonstrated that by increasing similarity of content, comparability of self-report and observational data are increased.

Often a subject is asked to make a global rating of social skill—a construct which possesses little meaning for him and for which he may have little understanding. Behavioral observers are given intensive training in rating this construct, and even after such training, reliability coefficients that are obtained often run in the moderate range (Curran & Mariotto, in press). Giving subjects training in rating the construct, providing them with anchor points, etc., may increase correspondence between self-report and observational ratings. Another alternative is to have subjects rate more specific behaviors which require less inference on their part. For example, Lowe and Cautela (1978) have developed a Social Performance Survey Schedule in which subjects rate the frequency with which they emit certain specific behaviors (e.g., disturbing noises such as burping, sniffling, etc). The level of inference required on the part of the subject in completing this scale seems to be much less than asking the subject to make a global rating of social skill. In one study, Lowe (1978) compared the self-report scores of psychiatric patients on the Social Performance Survey Schedule to nurses' ward ratings of social skill. A correlation of $r = .65$ was obtained between the nurses' ratings and the total score on the Social Performance Survey Schedule.

Most of our observational assessments of social skill are based on laboratory situations. The fidelity of our observational assessment may be improved in numerous ways, including improvement in rater training, sampling multiple situations, etc. (Curran & Mariotto, in press), but perhaps most important, by improving the representativeness of our laboratory situations to naturalistic settings. The question of the ecological validity of our laboratory-based situations will be addressed in the next section.

The Ecological Validity of Laboratory-Based Observational Procedures

The most commonly used observational assessment procedure for measurement of social skill is a laboratory-based simulated social situation in which subjects are asked to role-play as if it were a natural situation. Until recently, no data were available with respect to the representativeness of such role-plays of behavior in the natural environment. Recently, Bellack and associates have conducted two studies comparing the representativeness of indicators of social skill assertion in laboratory situations to more naturalistic settings. Specific behaviors were rated, such as eye contact, smiles, etc., which the investigators regard as components of social skill. In the first study, Bellack, Hersen, and Turner (1978) had observers rate the behavior of subjects on their laboratory simulation (Behavioral Assertiveness Test–Revised) and also had them rate subjects' behavior during a standard interview and a group-psychotherapy situation. In the second study, Bellack, Hersen and Lamparski (1979) compared skill performance of male and female subjects in simulated heterosexual-social situations to a more naturalistic situation involving interaction with a confederate in a waiting room. In general, there was a lack of correspondence between ratings in the simulated situations and ratings obtained in the more naturalistic settings. Curran (1978) had criticized the first-mentioned Bellack *et al.* (1978) study, stating that the lack of relationship could have been due to numerous factors, including differences with respect to the rating media, differences in raters, differences in behaviors rated, and the contextual choice of situations. Most of these same criticisms hold true for the second Bellack study.

Wessberg, Mariotto, Conger, Conger, and Farrell (1978) also compared the responses of subjects in two different simulated social interactions (a pizza parlor simulated date and an opposite-sex interaction in the student union) to their behavior in two waiting room situations. The subjects in the Wessberg *et al.* study were high- and low-frequency daters in a college population. The two simulated situations lasted for 4 min and the two waiting room periods lasted for 3 min. Videotapes were made of both the simulated situations and the waiting room situations. Trained raters then viewed these tapes and rated them on global measures of anxiety and skill. The correlation obtained in the Wessberg *et al.* study between observer's skill ratings of subject's performance across the two simulated situations was 0.71, and the correlation between the skill ratings across the two waiting periods was 0.73. More interestingly, the correlation between the skill rating in the first role-play to the first waiting period was 0.55 and to the second waiting period was 0.58. Likewise, the correlation between the skill ratings in the second role-play to those of the first waiting period was 0.48 and to the second waiting period was 0.51. In essence, then, the relationships between the global ratings of skill between the simulated situations and the naturalistic situations were almost as high as those established between the two naturalistic situations.

Although there may be numerous reasons why Wessberg *et al.* found good correspondence between simulated situations and naturalistic situations while Bellack and associates did not, two major reasons appear to be the different length of the interactions and the types of ratings. The simulated interactions in the Bellack studies were of extremely brief duration and did not involve a continual interaction between the subject and the confederate. In addition, in the Wessberg study, judges were asked just to make a global rating of skill, while in Bellack's studies, the judges were asked to rate specific behavior, such as eye contact. It appears reasonable that an extended interaction is more representative of what occurs in the natural environment and that global ratings of skill would better generalize across situations than ratings of more specific behaviors, such as eye contact, although this is an empirical question.

More studies need to be conducted assessing the degree of correspondence of behaviors rated in simulated interactions with those behaviors in more naturalistic situations. In addition, alternate ways of obtaining data with respect to ecological validity could be developed. One such innovation was used in the Wessberg *et al.* (1978) study.

Wessberg had his subjects judge the representativeness of their behavior in both the simulated situations and the waiting room situations for the perceived "realness" of these situations. The waiting periods were rated as significantly more "real life" than the role-plays and subjects reported their behavior as more representative of their everyday behavior in the waiting periods. Interestingly, the judges rated the skill performance of the subjects in the role-play as being significantly better than their performance in the waiting period. In addition, correspondence of the global skill ratings from the simulated situation to the behavior of the subjects in the more naturalistic situations were about as high as the correspondence of the skill ratings across both naturalistic situations. Another approach to obtaining data regarding ecological validity was demonstrated by Warren and Gilner (1978) in a study involving married or dating couples. All the subjects engaged in a simulated role-play task which was audiotaped. Their partners then listened to each other's taped role-plays and rated them with respect to how representative the behavior exhibited by the partner was to real-life situations. The ratings were generally high, indicating that the partners felt that the taped role-play behavior was representative of the subject's natural behavior.

Assessing Clinically Significant Change

Another issue which needs to be addressed is whether our assessment procedures are sensitive to clinically significant changes or merely reflect statistically significant changes. Arkowitz (1976) contrasted the effects he had with social skills training in a college analogue population with the success he had with a clinical population suffering from extreme heterosexual-social anxiety. The data on this clinical population indicated that only three of the patients showed significant improvement on most measures, three patients were slightly improved, while three other patients failed to complete their course of treatment. In addition, of the three patients who had shown significant improvement on most skill measures, two patients still reported some fear of dating and, in fact, had not dated. Arkowitz stated that "statistical changes in selective measures typically employed in mass outcome studies certainly reflect improvement but such improvement is far short of the desired levels of change needed to make a substantial difference in a person's life" (p. 8).

The problems associated with assessing clinically significant changes has been addressed by Kazdin (1977). Kazdin recommended two types of procedures which could assist in measuring clinically significant changes: social comparison and subjective evaluation. According to the social comparison criteria, behavior change can be viewed as clinically important if "the intervention has brought the clients' performance within the range of socially acceptable levels as evidenced by the clients' peer group" (p. 427). Of course, the problem here is finding an appropriate peer group. For example, if the targeted population is low-dating-frequency college students, and if the target behavior is an increase in dating frequency, then a high-frequency-dating group might be an appropriate normative population. However, in the case of psychiatric patients, do relatives of the patient serve as an appropriate normative group?

According to these subjective evaluation criteria, Kazdin states that a change can be regarded as clinically important if "the client's behavior is judged by others as reflecting a qualitative improvement on global ratings" (p. 427). Again, the difficulty exists in deciding which individuals or groups of individuals are relevant to make this subjective evaluation. In the case of unsocialized delinquents, should adolescents be making the decision, parole officers, or parents of the children? Of course, there is also the problem of whether the individuals making these subjective evaluations have the same understanding of the construct to be rated and whether their interpretations of qualitative improvements are comparable.

Other more objective measures can also be used to attempt to assess clinically significant changes. For example, Zigler and Phillips (1961) demonstrated that social competency within a psychiatric population is related to both length of hospitalization and to recidivism rates. In the Finch and Wallace (1977) study, a 3-month follow-up was conducted. Five of the eight patients receiving social skills training had been discharged from the hospital while only one of the eight control patients had been discharged. In the study by Monti *et al.* (1979), a psychiatrist who was blind to the treatment conditions of the study evaluated all patients at a 10-month follow-up. After a structured interview, the psychiatrist rated the patients on the Clinical Outcome Criteria Scale (Strauss & Carpenter, 1972). This scale measured aspects of an individual's functioning, such as whether he was gainfully employed, had been readmitted into the hospital, etc. Although the number of subjects contacted at this 10-month follow-up were few, the data collected were promising, indicating that those receiving social skills training appeared superior on many of the indices of social functioning.

Clearly, numerous factors in addition to a patient's social competency can affect variables such as recidivism rates (e.g., the death of a spouse, tolerance of a peer group, etc.), but in order to demonstrate the clinical utility of social skills training we must be able to indicate that such training results in a meaningful change in our subjects' functioning.

METHODOLOGICAL PROBLEMS

In this section, I wish to review certain methodological problems which I feel are especially pertinent to social skills training. I will address the problems of subject selection, follow-up data, and data analyses.

Subject Selection

The description of subject-selection procedures in many studies on social skills training is vague and unclear. In many cases it does not appear that any valid criteria were used other than convenience. In many cases it is unclear whether the selected subjects were actually performing inadequately in social situations, regardless of whether this poor performance was due to an actual deficit or the result of some inhibitory process. Often subjects are "nominated" as being eligible candidates for social skills training, but it is unclear what criteria the various individuals doing the nominating were using and whether these criteria were consistent across nominators. In those cases where subjects were assessed on purported measures of social skill prior to treatment, the lack of normative data still leaves it unclear whether these subjects were socially incompetent. Even if the subjects selected for a study are socially incompetent, they may be judged incompetent for different reasons; that is, one individual may have been nominated because of a low rate of positive behavior while another individual may have been nominated because of a high rate of obnoxious, antisocial behavior. The selected subjects may also be heterogeneous with respect to many other variables. For example, using the Liberman *et al.* (1977b) categories, individuals may be performing inadequately because of receiving, processing, or sending functions. The selected individuals may be experiencing rather isolated problems or they could be experiencing multiple difficulties. Arkowitz (1976) contended that if selected subjects are characterized by a single discrete problem, it follows that they will be more responsive to treatment than subjects with multiple problems.

I would now like to present descriptions of subject-selection procedures from several social skills training outcome studies with psychiatric populations. The selection of such descriptions was done on a random basis, but I feel they are representative of typical selection procedures and illustrate the nature of the subject-selection problem.

Weinman *et al.* (1972) utilized the following procedure. Two staff psychologists interviewed male patients on various wards of the hospital and reviewed their records. Selection was based on the following criteria: (a) diagnosis of schizophrenia; (b) minimum of 1 year of total hospitalization; (c) clinical impression of withdrawal and nonassertive behavior; (d) ability to respond to questions intelligently and (e) the absence of behavior problems that require close supervision. Eisler, Blanchard, Fitts, and Williams (1978) used the following criteria: 87 male psychiatric patients hospitalized at the Veterans Administration Medical Center were screened on 12 role-play tests adapted from the Behavioral Assertiveness Test–Revised. Those patients whose average scores were below a median (previously established norms were based on 60 psychiatric patients) were independently interviewed by a psychologist and psychiatrist, both with 5 years of diagnostic experience. The psychologists and psychiatrists, on the basis of their interview, determined the patients' psychiatric classification and eligibility for the study. Those patients who were found to be acutely psychotic or with organic brain syndromes were excluded from consideration. Goldsmith and McFall (1975) stated the following criteria: 41 male inpatients on a psychiatric ward at Chicago's West Side Veterans Administration Hospital were invited to participate as subjects. One refused and four were unable and/or unwilling to complete the initial assessment procedures. Finch and Wallace (1977) utilized the following criteria: the subjects were male schizophrenic inpatients residing in the same living unit at the Veterans Administration Hospital, Sepulveda, California. Nursing staff selected an initial pool of 29 patients who fullfilled the following criteria: diagnosis of schizophrenia, between 21 and 40 years of age, a minimum of 1 year of total hospitalization, clinical impression of depression or withdrawal with nonassertive behavior, ability to respond to questions and instructions cooperatively. These 29 patients were administered the Wolpe–Lazarus Assertiveness Questionnaire (1966) and were interviewed by the two therapists who conducted the assertion-training sessions. The 16 most nonassertive patients were selected for the study. Lomont *et al.* (1969) chose his subjects by the following criteria: (1) social anxiety judged to be an important part of the symptomatologies; (2) no evidence of organic brain damage; (3) no evidence of psychiatric thought disorders; (4) an IQ of at least 80; (5) able to read and write. Two staff clinical psychologists made the decision regarding symptomatology after receiving test data, including MMPI, Holtzman Inkblot Test, Ammon's Quick Test, Bender–Gestalt Test, and TAT scores.

It should be clear from these brief descriptions of selection procedures that subjects often constitute a fairly heterogeneous group both within and across studies. I would like to suggest that we use multiple screening procedures, such as nominations by experts who have been trained and have had ample opportunity to observe the skill level of patients, screening on both self-report and observational assessment procedures which have been established as valid and for which normative data exist, etc. Some attempt should be made to delineate subject characteristics such as whether he/she is regarded as socially inappropriate because of an excess of obnoxious, aggressive behavior or because of unassertive behavior, whether an actual deficit is probable or whether the inadequate performance is due to inhibition, whether it appears to be an isolated problem or the patient is suffering from multiple problems. It is clear from reading the literature that social skills training is a robust treatment that appears to produce statistically significant changes in a variety of populations. However, it is important for us adequately to describe our subject-selection procedures in order to facilitate comparison of results across studies. It is also essential for us to delimit our patient population if we are to study how treatment procedures may interact with subject characteristics.

Follow-Up Data

In Chapter 6 of this volume, Hersen reviewed (see Table 5) the results of clinical outcome studies utilizing social skills training with clinical populations. Of the 14 studies reviewed, only 6 studies contained a follow-up. Of those studies conducting follow-ups, the longest (Monti *et al.*, 1979) was of 10-month duration. In some of the studies where a follow-up was conducted, the investigators failed to use at follow-up all of the assessment procedures used pre- or posttreatment.

The goal of social skills training is to teach subjects new behaviors in order for them better to cope with problematic social situations and to increase their overall interpersonal effectiveness. Implicit in this goal is that the increase in interpersonal effectiveness is substantial and will result in meaningful significant changes in the subject's interpersonal behavior which will be maintained after treatment. The fact that most of the studies on social skills training do not include a follow-up period is a serious methodological deficit. Several issues regarding follow-up need to be discussed.

We must begin to develop a theoretical rationale regarding the appropriate length of time needed for a follow-up. Is a 6-month period too short or too long? If the goal of treatment is significantly to alter a subject's interpersonal style then a 6-month period may actually be too short a time span. Subjects may gradually start out utilizing their new skills in certain limited situations, and only after meeting some success will they then apply these new interpersonal skills in a more comprehensive fashion. Consequently, a 6-month period may be too short a time span. If, however, the intervention is brief and limited in scope and if the environment to which the subjects return does not reinforce these alternate ways of behaving, then a 6-month period may be too long a period to assess treatment effects adequately.

While it is hoped that the increase in interpersonal effectiveness developed in social skills training would be so substantive that environmental effects and changes would not seriously effect the maintenance of these changes, realistically, significant life changes may greatly effect the maintenance of the learned skills. For example, the death of a spouse may greatly affect whether an individual will maintain his treatment gains. Therefore, at follow-up investigators should attempt to evaluate the individual's current situational context in order to evaluate whether significant life changes have occurred in order better to understand those factors affecting treatment maintenance.

Another question to be resolved is the frequency of follow-up data collection. In most group-design studies, only one follow-up assessment is conducted; consequently, evaluation of treatment effects over several time periods cannot be made. In addition to sampling problems, the reactivity of one long-term follow-up assessment appears to be potentially greater than a series of repeated follow-up assessments. Multiple follow-up assessment is sometimes conducted in single-case studies, but here the frequency of follow-up assessment is very often less than assessment at either baseline or during treatment. The frequency of such follow-up assessment in single-case studies is often so few in number that real trends in the data cannot be specified.

Another problem that frequently occurs at follow-up assessment is the alteration of conditions of measurement. For example, in the case of observational data, the individual confederates used in the simulated role-plays or the set of judges making the observational ratings often have been changed. In one study, Royce and Arkowitz (1978) examined the effectiveness of social skills training in increasing friendship among same-sex peers. The pool of raters judging the performance of subjects at pre-, post-, and follow-up appears to have changed over these time spans. Although the results from most other dependent measures employed were consistent with the experimental predictions, the data from these observational ratings indicated no differences between the experimental and control groups. Lack of significant findings on the observational measures may have been due to a lack of consistency of raters over this time span.

Another issue which must be faced is the attrition problem which is especially prevalent at follow-up. In studies where follow-ups were conducted, often over 25% of the subjects completing treatment were not assessed at follow-up. There are basically two types of reasons for subject attrition. One is that the subject is not available (e.g., subject has moved, died, etc.) or the subject has not consented to be assessed at follow-up. The reasons for subject attrition should be noted when reporting the study. In addition, data analyses should be conducted in order to determine whether those subjects who are available at follow-up were different from those who were lost.

It is imperative that we conduct long-term follow-up on the effectiveness of social skills training; consequently, we need to pay more attention to the issues associated with follow-up assessment. While it may be interesting to demonstrate short-term changes in specific behaviors of our subjects, if social skills training is to remain a viable and active treatment strategy then long-term clinically significant changes in the interpersonal functioning of our subjects need to be documented.

Data Analyses

Most investigators in the social skills training literature use multiple dependent measures, often from multiple channels of assessment (self-report, observational, etc.). The usual procedure in these studies is for an investigator to submit each of the dependent measures to a univariate statistical comparison. Often a number of the dependent measures used in a particular study are highly correlated, and, therefore, separate univariate analyses will capitalize on chance. Multivariate analyses are recommended when a large number of dependent measures are used. If the dependent measures are highly intercorrelated, then a multivariate analysis will be a more conservative test. If, on the other hand, the dependent measures are not highly correlated, then the use of multivariate procedures may uncover differences between the groups which could be missed by separate univariate analyses. Kaplan and Litrownik (1977) noted that a treatment which has a minimal effect on two uncorrelated dependent variables may be shown to have a significant effect on the composite of the two measures. In effect, multivariate procedures combine dependent variables in a way that maximizes the differences between groups. Discriminant function analyses then can be used to determine which functions (i.e., weight composite of variables) best discriminates between the groups. These discriminant function analyses will assist us in understanding whether our treatments are affecting one or several processes.

A study by Twentyman and McFall (1975) is illustrative of the use of multiple dependent measures analyzed by univariate procedures. In this study, three self-report ratings of anxiety were obtained: a questionnaire, an anxiety rating taken after a simulated telephone call, and an anxiety rating taken after a forced-interaction test. Pulse-rate measures were taken before the simulated telephone call, during the forced interaction, and during a series of simulated social interactions. Observational ratings were taken during the simulated telephone call, the simulated social interaction, and the forced-interaction test. Observational ratings consisted of both global anxiety ratings and ratings of specific anxiety indicators, such as speech stammers. The subjects also kept a diary where they recorded a number of events such as total number of women interacted with in a specific time period. An avoidance measure was also obtained. This is an excellent example of the use of multiple outcome criteria from multiple channels of assessment, and as such provides a good example for other researchers to emulate. However, the data from these various dependent measures were analyzed by a series of *t*-tests when they should have been more appropriately analyzed by a multivariate analysis of variance. The use of multiple *t*-tests on highly correlated data greatly increased the probability of obtaining significant results by chance. If multivariate procedures had been used and significant results obtained, multiple discriminant functions could have been utilized to determine how many different processes were affected by the treatment conditions.

Other cases can be found in the literature where the use of multivariate procedures may be more appropriate than the univariate procedures generally conducted. For example, investigators (Hersen & Bellack, 1976) generally employ a large number of specific indicators of social skill, such as speech latency or eye contact, in their observational assessment. Rather than conducting separate univariate analyses on each of these measures, a multivariate analysis would be more appropriate.

Another area in which statistical procedures should be employed is in single-case studies. In most single-case designs, no statistics are employed and the reader is generally confronted with a visual display and asked to compare differences during baseline, treatment, and follow-up. Statistical procedures have been developed (Glass, Wilson, & Gottman, 1975) for $N = 1$ analyses which could be applied to social skills training research. It has been established (Jones, Weinrott, & Russell, 1978) that data from single-case studies, while appearing visually to demonstrate significant effects, may, in fact, be nonsignificant when subjected to a statistical analysis. Single-case studies featuring such procedures as return to baseline are superior in a control sense, but they, too, may be capitalizing on trends in the data.

Future Directions

In our research in the years to come, we must address the many methodological issues discussed in this chapter if social skills training is to remain a viable treatment alternative. We must examine closely those components of the social skills training "package" which are instrumental to its effect and especially study those procedures which promote transfer to naturalistic situations. We need to improve greatly the sophistication of our assessment methodology and we must study or restudy the assessment procedures and strategies derived from psychology's rich psychometric tradition. In our experiments, we must pay more attention to how we select our subjects, the composition of our subject population, the problem of subject attrition, and become more sophisticated with respect to our data analyses. However, the major problem which we must face squarely in the years to come is the definitional question: What is social skills and how can we best delimit and measure its components?

Let us pursue a brief analogy. If it were crucial for us to determine the precise time of day, our safest strategy would be to call the Bureau of Standards of Measurement. Unfortunately, no Bureau of Standards of Measurement exists in the social skills paradigm. If we could not call the Bureau of Standards, an alternative would be to query a number of individuals and reach some consensus. We have just undertaken a study of a similar fashion in order to arrive at some sort of consensual validation of the definition of social skills. The purpose of this investigation is to explore the degree of agreement across different experimental laboratories with respect to the social skill level assigned to various stimulus subjects. We have involved a number of prominent investigators' research teams in such an endeavor and have forwarded to each of these research teams copies of videotaped stimuli which contain the simulated role performance of a number of subjects. These stimulus tapes are representative of the most typical method of observational assessment within the social skill paradigm. More specifically, the stimulus interaction consists of a narrator describing a situation, a confederate issuing a prompt, followed by a response from the subject. We have asked each investigator to have the stimulus tapes rated by their most experienced raters in the same manner in which they typically rate stimulus tapes. For example, some investigators have their raters evaluate very specific indices of skill such as eye contact before making a global rating of overall skill. After they have completed their ratings in the usual manner, each team of raters will then be asked to rate each tape on an 11-point scale supplied by our research team. The data from this study should give us a better understanding of whether various investigators are indeed addressing the same construct. Such a study could also lead to the development of a library of tapes in which there is good consensual validation across raters, and, consequently, may be useful in training future raters and help insure comparability across investigators.

The other issue to be resolved is the specification of a methodology which will allow us to delimit those behaviors which comprise significant indicators of the construct of social skill. So far, those behaviors have been chosen mainly on an intuitive basis or by soliciting the opinion of several so-called experts. We need to develop methodological strategies in order to provide an empirical basis for the study of these components.

Duncan and Fiske (1977), in their book *Face To Face Interaction*, have outlined a research methodology and strategy which may prove useful in isolating significant components of social skill. They suggest a microanalysis of interpersonal interactions. They make useful suggestions with respect to the type of data to be collected, the conditions under which such data collection ought to occur, how the data should be processed and analyzed, and also provide a conceptual framework to guide the research process. They stress that the system to be used should be minimally inferential so that high reliability between observers can be obtained, wherein the observers function as instruments or machines. They stress that the conditions for data collection should be as naturalistic as possible and that the observers should be as unobtrusive as possible. They mention the importance of discovering and verifying sequential patterns which occur in any type of action sequence. Duncan has devised a statistical system called Crescat, whereby it is possible to identify complex sequential patterns in the data source. In their book, Duncan and Fiske examine two large-scale studies involving a microanalysis of face-to-face interactions. The second of these studies not only examines body motions, such as head nods, smiles, position of hands, etc., but also paralinguistic variables, such as pitch height, audible inhalations and exhalations, etc. No attempt was made in either of these studies to analyze language content, which I feel would be necessary in order to provide a firm grasp of the meaning of social skills. In their first study, Duncan and Fiske varied the stimulus characteristic of the confederate, and differences were found with respect to the microanalysis of the subject's behavior. Other situational contexts cues also need to be manipulated. The studies conducted by Duncan and Fiske are somewhat limited, and the data generated in these studies do not readily apply to the social skills area. However, many of their recommended procedures and strategies are transferable and some adaptations could be made.

Duncan and Fiske chose the behaviors to be observed on the basis of a number of considerations and criteria. The central criterion was that the acts should be recordable with a very high agreement between coders. Other criteria were: a relatively brief duration of the act, that the act be coded in terms of either presence or absence of the act, that the act be shown by many subjects and occur several times during the observational period, that the act presume relevance, and that the act be identified as a physical act *per se*. Although these appear to be reasonable criteria and should be considered, the number of such acts which met these criteria is enormously great. We need some strategy in order to reduce the number of possible acts which could be included for analyses.

A strategy employed by Royce and Weiss (1975) may prove useful. The Royce and Weiss study involved an examination of the parameters judges used in differentiating functional from dysfunctional married couples. Functional and dysfunctional couples participated in various problem-solving tasks and their performances were videotaped. Judges viewed these videotapes and were asked to sort the couples into either functional or dysfunctional categories. In addition, judges were asked to nominate the cues that they used in making this decision. A large number of cues were generated and then condensed. Operational definitions of these cues were constructed. Another group of judges then rerated the tapes using these operational cues. The investigators then performed a multiple-discriminant analysis using the data from these operationalized cues as an independent measure and the two groups of married couples as dependent measures. These investigators were able to determine which of these cues were important contributors in discriminating the groups. Of course, for these cues to be really useful, a cross validation would be necessary.

More recently, a similar strategy was applied in the social skills area (Conger, Wallander, Ward, & Mariotto, 1978). The investigators presented stimulus videotapes of males differing in heterosexual social skill and anxiety to 62 male and 73 female undergraduates. The undergraduate judges were asked to make global ratings of anxiety and skill after which they were requested to list the behaviors they had focused on in making the global ratings. In addition, they were asked what specific cues they had used in making judgments about anxiety, skill, or both constructs. The cues generated were analyzed for content and combined into a classification system by the investigators. The classification system was a hierarchical structure with four levels. The lowest level contained explicit cues (e.g., smiles), the next level contained cues of the same kind but less explicit (e.g., facial expressions), etc.

Results from the study indicated that the judges clearly discriminated levels of anxiety and skill. The cues generated by the judges were similar to variables under investigation in the field, but they emphasized conversational style and content to a much greater extent than they are currently weighted. According to the judges, some of the nominated cues were used mainly to judge skill, others to judge anxiety, with some cues used in the judgment of anxiety and skill. Female judges in comparison to male judges demonstrated a more discriminated use of skill cues and seemed particularly adept at specifying positive instances of skillful behavior. Conger *et al.* (1978) are in the process of determining the degree to which the cues generated by the judges match the behavior of the males in the stimulus videotapes. More studies need to be conducted to determine specific behavioral components used in evaluating social skill levels.

SUMMARY

In this chapter, an attempt was made to present some of the primary issues confronting investigators in the social skills area. The chapter began with a discussion of the lack of consensus regarding the definition of the construct of social skill. The definitional issue affects the content and process of our training programs, our assessment procedures, methodological strategies, and, indeed, the experimental questions addressed in our studies. Next, we looked at several parameters on which social skills training programs vary. Training programs differ with respect to content and process components, format, number of trainers, duration, modalities emphasized, molecularity versus molarity of the training, and the degree to which the training is individualized for a particular patient. Caution was urged in interpreting results across studies because of the potential lack of comparability of training programs. Selected assessment issues were then addressed, including the comparability, ecological validity, and clinical sensitivity of our assessment procedures. The next section of the chapter dealt with methodological problems. The problems reviewed were differences in subject-selection procedures, the absence of follow-up data, and inadequacies in data analyses. In the final section of the chapter, several research strategies were discussed that may prove useful in clarifying the definitional problem.

Methodological reviews in the research area are necessarily critical in tone, which may engender a feeling of pessimism in the reader. This is far from my own mood after completing this review. While cognizant of the fact that there are major problems in the social skills area, I retain a note of cautious optimism. My optimism is buoyed by my perception that investigators realize the complexities of the problems involved and are addressing these issues. A realization of these issues is instrumental to their solution.

REFERENCES

Arkowitz, H. *Clinical applications of social skill training: Issues and limitations in generalization from analogue studies.* Paper presented at the annual meeting of the Association for the Advancement of Behavior Therapy, New York, December 1976.

Argyle, M., & Kendon, A. The experimental analysis of social performance. In L. Berkowitz (Ed.), *Advances in experimental social psychology* (Vol. 3). New York: Academic Press, 1967.

Bandura, A. *Principles of behavior modification.* New York: Holt, Rinehart and Winston, 1969.

Barlow, D. H. Increasing heterosexual responsiveness in the treatment of sexual deviation: A review of the clinical and experimental evidence. *Behavior Therapy,* 1973, *4,* 655–671.

Bellack, A. S., Hersen, M., & Lamparski, D. Role-play tests for assessing social skills: Are they valid? Are they useful? *Journal of Consulting and Clinical Psychology,* 1979, *47,* 335–342.

Bellack, A. S., Hersen, M., & Turner, S. M. Role-play tests for assessing social skills: Are they valid? *Behavior Therapy,* 1978, *9,* 448–461.

Callner, D. A., & Ross, S. M. The reliability and validity of three measures of assertion in a drug addict population. *Behavior Therapy,* 1976, *7,* 659–667.

Cone, J. D. The relevance of reliability and validity for behavioral assessment. *Behavior Therapy,* 1977, *8,* 411–426.

Conger, A. J., Wallander, J., Ward, D., & Mariotto, M. J. *Peer judgments of heterosexual-social anxiety and skill: What do they pay attention to anyhow?* Paper presented at the meeting of the Association for the Advancement of Behavior Therapy, Chicago, November 1978.

Cox, R. D., Gunn, W. B., & Cox, M. J. *A film assessment and comparison of the social skillfulness of behavior problem and non-problem male children.* Paper presented at the meeting of the Association for the Advancement of Behavior Therapy, New York, December 1976.

Cronbach, L. J., Glaser, G. C., Nanda, A., & Rajaratnam, N. *The dependability of behavioral measures.* New York: Wiley, 1972.

Curran, J. P. Skills training as an approach to the treatment of heterosexual-social anxiety: A review. *Psychological Bulletin,* 1978, *84,* 140–157.

Curran, J. P., & Mariotto, M. J. A conceptual structure for the assessment of social skills. In M. Hersen, R. M. Eisler, & P. M. Miller (Eds.), *Progress in behavior modification,* Vol. 9. New York: Academic Press, in press.

Curran, J. P., Monti, P. M., Corriveau, D. P., Hay, L. R., Hagerman, S., & Zwick, W. R. *The comparability of social skill and social anxiety assessment procedures in a psychiatric population.* Unpublished manuscript, 1978.

Curran, J. P., Corriveau, D. P., Monti, P. M., & Hagerman, S. B. Self-report measurement of social skill and social anxiety in a psychiatric population. *Behavior Modification,* in press.

Duncan, S., & Fiske, D. W. *Face to face interaction: Research, methods, and theory?* New York: Wiley, 1977.

Eisler, R. M., Hersen, M., & Miller, P. M. Effects of modeling on components of assertive behavior. *Journal of Behavior Therapy and Experimental Psychiatry,* 1973, *4,* 1–6.

Eisler, R. M., Miller, P. M., Hersen, M., & Alford, H. Effects of assertive training on marital interaction. *Archives of General Psychiatry,* 1974, *30,* 643–649.

Eisler, R. M., Blanchard, E. B., Fitts, H., & Williams, J. G. Social skill training with and without modeling on schizophrenic and non-psychotic hospitalized psychiatric patients. *Behavior Modification,* 1978, *2,* 147–172. (a)

Eisler, R. M., Frederiksen, L. W., & Peterson, G. L. The relationship of cognitive variables to the expression of assertiveness. *Behavior Therapy,* 1978, *9,* 419–427. (b)

Farrell, A. D., Mariotto, M. J., Conger, A. J., Curran, J. P., & Wallander, J. L. Self- and judges' ratings of heterosexual-social anxiety and skill: A generalizability study. *Journal of Consulting and Clinical Psychology,* 1979, *47,* 164–175.

Feldman, M. P., & MacCulloch, M. J. *Homosexual behavior: Therapy and assessment.* Oxford: Pergamon, 1971.

Fiedler, D., & Beach, L. R. On the decision to be assertive. *Journal of Consulting and Clinical Psychology,* 1978, *46,* 537–546.

Finch, B. E., & Wallace, C. J. Successful interpersonal skills training with schizophrenic inpatients. *Journal of Consulting and Clinical Psychology,* 1977, *45,* 885–890.

Fischetti, M., Curran, J. P., & Wessberg, H. W. Sense of timing: A skill deficit in heterosocial-socially anxious males. *Behavior Modification,* 1977, *1,* 179–194.

Fishbein, M., & Azzen, J. *Belief, attitude, intentions, and behavior.* Reading, Mass.: Addison-Wesley, 1975.

Glass, G. W., Wilson, V. L., & Gottman, J. M. *Design and analysis of time series experiments.* Boulder: Colorado Associated University Press, 1975.

Goldfried, M. R., & Sobocinski, D. Effects of irrational beliefs on emotional arousal. *Journal of Consulting and Clinical Psychology,* 1975, *43,* 504–510.

Goldsmith, J. B., & McFall, R. M. Development and evaluation of an interpersonal skill-training program for psychiatric patients. *Journal of Abnormal Psychology,* 1975, *84,* 51–58.

Hersen, M., & Bellack, A. S. Social skills training for chronic psychiatric patients: Rationale, research findings, and future directions. *Comprehensive Psychiatry,* 1976, *17,* 559–580.

Hersen, M., & Bellack, A. S. Assessment of social skills. In A. R. Ciminero, K. S. Calhoun, & H. E. Adams (Eds.), *Handbook for behavioral assessment,* New York: Wiley, 1977.

Jones, R. R., Weinrott, M. R., & Russell, R. S. Effects of social dependency on the agreement between visual and statistical inference. *Journal of Applied Behavior Analysis,* 1978, *11,* 877–883.

Kaplan, R. M., & Litrownik, A. J. Some statistical methods for the assessment of multiple outcome criteria in behavioral research. *Behavior Therapy,* 1977, *8,* 383–392.

Kazdin, A. E. Assessing the clinical or applied importance of behavior change through social validation. *Behavior Modification,* 1977, *1,* 427–452.

Kraft, T. Social anxiety model of alcoholism. *Perceptual and Motor Skills,* 1971, *33,* 797–798.

Lentz, R. J., Paul, G. L., & Calhoun, J. F. Reliability and validity of three measures of functioning in a sample of "hard core" chronic mental patients. *Journal of Abnormal Psychology,* 1971, *78,* 69–70.

Liberman, R. P., Lillie, F., Falloon, I., Vaughn, C., Harper, E., Leff, J., Hutchison, W., Ryan, P., & Stoate, M. *Social skills training for schizophrenic patients and their families.* Unpublished manuscript, 1977. (a)

Liberman, R. P., Vaughn, C., Aitchison, R. A., & Falloon, I. *Social skills training for relapsing schizophrenics.* Funded grant from the National Institute of Mental Health, 1977. (b)

Libet, J. M., & Lewinsohn, P. M. Concept of social skills with special reference to the behavior of depressed persons. *Journal of Consulting and Clinical Psychology,* 1973, *40,* 304–312.

Lick, J. R., & Unger, T. E. The external validity of behavioral fear assessment. *Behavior Modification,* 1977, *1,* 283–306.

Lomont, J. R., Gilner, F. H., Spector, N. J., & Skinner, K. K. Group assertion training and group insight therapies. *Psychological Reports,* 1969, *25,* 463–470.

Lowe, M. *The validity of a measure of social performance in an inpatient population.* Unpublished manuscript, 1978.

Lowe, M. R., & Cautela, J. R. A self-report measure of social skill. *Behavior Therapy,* 1978, *9,* 535–544.

McFall, R. M., & Marston, A. R. An experimental investigation of behavioral rehearsal in assertion training. *Journal of Abnormal Psychology,* 1970, *76,* 293–303.

Miller, P. M., Hersen, M., Eisler, P. M., & Hilsman, G. Effects of social stress on operant drinking of alcoholics and social drinkers. *Behavior Research and Therapy*, 1974, *12*, 67–72.

Monti, P. M., Fink, E., Norman, W., Curran, J. P., Hayes, S., & Caldwell, A. The effect of social skills training groups and social skills bibliotherapy with psychiatric patients. *Journal of Consulting and Clinical Psychology*, 1979, *47*, 189–191.

Rehm, L. P., & Marston, A. R. Reduction of social anxiety through modification of self-reinforcement: An instigation therapy technique. *Journal of Consulting and Clinical Psychology*, 1968, *32*, 565–574.

Richardson, F. C., & Tasto, D. L. Development and factor analysis of a social anxiety inventory. *Behavior Therapy*, 1976, *7*, 453–462.

Royce, W. S., & Arkowitz, H. Multimodal evaluation of practice interactions on treatment for social isolation. *Journal of Consulting and Clinical Psychology*, 1978, *46*, 239–245.

Royce, W. S., & Weiss, R. L. Behavioral cues in the judgment of marital satisfaction: A linear regression analysis. *Journal of Consulting and Clinical Psychology*, 1975, *5*, 15–26.

Strauss, J. S., & Carpenter, W. T. The prediction of outcome in schizophrenia. *Archives of General Psychiatry*, 1972, *27*, 739–746.

Sylph, J. A., Ross, H. E., & Kedward, H. B. Social disability in chronic psychiatric patients. *American Journal of Psychiatry*, 1978, *134*, 1391–1394.

Trower, P., Yardley, K., Bryant, B. M., & Shaw, P. The treatment of social failure: A comparison of anxiety-reduction and skills-acquisition procedures on two social problems. *Behavior Modification*, 1978, *2*, 41–60.

Twentyman, C. T., & McFall, R. M. Behavioral training of social skills in shy males. *Journal of Consulting and Clinical Psychology*, 1975, *43*, 384–395.

Wallander, J. L., Conger, A. J., Mariotto, M. J., Curran, J. P., & Farrell, A. D. *An evaluation of selection instruments in the heterosexual-social anxiety paradigm*. Unpublished manuscript, 1978.

Warren, N. J., & Gilner, F. H. Measurement of positive assertive behaviors: The behavioral test of tenderness expression. *Behavior Therapy*, 1978, *9*, 178–184.

Watson, D., & Friend, R. Measurement of social-evaluative anxiety. *Journal of Consulting and Clinical Psychology*, 1969, *33*, 448–457.

Weinman, B., Gelbart, P., Wallace, M., & Post, M. Inducing assertive behavior in chronic schizophrenics: A comparison of socio-environmental, desensitization, and relaxation therapies. *Journal of Consulting and Clinical Psychology*, 1972, *39*, 246–252.

Wessberg, H. W., Mariotto, M. J., Conger, A. J., Conger, J. C., & Farrell, A. D. *The ecological validity of roleplays for assessing heterosocial anxiety and skill of male college students*. Unpublished manuscript, 1978.

Wood, M. A. *Structured learning, identical elements and subject differences in assertive behavior training with adolescents*. Unpublished manuscript, 1978.

Zigler, E., & Levine, J. Premorbid adjustment and paranoid-nonparanoid status in schizophrenia. *Journal of Abnormal Psychology*, 1973, *82*, 189–199.

Zigler, E., & Phillips, L. Social competence and outcome in psychiatric disorder. *Journal of Abnormal and Social Psychology*, 1961, *63*, 264–271.

Zigler, E., & Phillips, L. Social competence and the process-reactive distinction in psychopathology. *Journal of Abnormal and Social Psychology*, 1962, *65*, 215–222.

Index

9 780306 402333